Biointerface Characterization by Advanced IR Spectroscopy

Biointerface Characterization by Advanced IR Spectroscopy

Edited by,

C. M. Pradier
Y. J. Chabal

ELSEVIER
MASSON

Ámsterdam Barcelona Beijing Boston Filadelfia Londres Madrid
México Milán Múnich Orlando París Roma Sídney Tokio Toronto

Elsevier
Radarweg 29, PO Box 211, 1000 AE Amsterdam, The Netherlands
The Boulevard, Langford Lane, Kidlington, Oxford, OX5 1GB, UK

First edition 2011

Library of Congress Cataloging in Publication Data
A catalog record for this book is available from the Library of Congress

British Library Cataloguing in Publication Data
A catalogue record for this book is available from the British Library

ISBN: 978-0-444-53558-0

Printed and bound in the Great Britain
11 12 13   11 10 9 8 7 6 5 4 3 2 1

Contents

Preface

The interaction of biomolecules with inorganic materials is playing an increasingly more important role in medical and biomedical applications, such as implants or implanted devices (sensors, energy harvesting or sources), biosensors, and generally hybrid materials. Scientists from all disciplines, Physics, Chemistry, Biology, Mechanical/(bio)chemical Engineering, and Materials Science and Engineering, are getting engaged in the challenge of understanding and controlling the behavior of biomolecules in the vicinity of inorganic surfaces. For instance, several symposia have now been offered by the Materials Research Society (MRS) at its prominent Fall and Spring meetings. It is therefore not surprising that a number of characterization methods have been developed to monitor and quantify the presence of biomolecules at surfaces and their interaction with inorganic materials.

The more traditional methods originate from the Biology and medical field and involve the modification of biomolecules. They include fluorescent tagging, or use of specific reporter molecules. More recently, *surface scientists* have turned their attention onto the characterization of biomolecules at surfaces. For instance, there is now a division of ''Biomaterial interfaces'' in the AVS, one of the premier society primarily serving surface scientists, and a leader in the field of biomolecule/surface interaction, David Castner, was the 2010 AVS president. This section of the scientific community has thus brought to bear all the surface science techniques, including electron and ion spectroscopies (X-ray photoelectron spectroscopy, Auger electron spectroscopy, Rutherford Backscattering), imaging (Atomic Force microscopy, near field microscopy), optical spectroscopy (surface Plasmon resonance), and vibrational spectroscopy (Infrared and Raman). In fact, the interaction of biomolecules with solid surfaces constitutes an important part of the international conference on Vibrations at Surfaces, European Conference on the Applications of Surface Science (ECASIA) or even the European Conference on Surface Science (ECOSS).

Infrared spectroscopy is a remarkably versatile technique, ultra sensitive to molecular films but also adapted to the characterization of macromolecules on bulk or powder materials, using approaches best suited for the types of substrates. It is the most appropriate technique to probe the presence, formation and evolution of chemical bonds, i.e. the processes that are involved in biomolecule interaction with surfaces. It is therefore not surprising that many approaches, based on IR spectroscopy, have been devised to explore the interaction of biomolecules with solid surfaces. While complementary to other surface science techniques, FT-IR spectroscopy has the additional advantage of being compatible with a wide range of environments (vacuum, vapor, liquid) by using a variety of probing configurations (internal specular reflection for example). It can thus probe surfaces and interfaces in model or real « functional » conditions.

It is particularly timely to describe and review the progress made by infrared spectroscopists towards characterizing and understanding the interaction of biomolecules with solid inorganic surfaces in a single volume, so that a better assessment of the methods can be done and scientists in fields other than surface vibrational spectroscopy. It is also important that researchers using vibrational spectroscopy can be exposed to the challenges associated with the characterization of biomolecules and can bring their expertise to help the medical and biomedical field.

This book presents fundamental concepts associated with the characterization of biomolecules at surfaces in a variety of environments (vacuum, vapor, liquid), all centered on the use of infrared radiation as a probe. Each chapter focuses on a particular probing configuration and discusses both experimental issues/optimization schemes and scientific issues specific to using such an approach to study biomolecules. Examples are presented for the most relevant systems involving bio-related or biological interfaces, thus providing a practical advice for researchers in biomedical and chemical engineering and researchers using or optimizing bio-films or biosensors. For instance, optimization schemes are presented with caution on limitations of the technique, and an effort is made to address issues with in situ measurements (experimental limitations and spectral interpretation).

Theoretical concepts underlying the Fourier Transform Infrared (FTIR) reflection absorption spectroscopy (FT-IRRAS), also named Reflection Absorption Infrared Spectroscopy (RAIRS), technique based on a single the specular reflection at a metal surface are first presented in **Chapter I**. The selection rules associated with this special geometry and substrate conditions are described in detail. Examples of results, mainly concerning adsorption of small bio-related molecules, amino acids and small peptides on metal surfaces illustrate the advantage of the FT-IRRAS technique, and of applying it with polarisation modulation (PM-RAIRS or PM-IRRAS).

Chapter II focuses on the PM-IRRAS technique applied to air/water interfaces. Experimental setups are described, followed by examples of PM-IRRAS characterization of various biochemical systems like fatty acids, monolayer and multilayers of phospholipids pure or in interaction with antibacterial peptides, membraneous proteins, molecules and DNA.

Chapter III examines how the adsorption of biomolecules can be investigated using a single-bounce external reflection. While this may not be the most common geometry for semiconductor substrates, interesting properties associated with polarization dependence can be exploited to gain useful information about adsorbed molecular layers. This geometry is also convenient to study adsorption on both transparent and absorbing substrate.

Chapter IV considers similar adsorption systems as in Chapter III using in this case a single-transmission geometry. Such geometry is simple and versatile, with straightforward quantitative analysis, but requires transparent substrates preferentially polished on both sides. Emphasis is placed on the type of information that can be extracted from and typical issues associated with IR measurements of biomolecules attachment on both oxidized and oxide-free silicon surfaces.

In **Chapter V,** three specific modes of using IR spectroscopy for interfacial characterization are described, namely attenuated total reflection infrared (ATR-IR) spectroscopy, modulation excitation spectroscopy (MES) and vibrational circular dichroism (VCD). ATR-IR is shown to be a quantitative tool to study the adsorption and determine the orientation of molecular monolayers at solid-liquid interfaces. MES makes it possible to reduce the noise and therefore greatly increase the signal-to-noise ratio thanks to external signal modulation. This feature is particularly important to study weak dynamical changes at surfaces when there is strong background radiation or noise. VCD allows the selective detection of chirality of a layer by measuring the differential absorption of left and right circularly polarized light.

Chapter VI is devoted to the role of synchrotron radiation, focusing on the performance of the a 3rd generation machine. Such high brightness sources are ideal for IR spectro-microscopy, with spatial resolution better than 10 μm, enabling non invasive characterization and imaging of biomaterials on any type of solids with low photon energies. This chapter shows how synchrotron light is particularly well suited to characterizee substrate-molecule interactions in the Far-IR wavelength range.

Chapter VII describes the development of ATR, SEIRA, IRRAS and PM-IRRAS, based on IR single or multi reflection, for biosensing applications. These special configurations provide excellent sensitivity to low density of surface species, and can be used in situ in solution under flowing conditions. This chapter shows that the ability to distinguish metal carbonyls can enhance the sensitivity, thus providing a sensitive means to detect molecule interactions with High specificity.

Chapter VIII examines the characterization of non transparent materials by FT-IR with Fiber Evanescent Wave Spectroscopy (FEWS) that makes use of infrared transparent fibers. The chapter focuses on the use of chalcogenide fibers and their applications to biochemical sensing.

In **Chapter IX**, a new approach combining Atomic Force Microscopy (AFM) and infrared spectroscopy is presented; it provides high spatial resolution imaging by using a near-field photothermal technique coupled with specific infrared probes, thus going below the limits of optical resolution and allowing specific molecular identification. The physical principle of the technique is described, followed by some examples of applications in microbiology.

Chapter X presents a laser-based non-linear optical probe, Sum Frequency Generation (SFG), which is intrinsically sensitive to the interface between two centro-symmetric media. This technique is powerful for *in-situ* and time-resolved studies of adsorption, chemical evolution and molecular organization.

SFG spectroscopy is complementary to IR and Raman spectroscopy to study thanks to its sensitivity to interface species between two absorbing media. It has been successfully used to study lipid bilayers, protein conformation, biomolecule recognition as well as DNA pairing. The chapter also presents some prospects of the SFG technique towards ultra-fast time resolution studies and high resolution imaging of biointerfaces.

The aim of this book is to provide the reader with a detailed description of the currents IR-based methods to study biointerfaces, with a means to select the most appropriate approach for a given system, and with a guide for spectral analysis. Such a review of the latest developments in the IR techniques for biological interface characterization, illustrated by very recent examples, constitutes a useful reference tool for surface chemists, biochemists and advanced biotechnology engineers. This book should serve the biomedical community interested in exploring probes of adsorbed molecules that do not require tagging, the surface science community (including infrared spectroscopists) interested in biomedical applications, and students in biology, physics, chemistry, materials science or all disciplines in engineering interested in current, interdisciplinary challenges.

Contributors

Numbers in parentheses indicate the pages on which the author's contributions begin.

Souhir Boujday (167) UPMC, Universite Paris VI, Laboratoire de Reactivite de Surface, F-75005 Paris, France

Daniel Blaudez (27), CBMN, UMR5248 CNRS, Université Bordeaux, IPB

Bruno Bureau (217), UMR CNRS 6226 Sciences Chimiques, Groupe Verres et Céramiques, Université de Rennes I, Campus de Beaulieu, 35042 Rennes cedex, France

Thomas Bürgi (115), Département de Chimie Physique, Université de Genève, 30 Quai Ernest-Ansermet, 1211 Genève 4, Switzerland

Bertrand Busson (279), Laboratoire de Chimie Physique, Université Paris-Sud ; CNRS, Orsay, France

Sabine Castano (27), CBMN, UMR5248 CNRS, Université Bordeaux, IPB

Yves J. Chabal (83), University of Texas at Dallas, Department of Materials Science and Engineering Laboratory for Surface and Nanostructure Modification, 800 W. Campbell Road, RL 10 Richardson, TX 75080, USA

Alexandre Dazzi (245), Laboratoire de Chimie Physique, UMR8000, Université Paris-Sud, 91405 Orsay, FRANCE

Bernard Desbat (27), CBMN, UMR5248 CNRS, Université Bordeaux, IPB

Paul Dumas (145), SOLEIL Synchrotron, BP48, Saint Aubin, F-91192 Gif sur Yvette Céedex, France

Ryan Franking (57), Department of Chemistry, University of Wisconsin-Madison, 1101 University Avenue Madison, WI 53706, USA

Robert J. Hamers (57), Department of Chemistry, University of Wisconsin-Madison, 1101 University Avenue Madison, WI 53706, USA

Christian Humbert (279), Laboratoire de Chimie Physique, Université Paris-Sud ; CNRS, Orsay, France

Vincent Humblot (1, 145), Laboratoire de Réactivité de Surface, UMR CNRS 7197, Université Pierre et Marie Curie - Paris VI, 4 Place Jussieu, F-75252 Paris, France

Norman A. Lapin (83), BioSamPle Solutions, LLC, 116 Research Drive, Bethlehem, PA 18015

Pierre Lucas (217), Department of Materials Science and Engineering, University of Arizona, 4715 E. Fort Lowell Road, Tucson, AZ 85712, USA

Claire-Marie Pradier (1, 167), Laboratoire de Réactivité de Surface, UMR CNRS 7197, Université Pierre et Marie Curie - Paris VI, 4 Place Jussieu, F-75252 Paris, France

Clotilde Policar (245), Laboratoire des BioMole´cules, UMR 7203, de´partement chimie de l'ENS, 24 rue Lhomond, 75232, Paris Cedex 05

Rose Ruther (57), Department of Chemistry, University of Wisconsin-Madison, 1101 University Avenue Madison, WI 53706, USA

Michele Salmain (167), ENSCP, Chimie ParisTech, Laboratoire Charles Friedel, F-75005 Paris, France

Oliver Seitz (83), University of Texas at Dallas, Department of Materials Science and Engineering Laboratory for Surface and Nanostructure Modification, 800 W. Campbell Road, RL 10 Richardson, TX 75080 (USA)

Courtney Stavis (57), Department of Chemistry, University of Wisconsin-Madison, 1101 University Avenue Madison, WI 53706, USA

Xiaoyu Wang (57), Department of Chemistry, University of Wisconsin-Madison, 1101 University Avenue Madison, WI 53706, USA

Chapter 1

RAIRS under ultrahigh vacuum conditions on metal surfaces

V. Humblot[1] and C.-M. Pradier[2]
[1]*Laboratoire de Réactivité de Surface, UMR CNRS 7197,*
[2]*Université Pierre et Marie Curie - Paris VI, 4 Place Jussieu, F-75252 Paris, France*

SHORT HISTORY OF RAIRS

InfraRed Spectroscopy in the Reflection Absorption mode at grazing incidence (RAIRS) uses infrared light to excite the internal vibrations of molecules adsorbed on a reflective surface. The frequency of these vibrations is usually assigned to only one given chemical group in the adsorbate. From RAIRS spectra, one can determine the adsorption geometry and the adsorption site of the molecule.

Biointerface Characterization by Advanced IR Spectroscopy. DOI: 10.1016/B978-0-444-53558-0.00001-1

The historical development of RAIRS has been discussed by several authors [1–5] and came up after the infrared transmission work of Eischens *et al.* [6, 7] and Terenin *et al.* [8] in the 1950s. Eischens *et al.* used transmission infrared spectroscopy to study CO on supported Ni, Pd, and Pt. This technique is still widely used today, especially for studying adsorption on supported metals and dispersed catalysts. However, the use of well-defined single crystals, as model surfaces to address adsorption or reaction mechanisms, requires the experiments to be carried out in a reflection mode. RAIRS was first applied in 1959 by Pickering and Eckstrom [9] and Francis and Ellison [10] using multiple reflections. Although the multiple reflection technique was developed further, by using higher quality metal films, Greenler [11–13] first considered the possibility of single reflection experiments. The experimental RAIRS studies of CO on copper films by Pritchard and Sims [14] subsequently supported Greenler's findings.

HOW DOES IT WORK – THE PHYSICAL BASIS OF RAIRS

Classically, the general infrared selection rule states that for a vibration to be infrared active, it must induce a change of the electric dipole moment of the molecule. The intensity resulting from the transition between states ψ_i and ψ_f is defined by the following equation [15]:

$$I \propto \left| \psi_f^* E \mu_{fi} \psi_i \right|^2 \qquad (1)$$

where E is the electric field vector and μ_{fi} is the transition dipole moment.

Obviously, the intensity will be maximum when the electric field vector and the transition dipole moments are parallel and it is proportional to the square of the transition dipole moment [15].

For a molecule adsorbed on a metal surface, the interaction of the infrared radiation with the adsorbate dipole is influenced by the dielectric behavior of the metal, but the surface selection rule extends to molecules adsorbed in multilayers of thicker films, as illustrated by Poling *et al.* [16] with a 25 nm thick layer of Cu oxalate on copper.

The influence of a metallic substrate on absorption of infrared radiation by adsorbed molecules was originally examined by Francis and Ellison [10], and the classical electrodynamic macroscopic model was then developed by Greenler [11–13], leading to the definition of the optimum conditions for RAIRS experiments. Greenler showed, on the basis of an electrodynamic theory, first, that for metal surfaces only the component of the electromagnetic vector normal to the surface is active in exciting dipole-active vibrations [11], and second, that grazing angle incidence leads to the most intense electric field at the surface. Eventually, one single reflection at the surface was shown to be sufficient to achieve the desired sensitivity [12, 13].

Before considering the interaction of an infrared beam with an adsorbate-covered surface, one needs first to elucidate the interaction of a radiation with

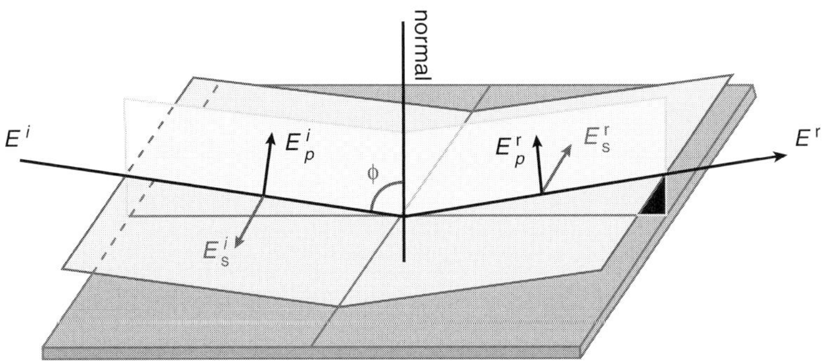

FIGURE 1 Schematic diagram of the s- and p-components of incident (E^i) and reflected (E^r) IR radiation.

a clean metallic surface, as discussed in more details below. This interaction can be described mathematically using the Fresnel equations [10]. The dipole selection rules will essentially derive from these equations. To understand the interaction between the IR radiation and the surface, the electric field vectors can be separated in two components; one perpendicular to the plane of incidence (s-component), E_s, and the other one parallel to the plane of incidence (p-component), E_p. Figure 1 shows the radiation components before and after reflection on a surface.

The complex index of refraction for each phase constituting the interface can be written as

$$\mathbf{n} = n - \imath k, \tag{2}$$

with $n^2 + k^2 \gg 1$, for light in the IR wavelength region reflecting on a metal surface; thus, the intensities of the p- and s-reflected light components, R_p and R_s, can be defined as follows:

$$R_p = \frac{(n - 1/\cos\phi)^2 + k^2}{(n + 1/\cos\phi)^2 + k^2} \tag{3}$$

$$R_s = \frac{(n - \cos\phi)^2 + k^2}{(n + \cos\phi)^2 + k^2} \tag{4}$$

$$\tan\Delta = \tan(\delta_p - \delta_s) = \frac{(2k\tan\phi\sin\phi)}{\tan^2\phi - (n^2 + k^2)}, \tag{5}$$

where δ_p and δ_s are the phase shift of the p- and s-light components upon reflection, respectively, and ϕ is the angle of incidence.

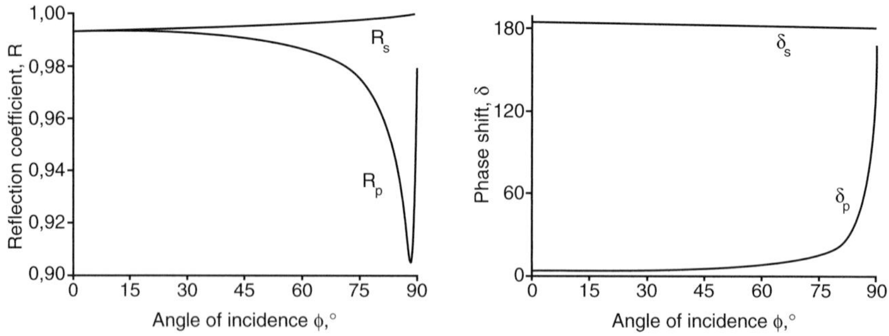

FIGURE 2 **The reflection coefficient (R) and the phase shift (δ) of the s- and p-components of the IR radiation on reflection from a metal surface as a function of the incidence angle (ϕ).**

Using appropriate values for n and k, typical of a highly reflecting metal surface ($n=3$ and $k=30$) [3], R_p, R_s and δ_p, δ_s can be calculated as a function of the incidence angle ϕ, as shown in Figure 2.

The electric field components of the incident, reflected, and refracted waves have to be considered, although the optical properties of the metal are such that most of the incident intensity is reflected, and the refracted wave contribution to the surface electric field is negligible [3]. If the amplitude of the incident electric field is $E^i\sin\theta$, where θ is an arbitrary phase, the electric field associated with the reflected wave is $E^i r\sin(\theta+\delta)$, where r is the reflection coefficient and δ is the shift phase. The resulting electric field at the surface is therefore given by

$$E = E^i[\sin\theta + r\sin(\theta+\delta)] \tag{6}$$

The phase shift (δ) of the s- and p-components of the infrared radiation reflected on a metal surface is shown in Figure 2. In the case of s-polarized light, equation (6) yields

$$E_s = E_s^i[\sin\theta + r_s\sin(\theta+\delta_s)] \tag{7}$$

Figure 2 shows that s-polarized light undergoes a phase shift close to 180° for all angles of incidence and $r_s \approx 1$; therefore, destructive interference occurs.

The incident and reflected beams lead to an electric field vector parallel to the surface close to zero; thus, vibrational modes with the vibrational moment change parallel to the surface will not be active. This is the main reason for the strict validity of the metal surface selection rule for dipole excitation of vibrational modes by infrared light on metal surfaces; *there is essentially no electric field parallel to the surface to interact with a transition dipole moment parallel to the surface.*

The p-component of the light, however, behaves differently; let us consider the two components of the incident and reflected electric fields, parallel, $E_p^{//}$, and normal $E_p^{\perp}$ to the surface. With a simple geometric approach, Figure 3 shows

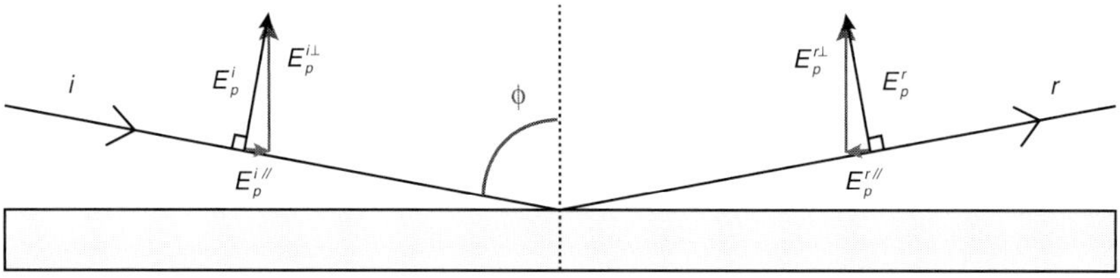

FIGURE 3 A schematic diagram of the magnitude of the p-component of the electric vector parallel ($E_p^{\|}$) and the perpendicular ($E_p^{\perp}$) to the surface as a function of the incidence angle ϕ.

that the reflected $E_p^{r//}$ and incident $E_p^{i//}$ are equal in magnitude but opposite in direction, and therefore cancel upon reflection on the surface.

In contrast, the reflected $E_p^{r\perp}$ and incident $E_p^{i\perp}$ are in the same direction and therefore lead to an increased field. The resultant electric field in the plane of incidence is given as follows:

$$E_p^{\perp} = E_p^i \sin\phi \left[\sin\theta + r_p \sin\left(\theta + \delta_p\right)\right]. \quad (8)$$

It is concluded that at angles close to grazing, incident p-polarized radiation can give rise to an enhanced (by a factor of 2 at maximum) electric field at the metal surface, in a direction normal to the surface. Greenler has shown that the optimum incidence angle for the RAIRS experiment is 80–88° depending on the metal [11]. Figure 4 shows the relative amplitude of the electric field perpendicular to the surface as a function of the incident angle, ϕ, for a typical

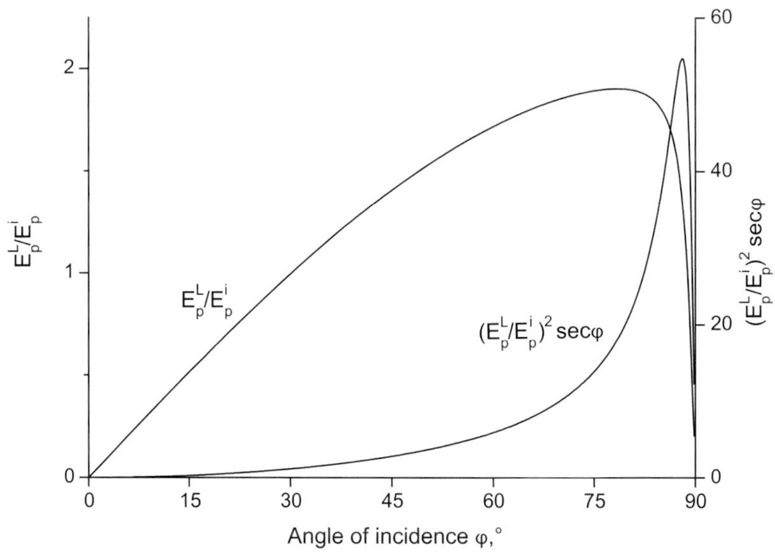

FIGURE 4 The relative amplitude ($E_p^{\perp}/E_p^i$) of the electric field perpendicular to the surface as a function of incident angle and the absorption intensity as a function of incident angle ϕ.

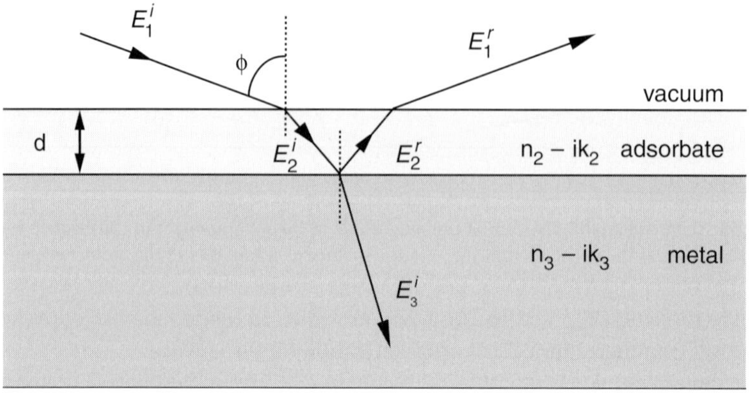

FIGURE 5 The classical three-phase model for an adsorbate layer ($n_2 - ik_2$) of thickness, d, on a metal ($n_3 - ik_3$), adapted from [4] copyright 1983 with permission from Elsevier.

metal, as well as the factor proportional to the differential reflectivity, ΔR

$$\Delta R = \frac{\left(\frac{E_P^{\perp}}{E_P^{i}}\right)^2 1}{\cos\phi}.$$

Greenler then extended the two-phase (vacuum/solid) system to a three-phase system, which includes a homogeneous isotropic layer adsorbed on the surface, characterized by an estimated complex dielectric constant [11]; the theoretical treatment has been greatly simplified by McIntyre and Aspnes [17]. Figure 5 displays a schematic diagram of the system under investigation. Greenler calculated the change in reflectivity at all angles of incidence using the Fresnel equations. It was found that an adsorbed layer does not make a significant difference in the main conclusion: *only light with a component of the electrodynamic vector normal to the surface will be absorbed, and the change in reflectivity for a metal is maximum at angles of incidence near to grazing* [11].

Incorporating an adsorbed layer also enables the calculation of an absorption function A, $A = \Delta R/R^{\circ}$, where $\Delta R = (R^{\circ} - R)$ is the reflectance in the three-phase system and R is the reflectance of the modeled adsorbate, see Figure 6.

The RAIRS experiment has been shown by Greenler to be most efficient when performed in a single reflection mode. This is due to the balance between the loss of intensity during reflection at grazing incidence and the increased absorption by the extra molecules available at each additional reflection [11]. It was also shown that the typical RAIRS spectrum of an adsorbate is directly comparable to the transmission spectrum [18].

It is also obvious, from the image dipole formalism, that only a vibration with a dipole moment perpendicular to the surface can be detected as illustrated by Figure 7. This is because the dipole parallel to the surface and its image are exactly opposite, whereas dipoles perpendicular to the surface reinforce each other.

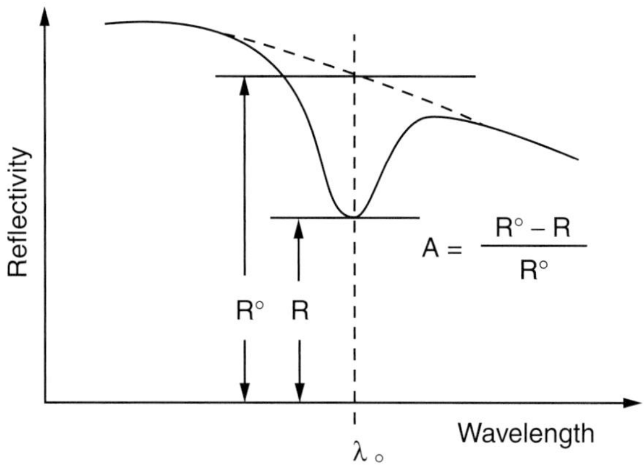

FIGURE 6 **Schematic diagram of the calculation of the absorption function $A = \Delta R/R^{\circ}$, where $\Delta R = (R^{\circ} - R)$ is the reflectance in the three-phase system with a nonabsorbing adsorbate and R is the reflectance of the modeled adsorbate, reproduced from [4] copyright 1983 with permission from Elsevier.**

In summary, theoretical evaluation of the RAIRS experiment, combined with experimental evidence, has shown the following:

1. Only the $p^{\perp}$ component of incident infrared radiation interacts with adsorbate vibrational dipoles; thus, only vibrational modes with a dipole component change perpendicular to the surface are active and can be observed.
2. The RAIRS experiment is most efficient near grazing incidence geometry.
3. The most reflective surfaces lead to the highest absorbance.
4. RAIRS is most efficient in a single reflection mode.

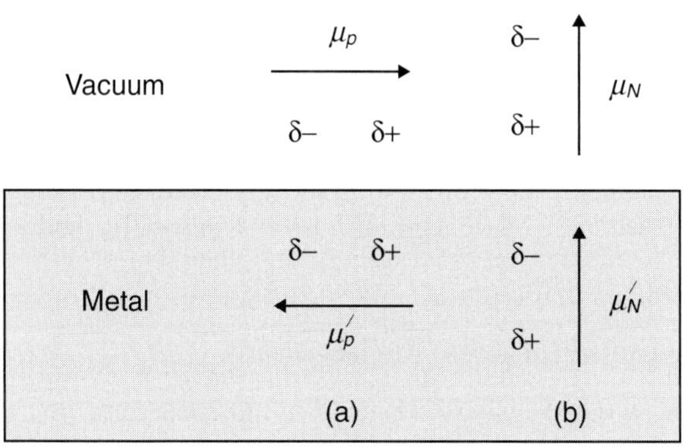

FIGURE 7 **The image dipole theory: parallel and perpendicular dipole moments.**

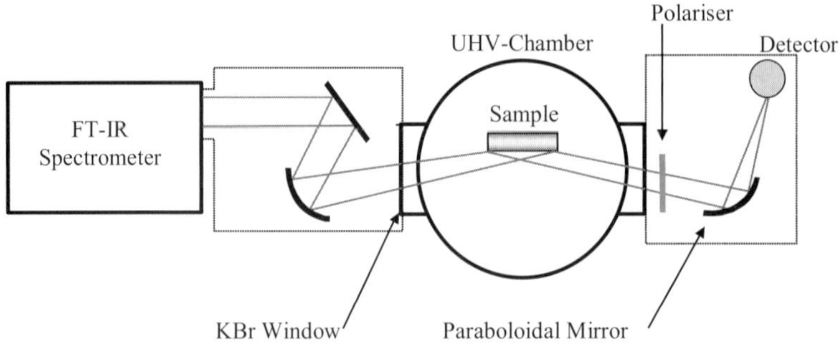

FIGURE 8 Experimental configuration of a UHV-RAIRS chamber coupled with a FT-IR spectrometer.

DESCRIPTION OF APPARATUS AND COUPLING WITH A UHV CHAMBER

The experimental configuration for an ultrahigh vacuum (UHV) RAIRS experiment is shown in Figure 8. A RAIRS experimental system has been developed for the first time in the mid-1980s by Chesters's group [19] and has become a standard UHV-RAIRS coupled system nowadays. The infrared beam is focused onto the sample surface at grazing incidence. The UHV chamber is equipped with KBr or ZnSe windows, which are IR-transparent. A semiconductor detector, usually a mercury–cadmium–telluride detector (MCT), then collects the resulting IR beam. The detector is cooled down by liquid nitrogen and its detection range is 4000–650 cm^{-1}. The infrared beam can be polarized prior to entry into the chamber, or on its outward path. For an optimized signal-to-noise ratio, the entire optical path length is purged with dry, CO_2-free air (dashed lines shown in Figure 8).

SELECTED EXAMPLES. WHY IS RAIRS SO SUCCESSFUL FOR THIN FILMS ON METAL SURFACES?

Vibrational spectroscopy of thin films adsorbed at metal surfaces provides direct information about the chemistry, the structure, the bonding, and the molecular orientation of the molecules constituting the adsorbed layer. Through few chosen examples, we will try to demonstrate the capability and resolution of RAIRS.

CO on pure metal surfaces: the beginning

One of the first reported analyses of adsorbed molecules on pure metal surfaces under UHV conditions was the adsorption of simple diatomic molecules. Due to the intensity of their dipolar moment, CO and NO were preferentially chosen.

CO adsorption has been characterized by RAIRS on a large ensemble of metals such as Pt, Pd, Cu, etc. A study of CO adsorption on a palladium surface will be presented in the next section. CO adsorption is still used for calibration of UHV FT-RAIRS systems (alignment and sensitivity), or to test the reactivity of a given surface.

CO/Pd(110)

CO adsorption on Pd(110) gave rise to a wide spectral range, where the ν_{CO} can be observed from ca. 2130 up to 1650 cm^{-1} [20–22]. Narrow regions have been identified, for various adsorption geometries of the molecule: 2130–2000 cm^{-1} corresponds to CO adsorbed on top of a metal atom, 2000–1880 cm^{-1} to a twofold bridging site, and finally 1880–1650 cm^{-1} for multiple coordinated CO, cf. Figure 9. Interestingly, the CO frequency is very sensitive to both the coverage and temperature. Adsorption at room temperature (300 K) yields a complex set of spectra, varying with the coverage, with bands observed from

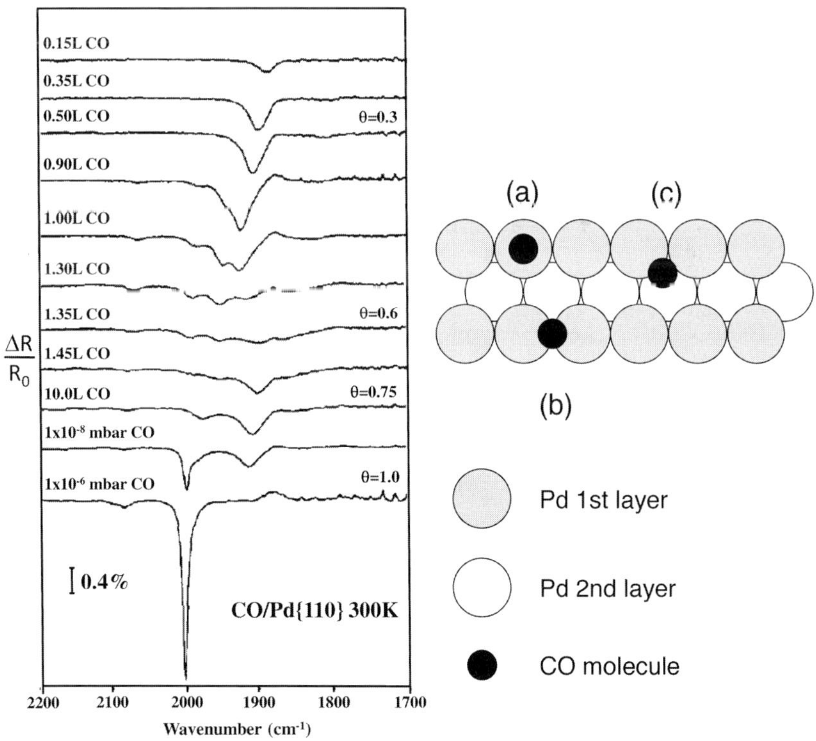

FIGURE 9 RAIRS spectra observed for CO adsorption on Pd(110) at 300 K, (a) CO molecules adsorbed on top sites, (b) CO molecules adsorbed on twofold short-bridge sites, and (c) CO molecules adsorbed multi-coordinated, adapted from [21, 24] copyright 1995 with permission from Elsevier.

1800 to 2050 cm^{-1} [20, 21]. At low coverage, $0.15 < \theta < 0.3$ L, the spectra show a single band that increases in intensity and frequency from 1882 to 1912 cm^{-1}, assigned to CO adsorbed on twofold short-bridged sites [23], Figure 9(b). For higher coverages ($\theta > 1$ L), the spectra exhibit again only one band at 2003 cm^{-1}, that is assigned to CO molecules adsorbed directly on top of a Pd atom, Figure 9(a). Eventually, at intermediate coverages ($0.3 < \theta < 1.0$) at least four broad bands co-exist ranging from 2068 up to 1861 cm^{-1} that can be assigned to molecules adsorbed on top (Fig. 9(a)), tilted asymmetric and symmetric twofold bridged sites (Figure 9(b)) [21, 22].

In contrast, when adsorption occurs at 180 K, the results show a more classical IR fingerprint with only one or two separate IR bands, cf. Figure 10. One can note, from the position and shift of the IR bands, that the adsorption site depends on the coverage and hence on the 2D organization of the admolecules and possible reconstruction of the palladium surface.

In order to clearly assign each IR band to the proper adsorption site, the surface reconstruction of Pd(110) should be taken into account. From LEED (low-energy electron diffraction) data, the 2D structure evolves from a (1×1),

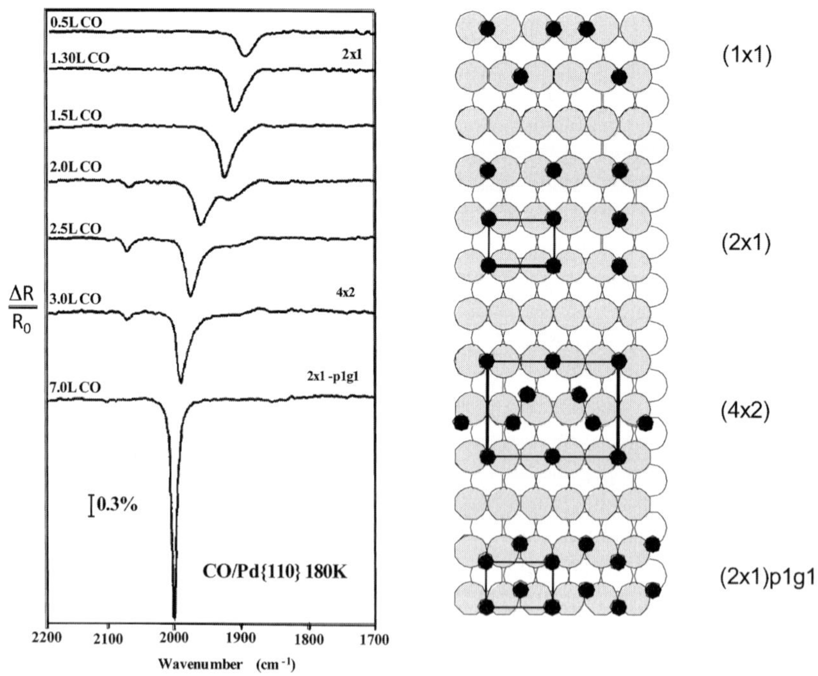

FIGURE 10 RAIRS spectra observed for CO adsorption on Pd(110) at 180 K [21]. Schematic representing the different adsorption site of CO molecules as function of coverage and 2D adlayer, adapted from [21] copyright 1990 with permission from Elsevier.

(2×1), (4×2) to eventually reach a (2×1)p1g1 arrangement, for around 4 L of CO (Fig. 10). Thus, Raval et *al.* [21] have assigned bands ranging from 1882 to 1912 cm^{-1} to symmetric twofold adsorbed CO molecules, and those at ca. 1950 cm^{-1} to asymmetric twofold species, due to surface reconstruction creating a near (111) surface. Eventually, they assigned the 1980–2000 cm^{-1} IR feature to tilted twofold bridged molecules, the usual on-top configuration ranging from 2068 to 2099 cm^{-1}[21].

Through this example, one can see that depending on the environment of the adsorption site, or of the possible surface reconstruction, a similar band can be shifted by several tens of wavenumbers, CO adsorption becoming a powerful tool to check surface reactivity/reconstruction.

Lysine and tartaric acid on Cu(110) single crystal surface; glutamic acid on Ag(110)

Some pioneer teams dedicated their work to the study of biomolecules, or bio-related molecules, on metal surfaces from thin films deposited in the liquid phase and analyzed *ex situ*. Among the vast literature, it is worth citing extensive studies from Lieberg's team dealing with small amino acids (AA) on gold and copper mainly. They have investigated glycine, histidine, or alanine and phenylalanine adsorption, by means of infrared spectroscopy and other surface science techniques [25–27].

Raval's group took great interest into studying adsorption of amino acids and carboxylic acids from spectroscopic and structural points of view up, with *in situ* UHV-RAIRS analyses during the exposure of molecules. The objective being of progressing in the field of enantioselective heterogeneous catalysis; they, in particular look carefully into the chirality of the molecules and how they can induce chiral properties to the surface [28, 29]. Worth citing examples are the early studies of the adsorption of methionine on gold and copper surfaces from liquid phase first, then studied in UHV environment, as well as the evaporation of glycine, alanine, proline, norvaline, or tartaric acid (presented below) on different single crystal surfaces under UHV conditions [24, 30–34].

Studies of single amino acid adsorption were the starting points of further studies involving di-peptides or tri-peptides, that is, several amino acids linked by in-between peptidic bonds. For instance, Barlow *et al.* [35] studied the adsorption of tri-alanine and tri-leucine on Cu(110). Vallée and co-workers [36–38] characterized adsorption of Gly-Pro, a combination of glycine (Gly) and proline (Pro), of GSH for glycine (Gly), cysteine (Cys), and glutamic acid (Glu), and of IGF for glycine (Gly), proline (Pro), and glutamic acid (Glu) on Au(111) and Au(110). Several examples of single amino acids, and small peptides adsorption on metal single crystals, under UHV conditions, are presented below.

Lysine on Cu(110)

Through this first example, we are turning toward molecules of biological interest; we will see the role of the amino acid radical (COOH/NH_2) in the adsorption processes, as well as the influence of the protonation/deprotonation phenomenon that leads to the presence of anionic, zwitterionic, or cationic chemical forms. A Cu(110) surface, exposed to a low pressure of lysine, leads to a series of spectra corresponding to the three ranges of surface coverage from the bare surface to adsorbed multilayers.

Low-coverage phase

Spectra (a) and (b), Figure 11, were recorded at the beginning of a continuous dosing regime mode ($P^\circ = 2 \times 10^{-9}$ mbar) at room temperature (300 K). From spectrum (b), one can isolate three main absorption bands in the 2000–1000 cm^{-1} region. The most intense one is observed at 1416 cm^{-1} and assigned to the symmetrical stretching vibration of a fully deprotonated carboxylic acid moiety, $\nu_{sym}^{COO^-}$. The two remaining weak absorption bands at 1624 cm^{-1} and 1328 cm^{-1} have been attributed to the $\delta_{asym}^{NH_3^+}$ antisymmetric deformation of a protonated amine group; the latter is the wagging mode of CH_2 of the alkyl chain, ω^{CH_2}.

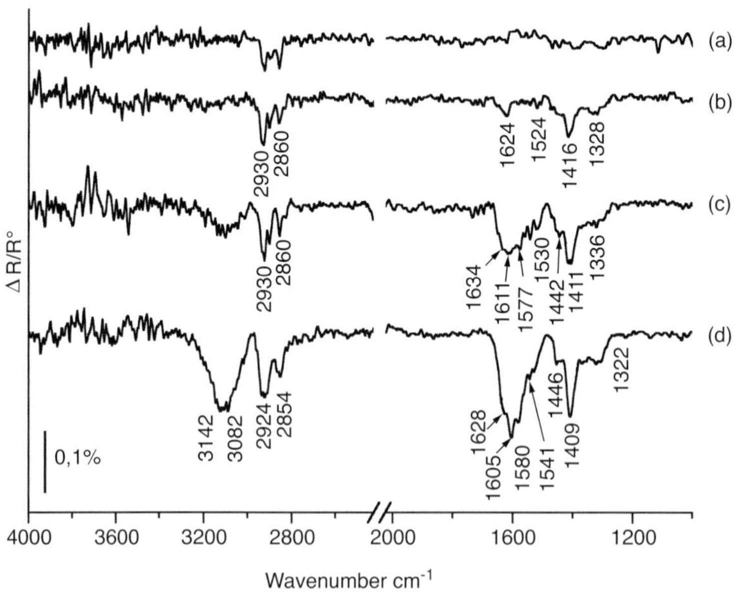

FIGURE 11 RAIRS spectra following the adsorption of lysine on Cu(110) at 300 K as a function of coverage. Sublimation exposure time (a) 5 min, (b) 15 min, (c) 22 min, and (d) 40 min, reproduced with permission from [39] copyright 2006 American Chemical Society.

The two bands, observed at 2930 cm^{-1} and 2860 cm^{-1}, are assigned, respectively, to the symmetric and antisymmetric stretching vibrational modes, $\nu_{asym}^{CH_2}$ and $\nu_{sym}^{CH_2}$. Eventually, a very weak, but still observable feature arises at 1524 cm^{-1}, attributed to the symmetric deformation mode of the amine group, $\delta_{sym}^{NH_3^+}$ [39].

The absence of any $\nu_{acid}^{C=O}$ (of the COOH group) stretching vibration around 1730 cm^{-1} and the presence of one peak at 1416 cm^{-1}, likely due to carboxylate stretch vibration, allow us to exclude the presence of lysine molecules under the forms $LysH_2^{2+}$ ($COOH/NH_3^+/NH_3^+$) [39]. On the basis of IR data, the distinction between NH_2 and NH_3^+ functionalities is rather difficult. The presence of a medium intensity feature at 1624 cm^{-1} was assigned mainly to a NH_3^+ group; however, the broadening of this band may suggest the contribution of some NH_2 moieties, contribution that will increase later on as the coverage increases. Based on these analyses, the main adsorbed species on the surface at this coverage is the zwitterionic Lys ($COO^-/NH_3^+/NH_2$) form, but the presence of a small fraction of $LysH^+$ ($COO^-/NH_3^+/ NH_3^+$) form cannot be excluded completely [39].

The adsorption geometry of the molecules at the surface can now be discussed on the basis of the RAIRS surface selection rules [40] applied to the observed vibrational modes. The presence of the $\nu_{sym}^{COO^-}$ vibration at 1416 cm^{-1}, and the absence of the $\nu_{asym}^{COO^-}$ vibration expected around 1600 cm^{-1} (inactive asymmetric stretch of the COO^- group) suggest that the two oxygen atoms are in symmetric positions with respect to the copper surface. Similarly to formates on Cu(110) [41, 42], it is proposed that the bonding of the molecules with the metal takes place via the oxygen atoms of the carboxylate group. The intensity of the $\nu_{sym}^{COO^-}$ vibration is, however, a factor of 10 lower than that observed for the equivalent vibration of formate on Cu(110) [42] thus suggesting that the carboxylate plane must be inclined toward the surface. Similarly, one can see that symmetric deformation $\delta_{sym}^{NH_3^+}$ is weak while the antisymmetric deformation shows a medium intensity band at 1624 cm^{-1}, implying that the NH_3^+ group lies parallel to the surface (C−N bond). Figure 12(a) shows a possible adsorption geometry for the lysine molecule at low surface coverage. Eventually, the contribution of the CH_2 groups of the alkyl chain shows a strong feature assigned to the symmetric stretching mode at 2930 cm^{-1} and a medium band at 1328 cm^{-1} attributed to the wagging mode. Hence, the alkyl chain adopts a near normal to the surface position generating a large dynamic dipole moment normal to the surface.

High-coverage phase

As the coverage of lysine is increased, spectrum (c) in Figure 11, one observes a number of new infrared features. The single band at 1624 cm^{-1} increases and broadens into a three-band contribution. The band assigned to the antisymmetric deformation of the amine group is slightly shifted to 1634 cm^{-1}, while two other

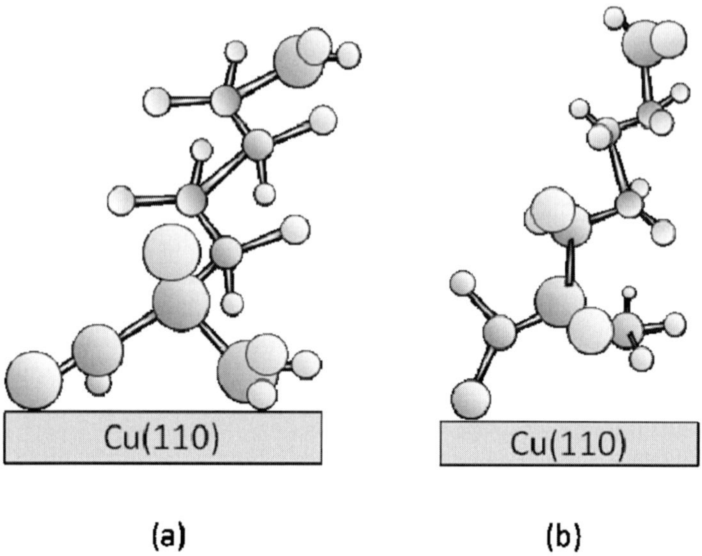

FIGURE 12 Orientation of L-Lysine on Cu(110) at 300 K, viewed along the surface at (a) low coverage: the "symmetric" geometry and (b) high coverage: the "asymmetric" geometry, reproduced with permission from [39] copyright 2006 American Chemical Society.

features at 1611 and 1577 cm^{-1} appear. The former band can be attributed to a δ^{NH_2} deformation while the latter is assigned to the antisymmetric stretching mode of the carboxylate group, $\nu_{asym}^{COO^-}$. This is the fingerprint of COO^- groups orientated with a sideway tilted orientation in which the two oxygen atoms are no longer held equidistant to the surface. In contrast, the symmetric stretching vibration $\nu_{sym}^{COO^-}$, at 1411 cm^{-1}, slightly varies, with the appearance of a shoulder peak at 1442 cm^{-1}, assigned to a δ^{CH_2} deformation mode. This behavior suggests that upon increasing coverage, a new type of adsorbed species appear on the Cu (110) surface. Details about the adsorption geometry of this new species may have been deduced by considering the vibrations of the functional group of the lysine molecules individually, in particular those of the carboxylate groups.

The contributions from the NH_3^+ groups also change with the appearance of a weak symmetric deformation mode at 1530 cm^{-1}, together with the antisymmetric mode $\delta_{asym}^{NH_3^+}$ at 1634 cm^{-1}. The proximal NH_3^+ group is likely now more parallel to the surface, with the C-N bond lying also parallel to the surface, see Figure 12(b). Although the positioning of the distal amine group and the carboxylate group is quite accurate, these RAIRS data do not allow the precise orientation of these upside functional groups to be determined, and, therefore the positions shown in the schematic Figure 12(b) are only one possible adsorption geometry.

Eventually, the spectrum (d) in Figure 11 shows the RAIRS data after 40 min of continuous dosing regime; there is no change in the band positions, but an increase of their intensity. All bands observed bear a strong similarity with the IR bands observed for the solid condensed phase of lysine [43, 44], suggesting that, at these high exposures, weakly perturbed physisorbed molecules are adsorbed, forming multilayers on the Cu(110) surface. It is worth noting that the broad absorption band around 3100 cm^{-1} is indicative of intermolecular hydrogen bonding between lysine molecules in the multilayer.

This behavior is often observed when multilayers of amino acid molecules are grown on copper surfaces at room temperature. [30, 34]

(R,R)-tartaric acid on Cu(110)

Tartaric acid (TA) is a molecule capable of existing in at least three different forms: the neutral bi-acid form; the monotartrate form, where one of the carboxylic acid groups has deprotonated to give a carboxylate moiety; and the bitartrate form, in which both acid groups have deprotonated. Furthermore, the molecule chemistry is complicated by the fact that protonated carboxylic acid groups are possibly involved in hydrogen bonds with other acid and alcohol groups, which potentially lead to a number of intermolecular interactions.

Infrared spectroscopy provides a way of distinguishing the different types of functional groups. The adsorption of tartaric acid on Cu(110) has been studied with a wide range of surface science techniques such as RAIRS, LEED, TPD (temperature programmed desorption), and STM (scanning tunneling microscopy) for example. The adsorption behavior of TA/Cu(110) under various coverage and temperature conditions reveals that the molecule/metal system yields to a complex and varied series of phases of adsorbed molecules [45, 46].

Room-temperature (300 K) adsorption phase

TA adsorption RAIRS data, recorded at low coverage and room temperature, are presented in Figure 13(b), and reveal bands at 1711 and 1437 cm^{-1}. The former band is correlated with a C=O group, while the latter is likely the symmetric ν_{COO-} vibration with the carboxylate groups. It, therefore, seems that both protonated acid and carboxylate groups are present in the adlayer; the authors concluded that every molecule in the adlayer exists as a monotartrate species in which one end retains the acid functionality and the other one deprotonates [45].

It is also reported that the carboxylate group strongly interacts with the metal surface as, for example, observed for the formate species [42], and must be responsible for the strong anchoring of the monotartrate species to the surface. Application of the infrared metal surface selection rule [40] allows the orientation adopted by this functionality to be determined; the presence of the symmetric ν_{COO-} vibration at 1437 cm^{-1} but the absence of the usually strong asymmetric ν_{COO-} vibration, expected in the 1550–1650 cm^{-1} range, implies that the two O atoms in the carboxylate group are equidistant from the surface.

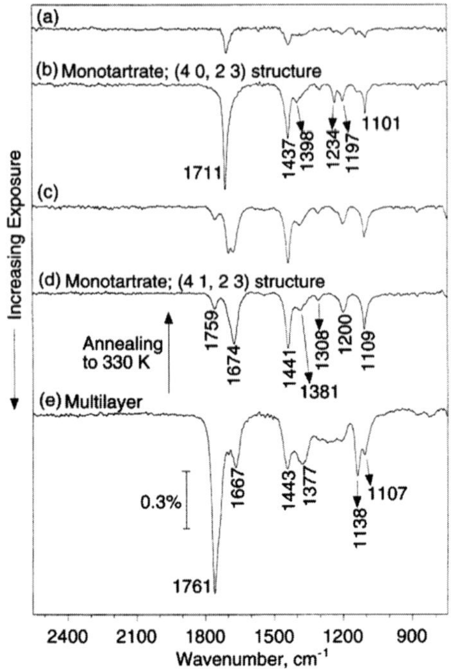

FIGURE 13 RAIRS spectra observed as a function of increasing coverage of *(R,R)*-tartaric acid adsorbed on Cu(110). Spectra shown in (a), (b), and (d) were obtained at 300 K while (c) and (d) were obtained at 350 K. Spectra (c) and (d) are not discussed here, reproduced with permission from [45] copyright 1999 American Chemical Society.

Turning to the acid end of the monotartrate species, the frequency of the $\nu_{C=O}$ vibration in the monotartrate species, 1711 cm^{-1}, much lower than that observed for monomeric carboxylic acids [45], is consistent with the presence of strong H-bonding within the adlayer. Therefore, the author concluded that the molecules are anchored to the surface via one carboxylate group, while the

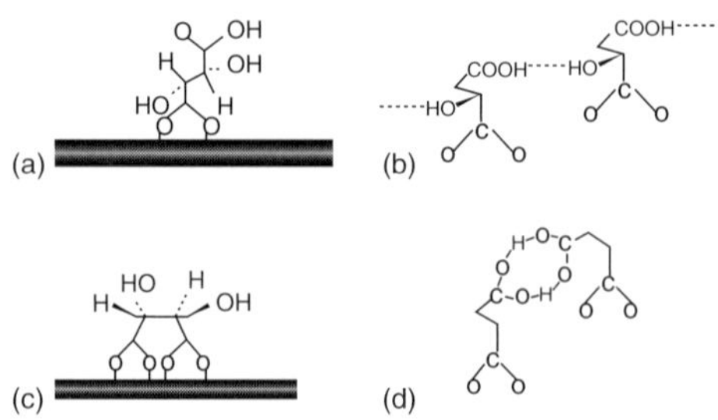

FIGURE 14 Schematic representation of the TA molecules adsorbed on Cu(110) under the monotartrate phase in (a) and the bitartrate phase in (c). (b) and (d) represent the H-bonding network of, respectively, linear acid-alcohol dimmers and cyclic acid dimmers, adapted with permission from [45] copyright 1999 American Chemical Society.

second one is pointing up from the surface, Figure 14(a), and the H-bond network could be described as presented in Figure 14(b), leading to strong intermolecular interactions (linear COOH-$OH_{alcohol}$ dimmers).

Eventually, when the coverage is increased up to a full monolayer, spectrum 13(e), the IR signature associated with this new phase displays slight changes: first, a downshift of the $\nu_{C=O}$ from 1711 to 1674 cm^{-1}, and, second, the appearance of a new band at 1759 cm^{-1} also associated with a $\nu_{C=O}$ vibration. The band at 1674 cm^{-1} indicates the presence of acid groups in H-bonded cyclic acid dimmers, Figure 14(d), while the latter is indicative of monomer-free acid groups, not involved in H-bonding [45]. These changes in the intermolecular interactions can be correlated to a new 2D organization of the molecules at the Cu(110) surface ((40,23)→(41,23)).

High-temperature (400 K) adsorption phase

The RAIRS data obtained for TA adsorption at 400 K are presented in Figure 15. The spectrum observed at low coverage (Fig. 15(a)) shows a complete absence of $\nu_{C=O}$ band in the 1700–1740 cm^{-1} region in contrast to the signals observed

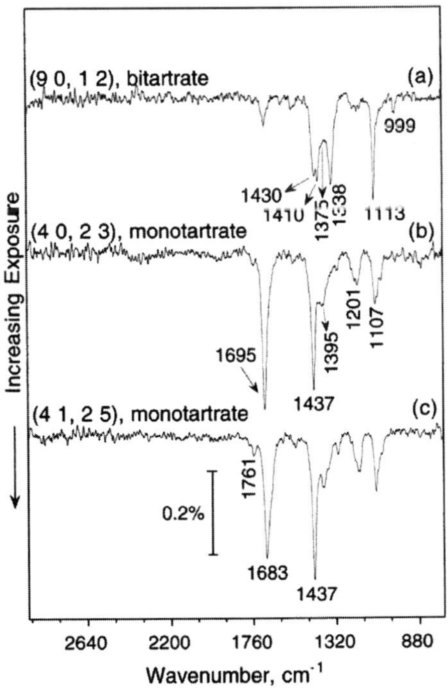

FIGURE 15 **RAIRS spectra observed as a function of increasing coverage of tartaric acid adsorbed on Cu(110) at 400 K, reproduced with permission from [45] copyright 1999 American Chemical Society.**

for the monotartrate phases (Fig. 13). Changes are also observed for the carboxylate band where the single band of the monotartrate form at 1437 cm^{-1} is replaced by two bands at 1430 and 1410 cm^{-1} (attributed to the coupling of the two identical COO^- oscillators on each molecule). Changes are also observed for the 1113 and 1338 cm^{-1} bands, which both increase and sharpen compared to the monotartrate phase at 300 K, Figure 13(d) at 1109 and 1308 cm^{-1}, respectively. These changes, corresponding to the δ_{C-H} and ν_{C-OH} vibrations, can be analyzed by considering the geometry of the bitartrate species (Fig. 14 (c)), in which the two carboxylate groups are bonded to the surface, leaving the C2–C3 bond almost parallel to the surface plane. In this geometry, the C–OH bond is directed toward the surface normal, while the C–H bond adopts a new orientation, closer to the surface parallel.

Glutamic acid on Ag(110)

As a final example for this section, we will present the work of Baddeley *et al.* [47], on the adsorption of (S)-glutamic acid (GA) on a Ag(110) single crystal surface. The GA molecule combines the amino acid functionality ($COOH/NH_2$) and the di-acid character, with a COOH group on the opposite position of the amino acid center. Interestingly, the presence of three functional groups might lead to strong intermolecular interactions driving the ordering of the adlayer at

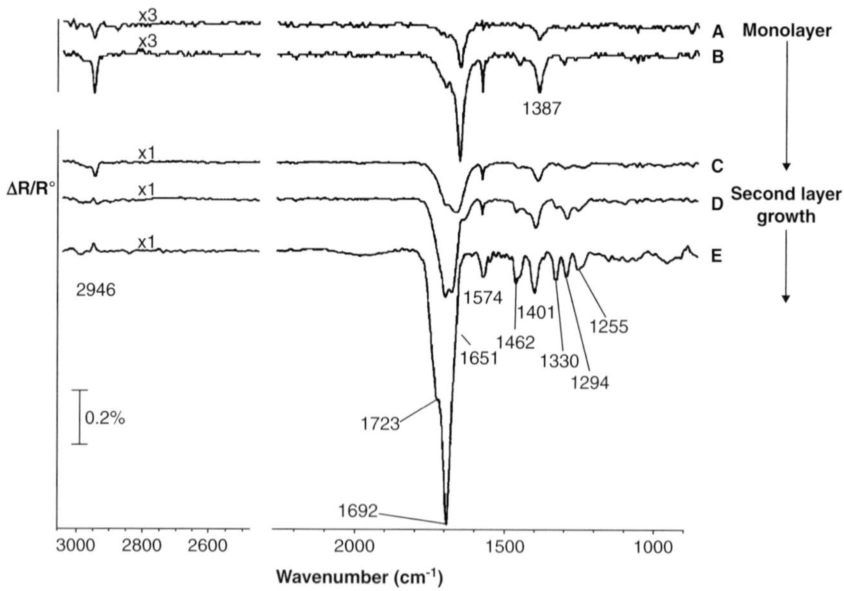

FIGURE 16 RAIRS spectra following the adsorption of glutamic acid on Ag(110) at 300 K as a function of coverage, from very low coverage up to multilayer regime, reproduced with permission from [47] copyright 2005 American Chemical Society.

the surface, in possible competition with the strength of the molecule-surface binding.

Figure 16 shows the RAIRS spectra obtained for the adsorption of GA at room temperature as a function of increasing coverage. At least two domains can be clearly identified, one a low coverage, corresponding to a sub-monolayer regime (spectra A and B), the second one, corresponding to the formation of multilayers. At low exposure, IR spectra are dominated by absorption bands at 1651, 1574, 1462, and 1387 cm^{-1}. These bands can be easily assigned to νC=O of acid groups, δNH_2, δCH_2, and $\nu^{sym}{}_{COO-}$, respectively, and suggest that the GA molecules are adsorbed as anionic moieties, where the acid group of the amino acid part is deprotonated into carboxylate, while the aliphatic acid group remains fully protonated. Moreover, such a low frequency for a νC=O stretching mode implies strong inter- or intra-molecular interactions, as observed in the previous tartaric acid example. Analyses of the IR bands and their intensity enabled the authors to propose a model for the adsorption geometry, where GA molecules are adsorbed rather flat at the Ag(110) surface, with all three functional groups interacting with the metal surface. In addition, it is worth noting that this low-coverage phase exhibits a 2D repetitive unit, evidenced by STM experiments, in which molecules strongly interact by hydrogen bonds.

When the coverage is increased (figure 16, spectra C to E), intensities of the previously observed IR features increase, and a set of extra RAIRS bands appear. This implies a change in the adsorption geometry of GA molecules (Fig. 16 spectrum C), with creation of close-range dimmers (additional νC=O at 1692 cm^{-1}). Eventually, when the coverage in increased further (spectrum D and E on Fig. 16), the IR signature now bears a close resemblance with the one obtained on crystalline GA powder, which is usually an indicator of a second physisorbed layer.

Bio-molecules (peptides) adsorbed on metal surfaces, studied by RAIRS

Adsorption of tri-alanine on Cu(110)

One of the first studies of poly-peptides adsorption on metal surfaces under the UHV condition was presented in 2001 by Barlow and co-workers [35], with the adsorption of tri-homo-amino acids, tri-alanine, and tri-leucine, on a Cu(110) single crystal surface.

When considering the adsorption of poly-peptides on a metal surface, not only one needs to consider their interaction with the metal surface, but also the conformation of the peptide itself. In this context, RAIRS will give a lot of information thanks to the analyses of IR bands, fingerprints of the conformation of peptides or their folding.

In the monolayer regime, Figure 17, Barlow *et al.* have identified three different adsorption phases for tri-alanine molecules. On the three spectra

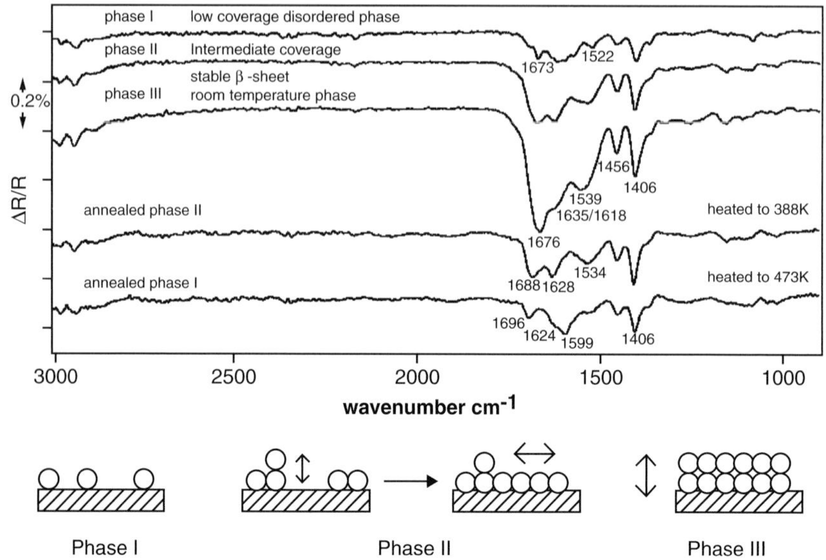

FIGURE 17 **RAIRS spectra of tri-L-alanine deposited on Cu(110) at 300 K under high-flux conditions; the adsorption geometries are reported for each of the three phases, reproduced and adapted with permission from [35] copyright 2001 American Chemical Society.**

acquired at room temperature, one notes the presence of the amide I and II IR signature, with bands present, respectively, in the 1640–1680 cm^{-1} and 1520–1540 cm^{-1} region, indicating that the molecules are adsorbed intact on the surface.

For the low coverage, phase I, in addition to the amide I and II bands, the vibrations associated with the carboxylate stretch vibrations at 1400 and 1616 cm^{-1} are observed, along with vibrations associated with terminal NH_2 groups, indicating that the molecules interact with the surface in a manner similar to L-alanine, via at least one amide bond. All the bands are, however, very sharp, evidencing weak H-bonds, thus very little interactions between tri-alanine molecules. The authors proposed that adsorption occurs randomly on the copper surface, as disordered isolated molecules, Figure 17 phase I scheme.

As the coverage of tri-alanine is increased, Figure 17 phase II spectrum, the adsorption peaks become broader, especially in the amide I region, indicating H-bonding between adjacent molecules. Intermolecular interactions may occur from either vertical or horizontal interactions, Figure 17 phase II scheme, suggesting that phase II is a transition state between the disordered phase I, and the compact bilayers of phase III.

This is confirmed by phase III spectrum, Figure 17, which is dominated by a strong and broad band at around 1676 cm^{-1}, assigned to interactions in between the amide C=O groups. In addition, this spectrum bears strong resemblance with IR spectra of tri-alanine crystalline solid composed of β-sheets layers. Thus, the authors concluded that phase III is constituted of stable bilayers with strong interlayer H-bonds, Figure 17 phase III scheme.

Adsorption of Gly-Pro, GSH (Glu-Cys-Gly), and IGF (Gly-Pro-Glu) on Au(110)

We now report the adsorption of a di-peptide, Gly-Pro and of two quite similar tri-peptides, GSH (Glu-Cys-Gly) and IGF (Gly-Pro-Glu), presented in Figure 18. The purpose of this work was to gradually increase the complexity of the adsorbed system and investigate the role of the SH group in the case of GSH. Each step was analyzed by PM-RAIRS and XPS data using single AA data from the step before. Thus, Gly-Pro was studied by comparing the data of glycine [30] and proline [33], IGF, after Gly-Pro and glutamic acid [48,49], and so on. The two tri-peptides, IGF and GSH, mainly differ by the central amino acid, cysteine in the case of GSH and proline in the case of IGF, Figure 18.

Gly-Pro on Au(110)

First, molecular Gly-Pro, a fragment of the IGF (Gly-Pro-Glu) molecule, was adsorbed on Au(110) [38]. The PM-RAIRS spectra of the Au(110) surface, recorded during Gly-Pro exposure ($P_{Dose}=1 \times 10^{-8}$ mbar), are shown in Figure 19(a). At low coverage (lower spectrum, 15 min of exposure), the peaks are not intense but well defined. All IR features increase with exposure time up to 30 min (middle spectrum), leading to the following main features: the most intense Gly-Pro-related band, observed at 1670 cm^{-1}, is attributed to the C=O amide I vibration, bands at 1735 cm^{-1} and 1357 cm^{-1} are assigned, respectively, to $\nu_{C=O}$ and ν_{C-O} of carboxylic acid groups of the Pro moiety [33; 50], and the

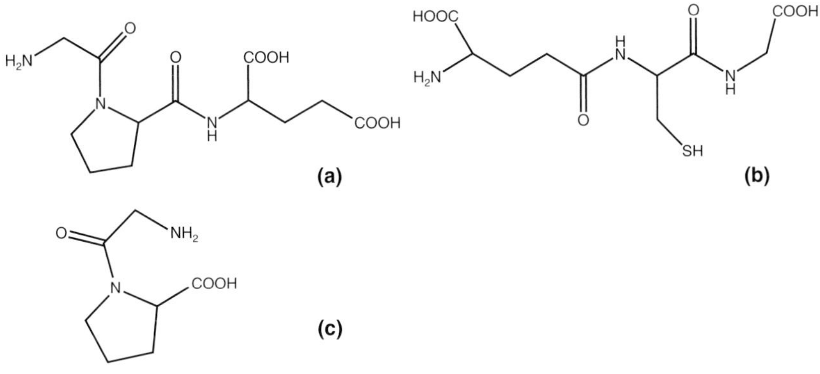

FIGURE 18 The three peptides molecules: (a) IGF (Gly-Pro-Glu), (b) GSH (Glu-Cys-Gly), and (c) Gly-Pro.

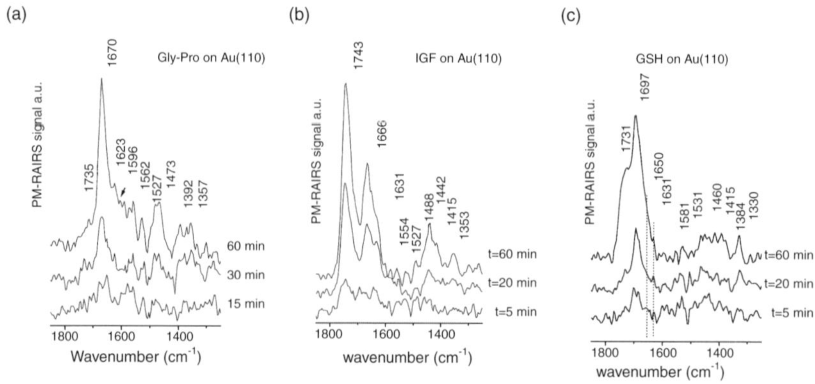

FIGURE 19 PM-RAIRS spectra recorded upon increasing exposure of Gly-Pro (a), IGF (b), and GSH (c), at 300 K on Au(110), reproduced with permission from [38] copyright 2009 American Chemical Society.

bands at 1596 and 1415 cm^{-1}, to the asymmetric and symmetric stretching vibrations of carboxylate groups [33]; this highlights that carboxylic acids and carboxylate groups are simultaneously present in the adsorbed layer. The weak absorption bands at 1623 and 1519 cm^{-1} may be assigned to the asymmetric and symmetric deformations of NH_3^+ of some protonated amine groups, and the band at 1562 cm^{-1} to the NH_2 deformation of the Gly fragment [30, 51]. Eventually, the band at 1477 cm^{-1} is attributed to the CH_2 scissor of the aliphatic ring of proline [33]. All expected vibrations of Gly-Pro (amide bands and functional goups) are identified on the spectra Figure 19(a), suggesting that Gly-Pro is adsorbed intact on the Au(110) surface.

When the coverage increases up to 60 min of exposure, the amide I vibration at 1670 cm^{-1} becomes predominant, with two shoulders at 1623 and 1596 cm^{-1}, attributed respectively to the asymmetric deformation of NH_3^+ and the asymmetric stretching of COO^-. Moreover, the stretching mode of C=O of carboxylic acid group at 1735 cm^{-1} has a very low intensity, suggesting that a majority of carboxylic acid groups are now deprotoned.

IGF on Au(110)

Figure 19(b) shows the PM-RAIRS spectra obtained as a function of increasing exposure of IGF on Au(110) at 300 K ($P_{Dose} = 1 \times 10^{-9}$ mbar). Clear IR bands can be observed from 5 min of exposure. As observed previously for the Gly-Pro molecules, the features can be assigned to the amide I and II vibrations and to vibrations from the constituent amino acid functionalities, indicating that IGF is adsorbed intact at the surface [38].

With increasing coverage, middle and upper spectra of Figure 19(b), the features at 1743 cm^{-1} ($\nu_{C=O}$) and at 1666 cm^{-1} (amide I) grow into two broad and very intense bands including other vibrations like the asymmetric deformation of NH_3^+ at 1631 cm^{-1}, and some weak absorption in the COO^- asymmetric stretching mode region (around 1600 cm^{-1}). In the 1500–1200 cm^{-1} region, all IR bands bear a close similarity with those observed for the Gly-Pro molecules, Figure 19(a). Like in the case of Gly-Pro adsorbed on Au (110), the presence of these features attributed to the carboxylic acid, carboxylate and NH_3^+ species on the PM-RAIRS spectra, shows that the zwitterionic forms of IGF predominate on the surface. Note that the PM-RAIRS spectra are all dominated by the C=O stretch of acid groups; the IGF molecule has two of them, one being likely partially deprotonated and the other one remaining under the COOH form.

From these data, it seems that the main differences between Gly-Pro and IGF are a slight different orientation of the molecules at the Au(110) surface and more importantly a difference growth mode, that will be discussed later.

GSH on Au(110)

Turning now to the last example, we will see differences induced by a different central fragment, proline in the case of IGF, and cysteine in the case of GSH. Figure 19(c) shows the PM-RAIRS spectra obtained for GSH on Au(110) from 5 min up to 60 min of exposure at 300 K ($P_{Dose}=1 \times 10^{-9}$ mbar) [36–38]. There is no change in band positions upon further GSH exposure, and all bands can be correlated to the functionalities of the GSH molecules. The PM-RAIRS spectra are dominated by a broad band at the 1650–1740 cm^{-1} region, with one main band at 1697 cm^{-1}, which likely includes two contributions: the free amide I signal and the $\nu_{C=O}$ band of carboxylic acid groups involved H-bonded intermolecular interactions with other GSH molecules. One also notes the presence of several other intense bands, at 1731 cm^{-1}, $\nu_{C=O}$ of free acid groups, 1650 cm^{-1}, and 1550 cm^{-1}, the amide I and II vibrations, and eventually the symmetric stretching vibration of the carboxylate group around 1415 cm^{-1}. These IR data reveal that the GSH molecules are present under their zwitterionic form at the Au(110) surface, with the carboxyl group of the Gly moiety fully protonated responsible for a strong H-bond network in between GSH molecules.

Gly-Pro, IGF, and GSH growth mode comparison

Kinetics of adsorption of the three peptides could be compared, by integrating the IR bands in a peptide-characteristic region, 1800–1600 cm^{-1}; it increases in the following order (cf. Fig. 20): Gly-Pro < GSH < IGF. This comparison does not take into account the orientations of the IR active groups that may enhance some signals and induce some errors in the correlation between IR intensities and number of adsobates. We however consider that, on an average, the spectral zone, considered in Figure 20, including signals from at least four dipole moments, $\nu_{C=O}$ (amide or acid groups), $\nu^{asym}{}_{COO^-}$, and δ_{NH2} provides a reasonable indication of the relative amounts of adsorbed molecules.

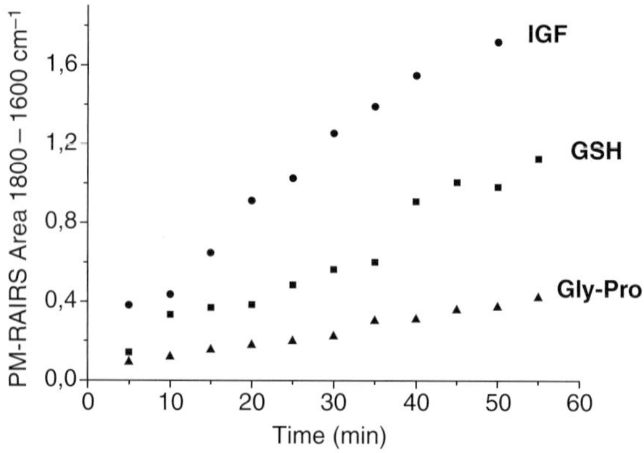

FIGURE 20 Evolution of the PM-RAIRS area in the 1793–1596 cm^{-1} region of GSH, IGF, and Gly-Pro on Au(110) under exposure, $P_{Gly\text{-}Pro} = 10^{-8}$ mbar, $P_{GSH} = P_{IGF} = 10^{-9}$ mbar , reproduced with permission from [38] copyright 2009 American Chemical Society

One notes that the total amount of organic material, after Gly-Pro adsorption, is significantly less than when adsorbing IGF or GSH, even at a pressure 10 times higher; this is a strong indication that an additional fragment, like on IGF or GSH, compared to Gly-Pro, induces a stronger interaction with the surface, likely because several centers of the molecule are binding with the gold surface.

Note that for these three systems, the evolution of the average thickness was also deduced from XPS calculations on the Au4f attenuation signal, and it showed exactly the same trend, of course achieving quantitative measurements is much simpler by PM-RAIRS than by XPS!

CONCLUSIONS

Through selected examples, it has been shown that RAIRS is a very powerful tool to characterize (bio)-molecules adsorption, binding mode, geometry, and kinetics, on metal surfaces.

In the case of very simple molecules, like CO, monitoring the frequency shift of a given vibration can help determining the adsorption site, top site, short-bridge, or long-bridge sites. One can even gain information on the positioning and orientation of a molecule with respect to the surface by looking at IR band intensities, by simply applying the metal surface selection rule of the RAIRS technique.

When turning to "more complex" molecules, such as amino acids or carboxylic acids (typically lysine and tartaric acid), RAIRS data can tell the experimentalist what is the chemical state of the adsorbed molecule, amine and carboxyl groups protonated or not for example, thanks to very distinct vibrational frequencies.

In the case of di- and tri-peptides adsorbed on gold surfaces, the main results were first the identification of all main IR characteristic bands, suggesting that the latter could be adsorbed under UHV conditions, without breaking; most important, crucial information could be obtained concerning the peptide surface assembling, with or without intermolecular interactions, as well as the kinetics of adsorption. RAIRS, performed under UHV conditions, thus appears as a very powerful technique, providing chemical, quantitative, information about adsorbates; opening a new way of investigating model biointerfaces, it appears to be very complementary to other more classical surface science characterization techniques.

REFERENCES

[1] Bradshaw, A.M.; Schweizer, E. *Spectroscopy of Surfaces.* (R. J. H. Clark and R.E. Hester (Eds.)), New York: Wiley 1988.
[2] Chabal YJ. *Surf. Sci. Rep.* 1988;8:211–357.
[3] Hayden, B.E. *Vibrational Spectroscopy of Molecules on Surfaces* (J.T. Yates, T.E. Madey (Eds.)) New York, 1987.
[4] Hoffmann FM. *Surf. Sci. Rep.* 1983;3:107.
[5] Hollins P, Pritchard J. *Prog. Surf. Sci.* 1985;19:275.
[6] Eischens RP, Francis SA, Pilskin WA. *J. Phys. Chem.* 1956;60:194.
[7] Eischens RP, Pilskin WA. *Advan. Catal.* 1958;10:1.
[8] Yaroslavsky NG, Terenin AN. *Dokl. Akad. Nauk. SSSR* 1949;66:885.
[9] Pickering HL, Eckstrom HC. *J. Phys. Chem.* 1959;63:51.
[10] Francis SA, Ellison AH. *J. Opt. Sci. Am.* 1959;49:131.
[11] Greenler RG. *J. Chem. Phys.* 1966;44:310.
[12] Greenler RG. *J. Chem. Phys.* 1969;50:1963.
[13] Greenler RG. *J. Vac. Sci. Tech.* 1975;12:1410.
[14] Pritchard J, Sims ML. *Trans. Faraday Soc.* 1970;66:427.
[15] Atkins PW. *Physical Chemistry. 1986.*
[16] Poling GW. *J. Colloid Interface Sci.* 1970;34:265.
[17] McIntyre J. DE, Aspnes DE. *Surf. Sci.* 1971;24:417.
[18] Greenler RG, Rahn RR, Schwartz JP. *J. Catal.* 1974;23:42.
[19] Chesters MA. *J. Electron Spectrosc. Relat. Phenom.* 1986;38:123–140.
[20] Raval R, Haq S, Blyholder G, King DA. *J. Electron Spectrosc. Relat. Phenom.* 1990;54/55:629.
[21] Raval R, Haq S, Harrison MA, Blyholder G, King DA. *Chem. Phys. Lett.* 1990;167:391.
[22] Raval R, Harrison MA, King DA. *Surf. Sci.* 1989;211/212:61–70.
[23] Chesters MA, McDougall GS, Pemble ME, Sheppard N. *Surf. Sci.* 1985;164:425.
[24] Raval R. *Surf. Sci.* 1995;331-333:1–10.
[25] Carlsson C, Liedberg B. *Mikrochimica Acta* 1988;1:149–152.
[26] Liedberg B, Carlsson C, Lundström I. *J. Colloid Interface Sci.* 1987;120:64–75.
[27] Liedberg B, Lundström I, Wu CR, Salaneck WR. *J. Colloid Interface Sci.* 1985;108:123–132.
[28] Baiker A. *J. Mol. Catal.* 1997;115:473.
[29] Webb G, Wells PB. *Catal. Today* 1992;12:319.
[30] Barlow SM, Kitching KJ, Haq S, Richardson NV. *Surf. Sci.* 1998;401:322–335.
[31] Barlow SM, Raval R. *Surf. Sci. Rep.* 2003;50:201–342.

[32] Cooper E, Krebs F, Smith M, Raval R. *J. Electron Spectrosc. Related Phenom.* 1993;64/65:469–475.
[33] Mateo Marti E, Barlow SM, Haq S, Raval R. *Surf. Sci.* 2002;501:191–202.
[34] Williams J, Haq S, Raval R. *Surf. Sci.* 1996;368:303–309.
[35] Barlow SM, Haq S, Raval R. *Langmuir* 2001;17:3292–3300.
[36] Vallée A, Humblot V, Méthivier C, Pradier C-M. *Surf. Sci.* 2008;602:2256–2263.
[37] Vallée A, Humblot V, Méthivier C, Pradier C-M. *Surf. Interface Anal.* 2008;40:395–399.
[38] Vallée A, Humblot V, Méthivier C, Pradier C-M. *J. Phys. Chem. C* 2009;113:9336–9344.
[39] Humblot V, Méthivier C, Pradier C-M. *Langmuir* 2006;22:3089–3096.
[40] Greenler RG, Snider DR, Witt D, Sorbello RS. *Surf. Sci.* 1982;118:415–428.
[41] Hayden BE, Prince K, Woodruff DP, Bradshaw AM. *Surf. Sci.* 1983;133:589–604.
[42] Humblot V, *et al.* *Surf. Sci.* 2003;537:253–264.
[43] Barth A. *Prog. Biophys. Molecular Biol.* 2000;74:141–173.
[44] Rahmelow K, Hubner W, Ackermann T. *Anal. Biochem.* 1998;257:1–11.
[45] Ortega Lorenzo M, Haq S, Murray P, Raval R, Baddeley CJ. *J. Phys. Chem. B* 1999;103:10661.
[46] Ortega Lorenzo M, Humblot V, Murray P, Baddeley CJ, Haq S, Raval R. *J. Catal.* 2002;205:123.
[47] Jones TE, Baddeley CJ, Gerbi A, Savio L, Rocca M, Vattuone L. *Langmuir* 2005;21:9468–9475.
[48] Trant, A.G.; Jones, T.E.; Noakes, T.C.Q.; Bailey, P.; Baddeley, C. *J. Surf. Sci.* 604 300–307.
[49] Jones TE, Urquhart ME, Baddeley CJ. *Surf. Sci.* 2005;587:69–77.
[50] Jones TE, Baddeley CJ. *Surf. Sci.* 2002;513:453–467.
[51] Mateo Marti E, Methivier C, Pradier CM. *Langmuir* 2004;20:10223–10230.

Chapter 2

PM-IRRAS at liquid interfaces

Daniel Blaudez, Sabine Castano and Bernard Desbat
CBMN, UMR5248 CNRS, Université Bordeaux, IPB

INTRODUCTION

Since the theoretical and experimental works of Greenler in the 1960s [1, 2] it is well known that the analysis of ultrathin films deposited on a metallic substrate by infrared reflection absorption spectroscopy (IRRAS) is readily realizable, thanks to the surface electric field enhancement. Moreover, surface electric field anisotropy gives a selection rule that allows easy determination of the orientation of the molecular groups. On the other hand, when the substrate is dielectric, surface field enhancement does not exist anymore and the surface electric field has components both on the surface plane and perpendicular to this plane. These features result in much weaker and less readable IRRAS spectra.

For the analysis of monolayers spread at the air–water interface, problems are still increased because of the very strong absorptions occurring in the liquid

Biointerface Characterization by Advanced IR Spectroscopy. DOI: 10.1016/B978-0-444-53558-0.00002-3

subphase and, especially, in the surrounding water vapor environment. Obtaining good IRRAS spectra of spread monolayers is possible but requires long acquisition duration and accurate control of vapor pressure during sample and reference (uncovered water) spectrum recordings. To overcome these problems, a differential IR reflectivity technique (PM-IRRAS) has been developed in the early 1990s. This method previously used with metallic substrates had allowed an increase in the detection of the surface absorptions by several orders of magnitude and provided in a few minutes, without purging the sample environment, very high quality spectra. Some effort has thus been made to find the best experimental conditions to apply PM-IRRAS to dielectric substrates and, in particular, to the air–water interface. Quantitative analysis and modeling of the detected signal have revealed that there is also a surface selection rule in the case of a dielectric substrate, which allows differentiation of transition moments according to their orientation with respect to the surface. The first use of PM-IRRAS for the study of a Langmuir monolayer at the air–water interface was reported in 1993 [3] and its potential for the study of biosystems has been rapidly recognized.

PRINCIPLE, THEORY AND EXPERIMENTAL SETUP

PM-IRRAS signal

Basically, PM-IRRAS combines Fourier transform mid-IR reflection spectroscopy with fast polarization modulation of the incident beam (ideally between p- and s-linear states) and with two-channel electronic and mathematical processing of the detected signal. With a metallic substrate, the obtained signal is simple and corresponds to a normalized differential reflectivity:

$$\frac{\Delta R}{R} = J_2(\phi_0)\frac{R_p - R_s}{R_p + R_s} \tag{1}$$

R_p and R_s are the polarized reflectivities of the sample and J_2 is the second-order Bessel function of the maximum dephasing ϕ_0 introduced by a photoelastic modulator (PEM).

With a dielectric substrate, the signal is much more complicated because of the significant difference between the values of the polarized reflectivities. At the detector output, the signal can be electronically split into a first part carrying only the intensity modulation at frequencies ω_i introduced by the interferometer:

$$I_+ = C_+[(R_p + R_s) \pm J_0(\phi_0)(R_p - R_s)]\, I_0(\omega_i), \tag{2}$$

and a second part

$$I_- = C_-[J_2(\phi_0)(R_p - R_s)]\, I_0(\omega_i)\cos(2\omega_m t). \tag{3}$$

Here C_+ and C_- are constants accounting for amplifications during the two-channel electronic processing, J_0 is the zero-order Bessel function and $2\omega_m$ is the fixed frequency of the polarization modulation imposed by the PEM. The sign $\pm$ depends on the direction chosen for the entrance polarizer.

After demodulation, the ratio of these two parts gives the PM-IRRAS signal:

$$S = \frac{C_{-}\,[J_2(\phi_0)(R_p - R_s)]}{C_{+}\,[(R_p + R_s) \pm J_0(\phi_0)(R_p - R_s)]}. \tag{4}$$

A dielectric substrate (such as a water subphase) has a large and frequency-dependent contribution to the signal and, consequently, comparison with the bare substrate spectrum is always necessary to extract very weak absorption bands of the deposited monolayer.

Hence, PM-IRRAS spectra of covered $S(d)$ and uncovered $S(0)$ substrate have to be recorded. When the two spectra are obtained under exactly the same experimental conditions, they superimpose quasi-perfectly and present the J_2 envelope. In order to eliminate the J_2 dependence, a final computation is made to obtain the normalized difference PM-IRRAS spectrum:

$$\frac{\Delta S}{S} = \frac{S(d) - S(0)}{S(0)}. \tag{5}$$

Signal modeling: optimum angle of incidence and surface selection rule

Optimization of the PM-IRRAS detection of the surface absorptions of an ultrathin film deposited on a dielectric surface is nontrivial. Indeed, depending on the angle of incidence of the infrared beam, R_p may be very different from R_s and equation (4) cannot be simplified as it is the case for metallic substrates. Moreover, as the surface electric field has components both in the surface plane and perpendicularly to this plane, a vibrational mode is detected, whatever the orientation of its associated transition moment.

Optimum angle of incidence

Using a general software program developed for modeling the reflectivity of absorbing layered systems, the $\frac{\Delta S}{S}$ response of an uniaxial monolayer deposited on a semi-infinite substrate (with the refractive index $n_s = 1.3$) is calculated. The monolayer has two absorption bands: one with its transition moment parallel (x, y) to the surface and the other with its transition moment perpendicular (z) to the surface. Calculated behavior patterns of $\frac{\Delta S}{S}$ versus angle of incidence are presented in Figure 1.

The results in Figure 1 show that $\frac{\Delta S}{S}$ values corresponding to (x, y)- and (z)-oriented transition moments have opposite signs, which switch when the incidence passes through the Brewster angle θ_B. Consequently, (x, y)- and (z)-oriented absorptions will always have opposite signs on a PM-RRAS spectrum. Additionally, as $\left|\frac{\Delta S}{S}\right|$ is larger for (x, y)-oriented moments than for (z)-oriented moments (in a ratio of ca. 5:3), detection of the in-plane monolayer

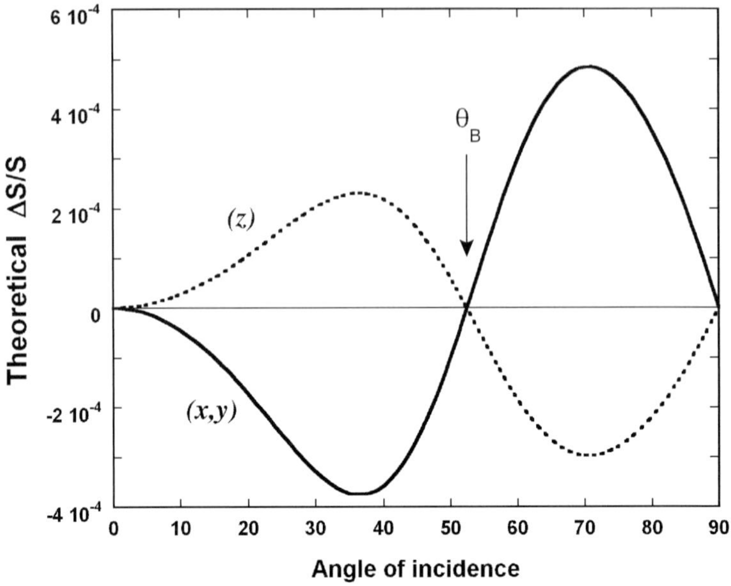

FIGURE 1 **Behavior versus angle of incidence of the theoretical PM-IRRAS $\frac{\Delta S}{S}$ signal from a uniaxial monolayer having an absorption band with its transition moment oriented along the surface (full line) or perpendicular to the surface (dotted line). Refractive index of the substrate is $n_s = 1.3$.**

absorptions is more efficient. Figure 1 also shows that there are two angles of incidence (36° and 71°) that maximize the signal, whatever the orientation of the transition moment. Although these two maxima have theoretically almost the same values, it is found experimentally that the second maximum at 71° gives a much better signal-to-noise (S/N) ratio (10) than the first one at 36° (S/N = 2.4). This difference is due to the fact that the polarized reflectivities R_p and R_s have much weaker values at 36°. This makes the denominator in equation (4) very small and, consequently, $\frac{\Delta S}{S}$ much more sensitive to noise.

The optimal value of the angle of incidence exclusively depends on the substrate refractive index n_S. It shifts to higher values when n_s increases. For usual dielectric substrates (n_s values between 1.3 and 1.7), the angle of incidence has to be set between 70° and 75°. When using intrinsic semiconductors (e.g., silicon or germanium) for which the Brewster angle is close to 75°, the second maximum corresponds to a grazing incidence (about 85°) and is not appropriate because the convergent beam has a very large angular distribution and it spreads on the sample over a very large area. It is preferable to set the angle of incidence at the first maximum value (about 55°).

When monolayer absorptions are very weak, Saccani *et al.* [4] have shown that the sensitivity of PM-IRRAS can be significantly increased by a very simple way. The enhancement of the absorption band intensity is obtained by reducing the signal arising from the substrate. This is achieved by adding a polarizer or a tilted plate after the sample in order to balance the polarized reflectivities of the sample. As a consequence, the contribution of the ultrathin film to the PM-IRRAS signal is increased relative to that of the substrate. An enhancement factor of about 10 has been obtained on monolayer films spread at the air–water interface.

Surface selection rule

The up (down) sense of a band relative to the baseline and its intensity are governed by a surface selection rule dependent on the orientation of the transition moment. Setting the angle of incidence at the second maximum, a transition moment lying in the surface plane gives rise to an absorption band oriented positively with respect to the baseline while a transition moment perpendicular to the surface gives rise to an absorption band oriented negatively. Between these two extreme cases, a general orientation of the transition moment leads to competitive positive and negative contributions. In Figure 2 is presented the calculated PM-IRRAS spectrum of an absorption band with an orientation of the transition moment varying from parallel ($\theta = 0^\circ$) to perpendicular ($\theta = 90^\circ$) to the surface normal. The substrate is a water subphase in the present case.

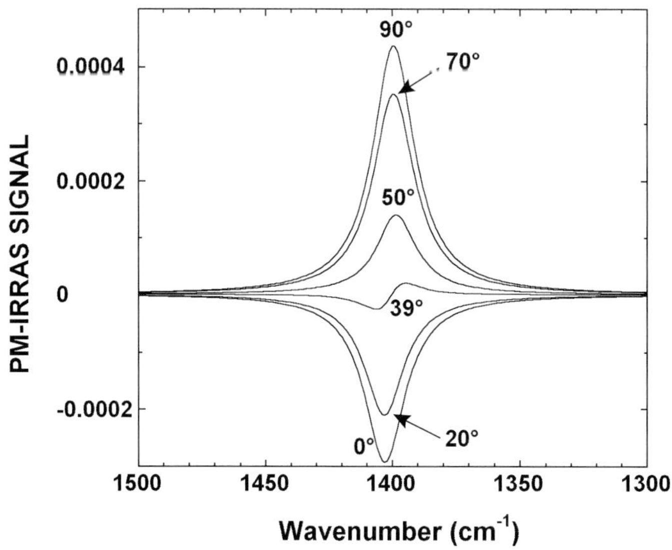

FIGURE 2 **Calculated PM-IRRAS spectrum of an absorption band located at 1400 cm^{-1} with an orientation of the transition moment varying from parallel ($\theta = 0^\circ$) to perpendicular ($\theta = 90^\circ$) to the surface normal.**

The absorption band is negative when the transition moment is along the surface normal ($\theta = 0°$) and decreases in intensity as θ increases. At $\theta = 39°$, the positive and negative contributions balance but do not perfectly cancel because of the dispersion effect associated with the monolayer absorption (TO–LO (transverse optic–longitudinal optic) splitting, see below). This particular angle, which depends on the substrate refractive index is referred to as the vanishing angle. Beyond $\theta = 39°$, the positive contribution becomes predominant and increases until reaching a maximum when the transition moment is in the surface plane ($\theta = 90°$).

Figure 3 presents the value of the vanishing angle as a function of the substrate refractive index n_s, in a range covering the main dielectric materials used as IR substrates. Not surprisingly, the vanishing angle increases regularly as n_s increases because the perpendicular surface component of the incident electric field becomes more and more predominant.

To sum up, the intensity of an absorption band does not depend only on the oscillator strength of its transition moment, but also on its orientation. This last dependence is complex and spectral simulation is necessary to draw quantitative information on molecular orientation in ultrathin films deposited on dielectric substrates.

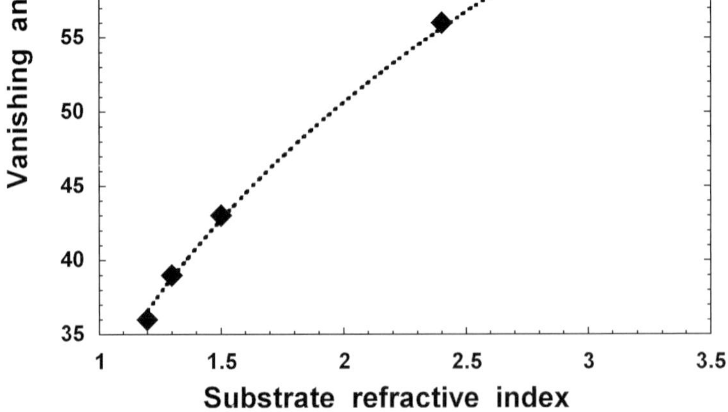

FIGURE 3 **Calculated behavior of the vanishing angle versus n_s.**

Consequence of the TO–LO splitting on the PM-IRRAS bandshape

In specific cases, strong polar vibrational modes present a derivative-like shape in PM-IRRAS spectra. This can be understood considering the infrared optics of condensed layers observed under grazing incidence and the TO–LO splitting of the IR modes [5–9]. Depending on the infrared beam polarization, the electric field generated at a dielectric interface reveals the TO component or the LO component of a polar vibrational mode. The s-polarization produces an electric field in the interface plane (y direction) and only the TO component can be observed. The p-polarization produces an electric field in the plane (x direction) which not only excites the TO component, but also generates a perpendicular electric field (z direction) in which intensity increases with the angle of incidence, and which yields the LO component. Except for some very special cases, the LO component maximum (LO mode) has a higher wavenumber than the TO component maximum (TO mode). The wavenumber difference between the two modes is directly proportional to the intensity of the vibration mode. The difference can be very strong for ionic or polar compounds (e.g., 150 cm^{-1} for SiO_2 stretching), very small for weak absorptions (about 0.2 cm^{-1} for CH stretching) and medium for strong bands in organic compounds (e.g., 5 to 12 cm^{-1} for the Amide I band of proteins).

According to the PM-IRRAS surface selection rule, the LO mode in the normal direction (z) gives a negative band and the TO mode in the (x, y) plane provides a positive band. The sum of the two bands produces a band with a derivative-like shape because wavenumbers of the two modes do not coincide. It is also easy to demonstrate that the negative minimum and the positive maximum of the resulting band do not coincide exactly with LO and TO mode positions but are shifted at higher and lower frequencies, respectively. Such a derivative-like shape has been observed in several studies on the Amide I band [10], on the carboxylate symmetric stretching [11] (see Fig. 6), and on the phosphate asymmetric stretching [12] as shown in Figure 4.

Finally, it is important to note that this TO–LO splitting is also present in conventional IRRAS spectra of monolayers deposited on dielectric substrates when spectra are registered with a p-polarized electric field. On metallic substrates, the infrared beam produces an electric field perpendicular to the surface which only probes the LO component. As a consequence, the Amide I band of proteins detected with IRRAS or PM-IRRAS on metals are always located at higher frequency than the band obtained for the same protein with a direct absorption technique [13]. This has to be kept in mind when the Amide I band position is used to determine the secondary structure of a given protein.

Experimental setup

The optical setup is based on a commercial FTIR spectrometer with two-channel real time processing capability (Fig. 5). At the output of the interferometer, the parallel beam (modulated in intensity at frequencies ω_i) is deflected out of the

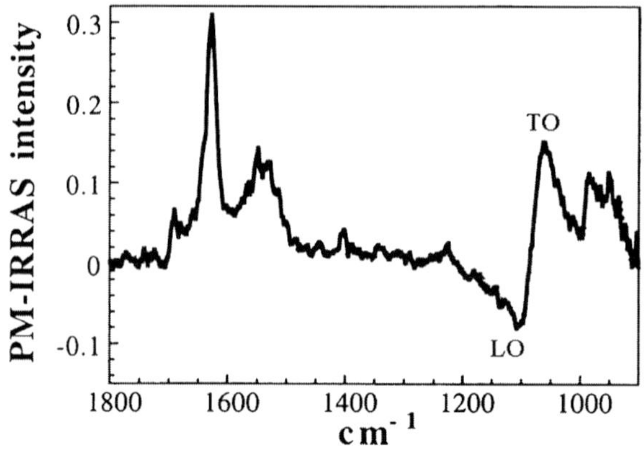

FIGURE 4 PM-IRRAS spectrum of a mixed DPPC/DPPS Langmuir monolayer in the presence of penetratin peptide. The phosphate asymmetric stretching band located around 1070 cm^{-1} presents a TO–LO splitting.

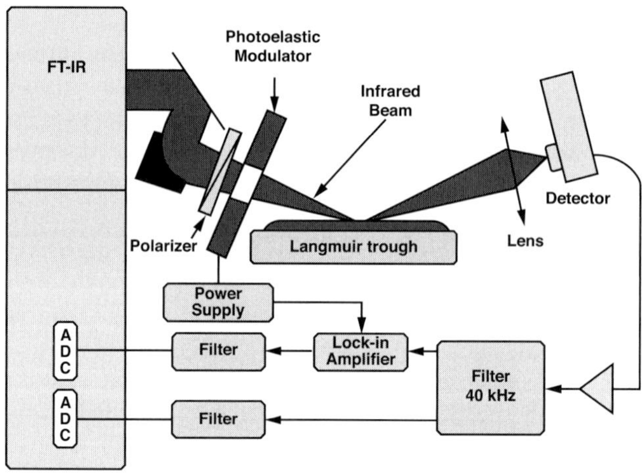

FIGURE 5 Schematic of the PM-IRRAS setup at the air–water interface.

spectrometer box and focused on the sample, after passing through a linear polarizer and a ZnSe PEM which superimposes a polarization modulation at a fixed frequency $2\omega_m$ (this frequency usually ranges from 50 to 100 kHz). After reflection on the sample at an optimum angle of incidence for the used substrate, the light beam is focused on a sensitive photoconductor or photovoltaic HgCdTe detector.

The two-channel PM-IRRAS processing relies on the existence of a difference of at least one order of magnitude between the frequency of the polarization modulation and those of the intensity modulation, which are typically in the 0.1–3 kHz range for the mid-IR spectral domain. Using electronic filtering, it is very easy to split the component of the detected signal having double modulation from that which has simply an intensity modulation. Then, after demodulation of the doubly modulated component by a lock-in amplifier having a very low time constant (0.1 ms), one obtains two interferograms, which are processed and Fourier transformed as usual in FTIR spectroscopy. Finally, a ratio is found between the two components in order to achieve the PM-IRRAS signal.

It is worth noting that owing to the fast polarization modulation and to the identical and precise real time acquisition and processing of the two interferograms, this ratio is, in principle, extracted from all the polarization-independent signals such as strong water vapor absorptions and from all the instrumental drifts and fluctuations. Moreover, the wavelength dependence of the polarization modulation imposes on the PM-IRRAS signal a wavelength-dependent efficiency, which depends both on the maximum position function of the modulator setting and on the zero- and second-order Bessel functions. Generally, spectra are recorded with a spectral resolution of 8 cm^{-1} and by co-addition of 600 scans, which corresponds to an acquisition time of ca. 10 min.

APPLICATIONS

Lipids

Fatty acids, interfacial pH

The first example presented in this section deals with the fundamental issue of the effect of cations on the dissociation of an acidic Langmuir monolayer and, following, the supposed significant difference between the pH at the interface with respect to the subphase pH. Using PM-IRRAS, it is possible to study quantitatively the dissociation of a fatty acid Langmuir monolayer as a function of the subphase pH by measuring the integrated intensities of the C=O and COO^- stretching vibration bands at 1700 and 1530 cm^{-1}, respectively.

Figure 6 shows the PM-IRRAS spectra of the monolayer recorded along a pH range covering the entire dissociation process. The spectral shift between the two absorption bands allows unambiguous discrimination between acid and salt species and thus, accurate determination of the ionized fraction of molecules (Fig. 7).

The balance between the acid and salt species in the monolayer is obtained for a pH subphase (half-neutral pH), which is different from the arachidic acid pK_a and strongly varies with the counterion. Comparing the experimental results with the purely electrostatic Gouy–Chapman (GC) theory, specific behavior of each counterion can be described (see Table 1). These results have been discussed elsewhere in details [11]. We can just note that the cations, which reduce the most the apparent pK_a of arachidic acid (i.e., Cd^{2+} and Al^{3+}) are also the most toxic for biological membranes.

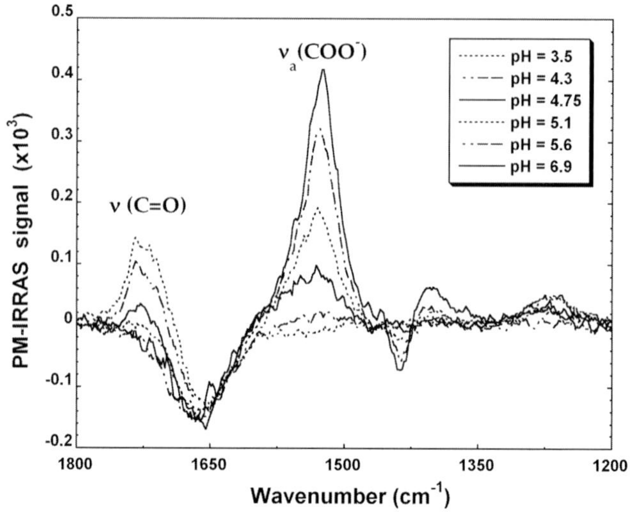

FIGURE 6 **PM-IRRAS spectrum of an arachidic acid monolayer spread onto a water subphase containing $CdCl_2$ as a function of the subphase pH.**

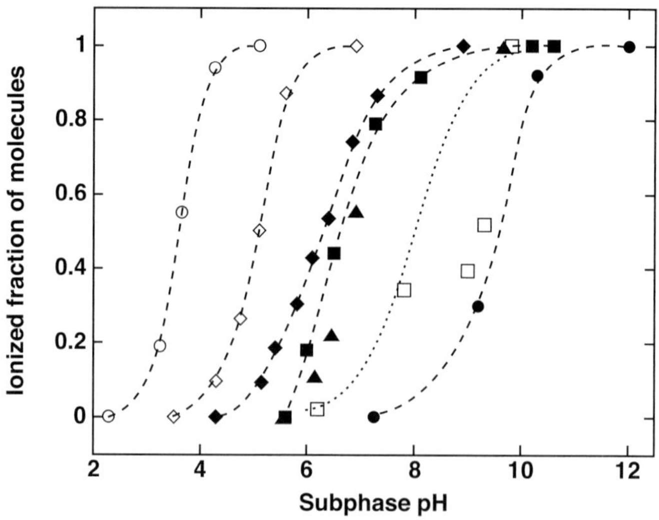

FIGURE 7 **Ionized fraction of molecules as a function of the subphase pH for several countercations (Al^{+++}(○), Cd^{++}(◇), Ca^{++}(♦), Mg^{++}(▲), Li^{+}(■), Na^{+} (□) and $(CH_3)_4N^{+}$(●)).**

TABLE 1 Theoretical and experimental values of the subphase pH at half-neutrality for various counterions.

Half-neutral pH	$(CH_3)_4N^+$	Na^+	Li^+	Mg^{2+}	Ca^{2+}	Cd^{2+}	Al^{3+}
GC theory	9.25	9.25	9.25	7.68	7.67	7.65	7.38
Experimental	9.6	9.2	6.6	6.85	6.3	5.1	3.6

Bi and multilayers at the air–water interface

Compression beyond the collapse of phospholipid monolayers on a modified Langmuir trough results in the formation of three-dimensional structures. It is important to determine if these structures are composed of stable multilayers or other more complex organized systems (such as aggregates or micelles). Indeed, stable multilayers are relevant new models for studying the properties of biological membranes and for understanding the nature of interactions between membranes and peptides or proteins. To address this issue, Saccani *et al.*[14] have studied the collapse of 1,2-dimyristoyl-*sn*-glycero-3-phosphocholine (DMPC), 1,2-di[*cis*-9-octadecenoyl]-*sn*-glycero-3-[phospho-L-serine] (DOPS), and 1,2-di[*cis*-9-octadecenoyl]-*sn*glycero-3-phosphocholine (DOPC) monolayers using complementary techniques: isotherm measurements, Brewster angle microscopy (BAM), and PM-IRRAS.

The PM-IRRAS spectra of DMPC at the air–water interface are presented in Figure 8. After compression beyond the collapse, the intensities of the asymmetric ($\nu_{as}CH_2$) and symmetric ($\nu_s CH_2$) stretching bands, located at 2920 and 2850 cm^{-1}, respectively, are multiplied by a factor of 1.6 with respect to the intensities measured on the monolayer spectrum. This value, slightly inferior to 2, shows that the trough surface is not totally covered by a bilayer and that monolayer domains subsist. Furthermore, the PM-IRRAS results are in agreement with the BAM results and mean that fluid DMPC bilayers are partially formed at high surface pressure. Moreover, the small shift of the CH_2 bands to low frequencies for the system beyond the collapse indicates that the DMPC bilayer is slightly more rigid than the monolayer.

The main difference between the two infrared spectra at the air–water interface, before and after collapse, comes from the apparition at high pressure of a well-defined positive band around 2950 cm^{-1} corresponding to the asymmetric methyl stretching vibration ($\nu_{as}CH_3$) and a negative $\nu_s CH_3$ band located at 2870 cm^{-1} characteristic of the symmetric methyl stretching vibration. According to the PM-IRRAS selection rule the positive and negative signs of these two bands, respectively, suggest a methyl axis orientation more vertical in the bilayer than in the monolayer. These additional interactions at the methyl extremities suggest a model where the alkyl chains of each layer in the bilayer

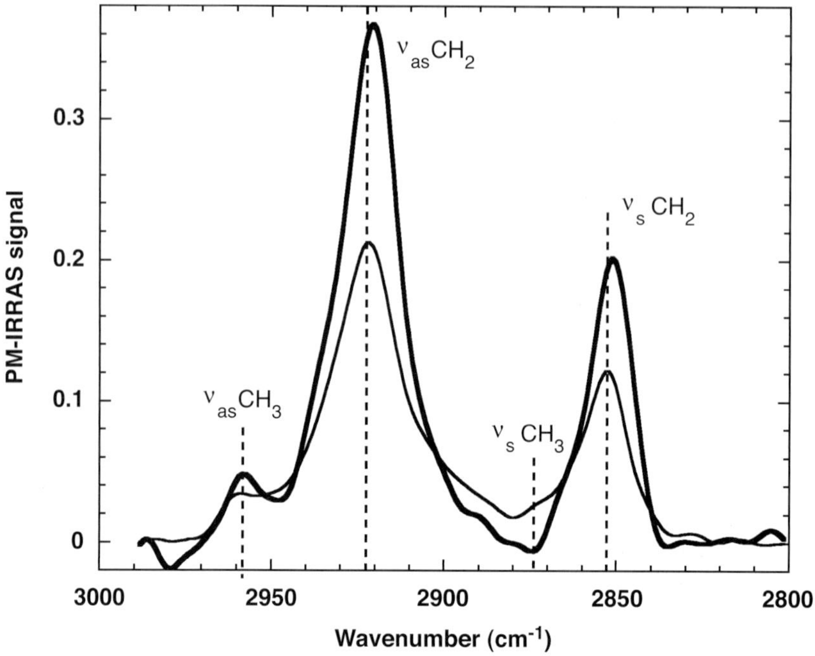

FIGURE 8 PM-IRRAS spectrum of a DMPC monolayer before collapse (30 mN/m) (thin solid line), and a DMPC monolayer compressed beyond the collapse (51 mN/m) (thick solid line).

are in contact. This tail-to-tail contact organization seems to be more stable than the organization where the polar headgroups of the top layer are in interaction with the alkyl chains of the bottom layer. Obviously, the tail-to-tail contact organization supposes that a hydration layer exists at the polar head level to minimize repulsion between the polar heads of the top layer and the air. This hypothesis was confirmed by molecular dynamics simulations [14].

In the same article, the authors have clearly shown that DOPS (a charged phospholipid) stabilizes in complete three-layer systems. In this case, a monolayer in contact with water is topped by a bilayer with a head-to-head contact organization.

Peptides/interaction peptide membrane

Antibacterial LK peptides

Currently a great deal of interest is shown in antibacterial peptides, so-called nature's antibiotics, which seem to be promising to overcome the growing problem of antibiotic resistance. The design of novel peptides with antimicrobial activities requires the understanding of their action mechanism on membranes. Most of the time such peptides are amphipathic and they have a positive net charge

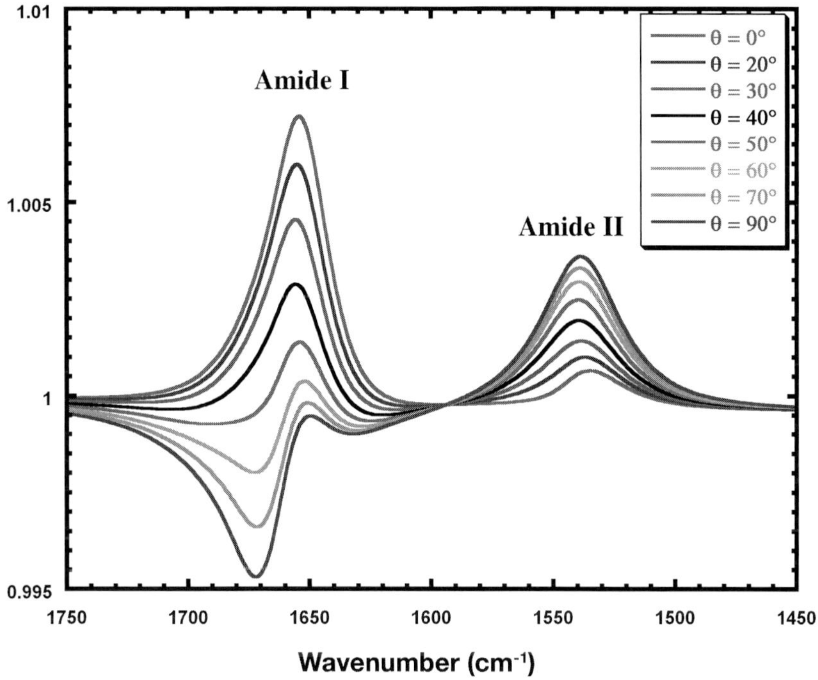

FIGURE 9 **Calculated PM-IRRAS spectrum in the Amide I and Amide II spectral region of an α-helix as a function of the helix axis orientation with respect to the surface plane.**

given mainly by the lysine residue. The helix structure seems to be the main folding adopted but it is not a general rule and the repartition of the charged group in the hydrophobic residues is variable. Therefore, many studies have been conducted, among others by PM-IRRAS, on lipid monolayers to determine the orientation, the secondary structure and the affinity for the interface of these peptides. To better monitor the behavior of these peptides a preliminary study was conducted to determine the shape of the PM-IRRAS spectra following the orientation, on the water surface, of the two main secondary structures, that is, α-helix and antiparallel β-sheet [15]. Indeed, the anisotropic absorption of these structures cannot be assigned to only one direction for the transition moment, and each amide I and amide II bands should be defined in the three space directions x, y, z [16]. The calculation is simpler for the α-helix because of its higher symmetry and therefore only the tilt angle θ of the helix with respect to the surface plane has to be varied. The data presented in Figure 9 show that the PM-IRRAS spectrum is very sensitive to the helix axis orientation with respect to the surface.

When the helix lays flat at the surface ($\theta = 0°$), a strong positive Amide I band (TO component) and a weak positive Amide II band are expected. When the helix is perpendicular to the water surface ($\theta = 90°$), a strong negative

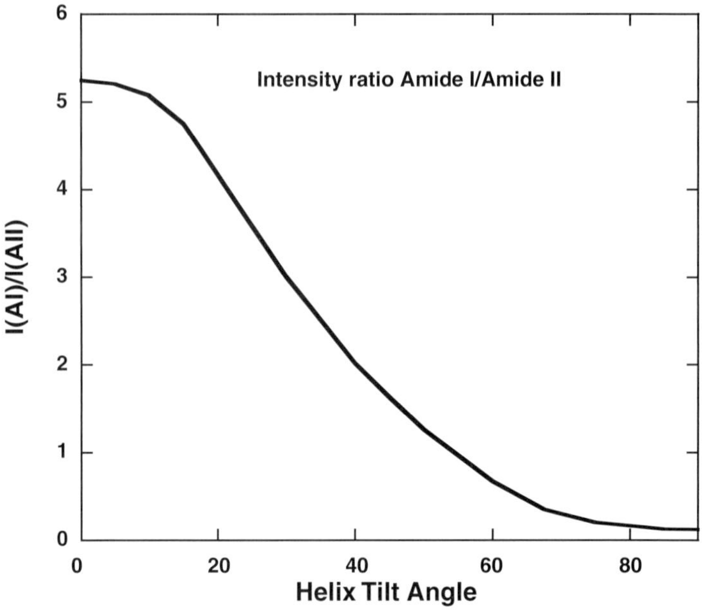

FIGURE 10 **Calculated intensity ratio between the Amide I and Amide II absorption bands in a PM-IRRAS spectrum as a function of the helix tilt angle with respect to the surface plane. The calculation of the ratio has been done using the integrated intensity of the positive component of the Amide I band and the integrated intensity of the Amide II band.**

Amide I, shifted at higher frequency (LO component) and a strong positive Amide II band are expected. For intermediate values of the tilt angle the Amide I band shows a derivative-like shape because the projection of the Amide I transition moment along the surface normal gives a LO component at high frequency, whereas the projection of the transition moment in the surface plane gives a TO component at low frequency. These shape modifications show that the ratio intensity of the Amide I and Amide II bands in a PM-IRRAS spectrum is a very sensitive parameter to follow the α-helix orientation (see Fig. 10) [17].

The antiparallel β-sheet secondary structure is more complicated to describe because it does not own an axial revolution symmetry. The spectral simulation should be done taking into account two rotation angles: the angle between the direction of the peptide chains and the normal surface (angle θ) and the rotation angle of the β-sheet plane around its revolution axis (angle ψ). The calculated PM-IRRAS spectra of the most characteristic orientations are reported in Figure 11.

These few examples show the variety of shapes, which can be encountered on a PM-IRRAS spectrum of a β-sheet according to its orientation. However, some "basic" orientations can be recognized rather easily.

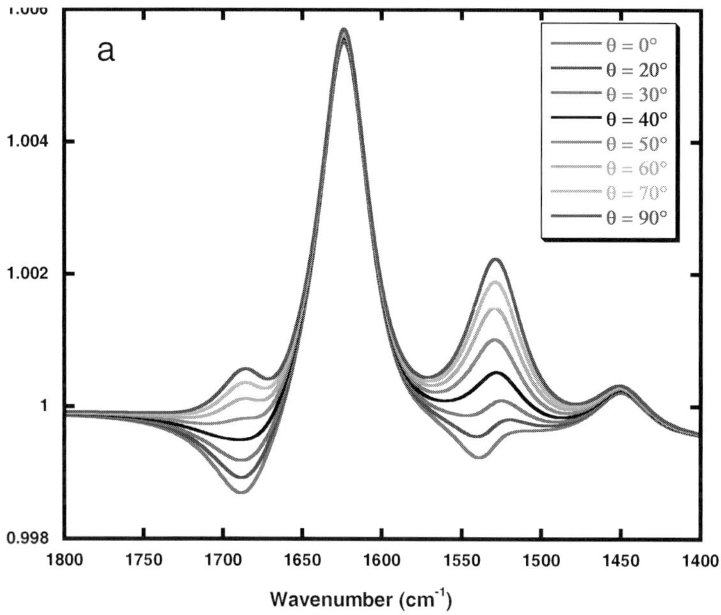

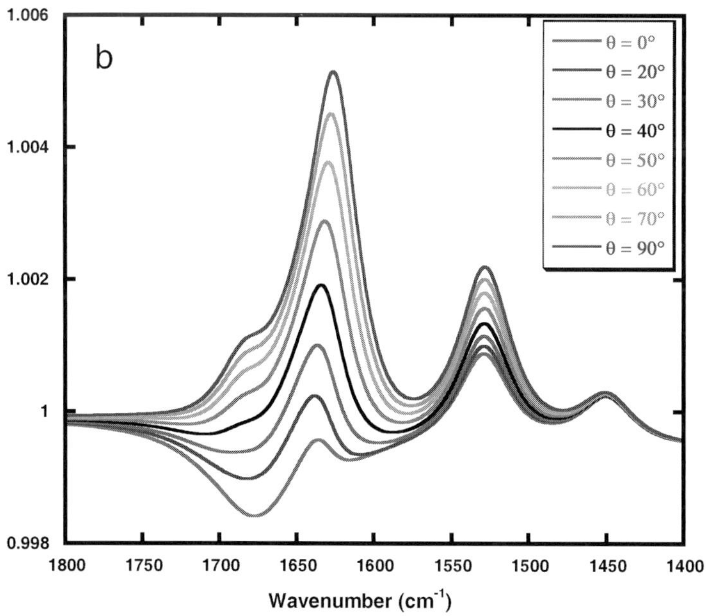

FIGURE 11 **Calculated PM-IRRAS spectrum in the Amide I and Amide II spectral region of an antiparallel β-sheet as a function of the rotation angles θ and ψ.**

(a) $\psi = 0°$ and θ varies from 0° to 90°.
(b) $\psi = 45°$ and θ varies from 0° to 90°.
(c) $\psi = 90°$ and θ varies from 0° to 90°.

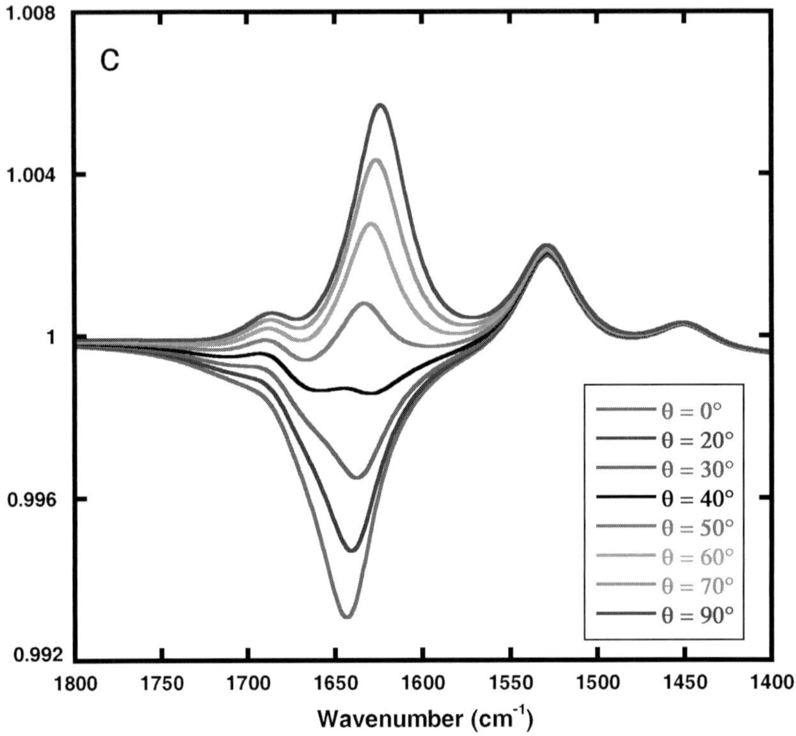

FIGURE 11 *(Continued)*

- When the β-sheet lays flat at the interface ($\psi = 0°$ and $\theta = 90°$), the spectrum presents positive Amide I and II bands with relative intensities identical to that measured on an isotropic absorbance spectrum.
- When the β-sheet plane is perpendicular to the interface with the carbonyl parallel to the interface plane ($\psi = 0°$ and $\theta = 0°$), the spectrum presents a strong positive Amide I band (≈1625 cm^{-1}) and weak negative components for the Amide I' (≈1690 cm^{-1}) and Amide II (≈1540 cm^{-1}) bands.
- When the β-sheet plane is perpendicular to the interface with the carbonyl perpendicular to the interface plane ($\psi = 90°$ and $\theta = 0°$), the spectrum presents a strong negative Amide I band shifted at higher frequency (≈1640 cm^{-1}) and two positive Amide I' and Amide II bands.

On the basis of these calculations, secondary structure and orientation of linear synthetic amphipathic L_iK_j peptides were investigated *in situ* by PM-IRRAS spectroscopy either for the peptides alone at the air–water interface or for the peptides inserted in a phospholipid monolayer [18, 19]. In brief, for the $L_iK_{j(i=2j)}$ series designed with a 2/1 L/K ratio to generate ideally amphipathic putative α-helices, PM-IRRAS spectra of pure peptide films at the interface or inserted into a DMPC monolayer clearly showed structural changes within the peptide series. The peptides shorter than 12 residues adopt a quasi pure β-sheeted secondary structure oriented mostly parallel to the interface. The longer peptides ($n \geq 15$) were fully in α-helix and oriented in the plane of the interface. The 12-residue peptide gives a more flexible molecule which adopts a β-sheet structure at the interface and an α-helix structure inserted in a DMPC monolayer. The peptides of the $(KL)_mK$ series, designed with a strict twofold periodicity in the sequence to generate ideally amphipathic β-strands, form films at the air–water interface and their *in situ* PM-IRRAS spectra display similar shapes, whatever the environment (pure peptide film or peptides inserted into a compressed DMPC monolayer). As expected, they fold into antiparallel β-sheets oriented mostly parallel to the interface. These works have been completed by the study of another type of LK peptides with primary amphipathy, that is, where the hydrophobic and hydrophilic residues were separated in the sequence, giving L_iK_j or K_iL_j compounds. These peptides were named diblock peptides. We particularly studied the $L_{10}K_5$ and K_5L_{10} peptides, which have inverse sequences. Each peptide shows the same aptitude to be folded in α-helix parallel to the interface whatever the surface pressure. On the other hand, a monolayer composed of $L_{10}K_5$ and K_5L_{10} in a 1/1 ratio shows a completely different behavior. The PM-IRRAS spectra show that α-helix is still the predominant secondary structure but the orientation strongly varies as the surface pressure π increases (Fig. 12). When $\pi = 22$ mN/m, spectral calculations show that the experimental Amide I/Amide II intensity ratio corresponds to a helix axis parallel to the interface. As π increases, the ratio decreases indicating that the helix progressively straightens up. At the highest surface pressure, the tilt angle with respect to the surface plane can be estimated to about 50°.

Inserted in a DMPC monolayer at the air–water interface, the diblock $L_{10}K_5$ and K_5L_{10} peptides also show different behavior if they are present as a mixture or not (Fig. 13). The PM-IRRAS spectrum of one diblock in interaction with the phospholipid monolayer shows a broad Amide I band centred at 1658 cm^{-1} and a very weak Amide II band (top curve in Fig. 13). These observations indicate that the peptide is essentially present as α-helices lying flat at the interface. With a 1/1 mixture, the amide bands are very sharp and well defined (bottom curve in Fig. 13). The Amide I band located at 1654 cm^{-1} shows a pure α-helix secondary structure and the Amide I/Amide II intensity ratio, close to 1.25, indicates a helix axis average tilt angle of 50° with respect to the normal to the interface.

These results show the PM-IRRAS possibilities to determine easily both the secondary structure and the orientation of peptide in interaction with a lipid monolayer.

Gramicidin

Many PM-IRRAS experiments have been performed on natural antibacterial compounds. In 1999, Ulrich *et al.* [20] have applied the method to gramicidin A,

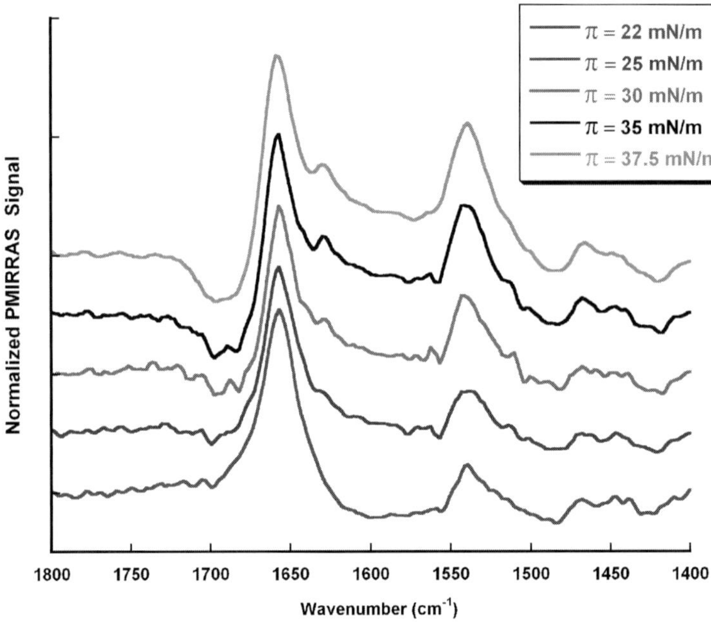

FIGURE 12 PM-IRRAS spectrum of a $L_{10}K_5$ / K_5L_{10} mixed monolayer (1/1 ratio) at the air–water interface, as a function of the surface pressure, 22mN/m bottom curve, 37.5 mN/m top curve.

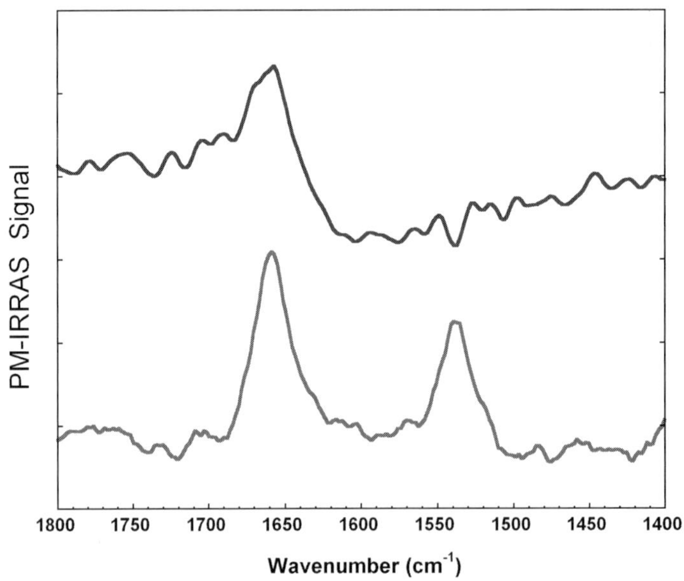

FIGURE 13 PM-IRRAS spectrum of a K_5L_{10} / DMPC monolayer (top curve) and of a $L_{10}K_5$ / K_5L_{10} / DMPC monolayer (bottom curve) at the air–water interface.

a natural antibacterial that was the first manufactured antibiotic to be commercialized. This peptide is formed of alternate L and D residues, which gives it the possibility of adopting multiple and original single- and double-stranded secondary structures. Monolayers of gramicidin A, pure and in mixtures with DMPC, were studied *in situ* at the air/H_2O and air/D_2O interfaces. Simulations of the entire set of Amide I absorption modes were also performed, using complete parameter sets for different conformations based on published normal mode calculations. The authors deduced that the secondary structure of gramicidin A in the pure monolayer and in a DMPC monolayer could be assigned to a β-6.3 double helix. A marked dependence of the peptide orientation on the applied surface pressure was observed for the mixed lipid–peptide monolayer and at low pressure the helix lays flat on the surface, whereas at high pressures the helix was oriented almost perpendicularly to the surface. In 2002, a complementary work has been performed by Lavoie *et al.* [21] on pure monolayer of valine gramicidin A. They also found the propensity of this peptide to be structured in α-helix, but their infrared simulations suggest a β-5.6 double helix almost vertical at high surface pressure, in agreement with the thickness found for the monolayer by X-ray reflectivity. In addition, the comparison between PM-IRRAS spectra recorded on H_2O and D_2O subphases and calculated spectra, and X-ray reflectivity data analysis show that a structural modification of the interfacial water occurs when the monolayer of valine gramicidin A is present.

PROTEINS/INTERACTION PROTEIN MEMBRANE

Membraneous proteins

PS II core complex and rhodopsin

Stabilization and crystallization of membrane proteins is an important issue in view to determine their atomic structure by X-ray diffraction. Among the methods proposed to obtain protein crystals, one used the possibility of forming an organized protein monolayer on an aqueous buffer surface. However, it soon became apparent that application of this method was only possible if one could control the state of the protein and also determine the parameters which allow us to maintain stable its secondary structure. For this kind of study PM-IRRAS constitutes a relevant tool as it allows us to give accurate information on the secondary structure within a relatively short time scale (a few minutes or even less) thanks to its differential nature.

A few years ago, two PM-IRRAS studies have been achieved on membraneous proteins, photosystem II core complex (PS II CC) and rhodopsin [22, 23]. In the first one, the state of PS II CC in monolayer at the gas (nitrogen)–water interface was investigated using PM-IRRAS and X-ray reflectivity techniques. Two approaches for preparing and manipulating the monolayers were examined and compared. First, PS II CC molecules were compressed immediately after spreading at the interface at an initial surface pressure of 5.7 mN/m. In the second approach, the monolayer was incubated for 30 min at a surface pressure of 0.6 mN/m before compression.

In the first approach, the protein complex maintained its native α-helical conformation upon compression, and the secondary structure of PS II CC was found to be stable for 2 h (Fig. 14).

The second approach resulted in films showing stable surface pressure below 30 mN/m and the presence of large amounts of ß-sheets, which indicated denaturation of PS II CC (Fig. 15). Above 30 mN/m, those films suffered surface pressure instability, which had to be compensated by continuous compression. This instability was correlated with the refolding of new α-helices in the film.

To probe the process of denaturation when PS II CC was spread at low surface pressure, PM-IRRAS spectra were measured immediately after spreading and compared with the one obtained after a 30 min incubation time at a surface pressure of 0.6 mN/m (Fig. 16). To obtain a band position that would represent the initial state of PS II CC right after spreading, the number of scans was reduced to 20, which corresponded to a measuring time of 1 min. The spectrum obtained

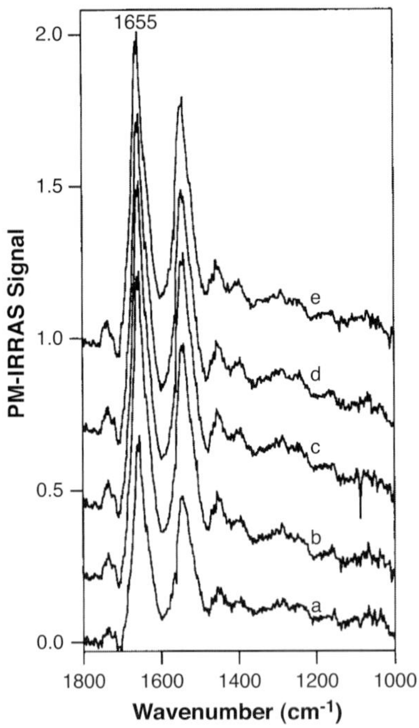

FIGURE 14 **PM-IRRAS spectra of PS II CC monolayers at the nitrogen–water interface. The initial surface pressure was set at 5.7 mN/m. Compression was started immediately after spreading of PS II CC. Surface pressure was maintained at 20 mN/m. Spectra were measured at 15 min (a), 30 min (b), 45 min (c), 75 min (d), and 120 min (e) after reaching 20 mN/m.**

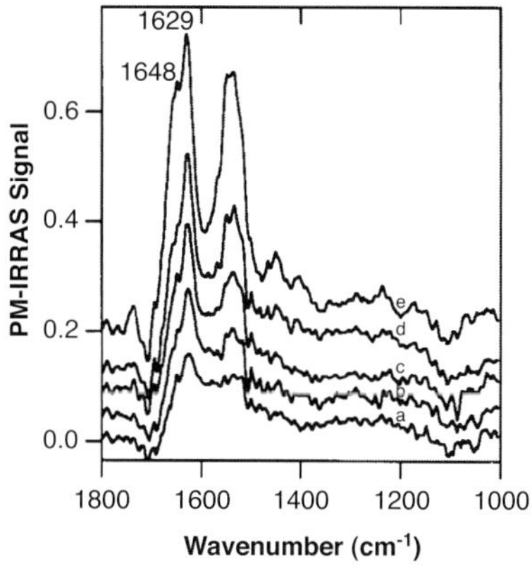

FIGURE 15 PM-IRRAS spectra of PS II CC monolayers at the nitrogen–water interface. Initial surface pressure was set at 0.6 mN/m. Compression was started 30 min after spreading of PS II CC. Spectra were measured at 0.6 mN/m (a), 10 mN/m (b), 20 mN/m (c), 30 mN/m (d), and 40 mN/m (e).

(spectrum *a*, Fig. 16), although noisy, shows a distinct Amide I band maximum at approximately 1654 cm^{-1}, indicating that 1 min after spreading, the protein complex still retained its α-helical structure. However, after 30 min of incubation at 0.6 mN/m (spectrum *b*, Fig. 16), the maximum was shifted to 1631 cm^{-1}, the characteristic band position of β-sheets. This indicates that the incubation time at low surface pressures is a critical factor in the denaturation of proteins at the interface as is the value of the initial spreading surface pressure.

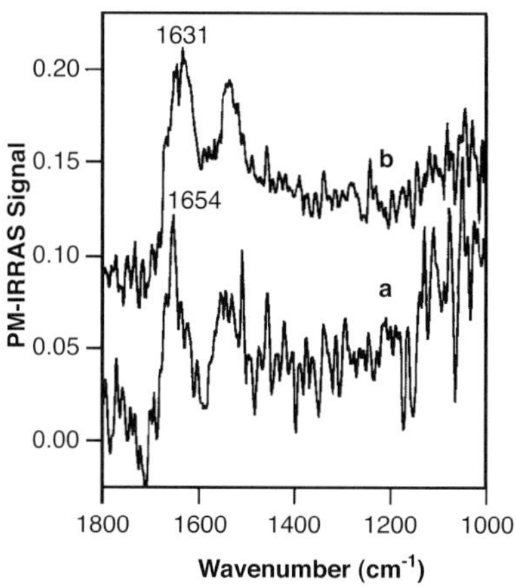

FIGURE 16 PM-IRRAS spectra of PS II CC monolayers at the nitrogen–water interface. Surface pressure was set at 0.6 mN/m and the film was not compressed. Spectra were recorded (a) immediately after spreading (1 min acquisition time) and (b) 30 min after spreading.

In the case of rhodopsin (a member of the large family of G-protein coupled receptors), PM-IRRAS has evidenced that its secondary structure can be retained at the gas–water interface only when the experiment is performed at low temperature (4°C).

For the two proteins, PM-IRRAS experiments have also shown that the Langmuir method allows purification during the compression. Indeed, the authors observed that the octaglucosyl surfactant necessary for protein extraction dissolved in the buffer subphase during the compression. Altogether, these studies provide direct structural and molecular information on membrane proteins when spread in monolayers at the gas–water interface.

Other systems

Molecular antibiotic

The topic of antibiotic systems has been also explored by Shanmuk *et al.* [24] by crossing the two-dimensional (2D) correlation method with PM-IRRAS. The technique was applied to the 1,2-dipalmitoyl-sn-glycero-3-phosphatidic acid (DPPA)/tetracycline hydrochloride (TC) system to gain a better understanding of the surface pressure-induced effects on the interaction between the phospholipid and the antibiotic at the air–water interface.

Statistical correlation spectroscopy using 2D IR, $\beta(v)$ (asynchronous cross-correlation) and $k(v)$ (synchronous cross-correlation) correlation analyses were performed on six spectra recorded at 3, 5, 10, 20, 30, and 40 mN/m. Conventional 2D IR correlation maps revealed strong correlation behavior between the vibrational modes of the lipid and the antibiotic. $\beta(v)$ correlation plots provided information about the relative rates of occurrence of the coupled responses noted in the conventional 2D IR plots and were able to discriminate the behavior of the four TC cycles (A,B,C,D), Scheme 1. These calculations indicated that molecular reorientation occurs at lower surface pressures for the modes in Ring A than for the modes in Ring C.

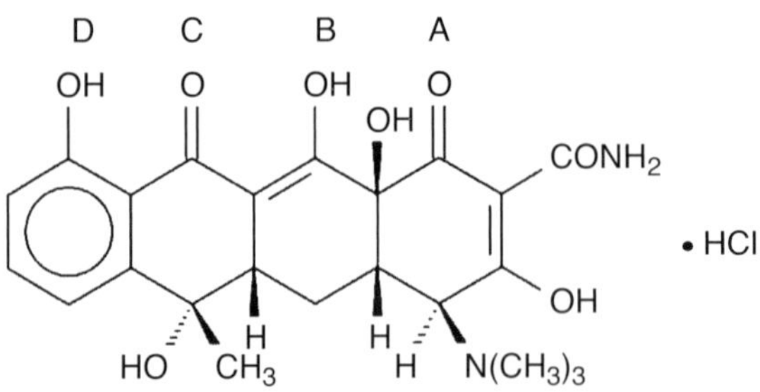

SCHEME 1 Indexation of tetracycline cycles.

A new model-dependent 2D correlation method, exponential *k(v)* correlation analysis, confirmed the results obtained with the previous correlation methods, that is, the lipid–antibiotic interactions occurred in a bimodal fashion, depending upon surface pressure. Initial interaction between the tetracycline molecule and the DPPA molecule occurs at low surface pressures, primarily between Ring A of the tetracycline molecule and the lipid headgroup region. However, with increasing surface pressure the mode of interaction changes and the strongest interaction at high surface pressures occurs between Ring C of tetracycline and the DPPA headgroup. This work clearly shows that PM-IRRAS efficiency can be improved by carrying out statistical calculation on data.

DNA–cationic lipids interactions

The lipid bis(guanidinium)-tris(2-aminoethyl)amine-cholesterol (BGTC) is a cationic cholesterol derivative bearing guanidinium polar headgroups which displays high transfection efficiency *in vitro* and *in vivo* when used alone or formulated as liposomes with the neutral colipid 1,2-di-[*cis*-9-octadecenoyl]-*sn*-glycero-3 phosphoethanolamine (DOPE) [25–29]. Since transfection may be related to the structural and physicochemical properties of the self-assembled supramolecular lipid–DNA complexes, we used the Langmuir monolayer technique coupled with PM-IRRAS to investigate DNA-BGTC/DOPE interactions at the air–water interface in mixture conditions used for transfection assays: 3/2 BGTC/DOPE molar ratio and an excess of positive charge of BGTC [30].

DNA interactions with the BGTC/DOPE (3/2) monolayer (20 mN/m) at the air/water interface

When DNA is injected into the subphase of a preformed BGTC/DOPE monolayer at $\pi = 20$ mN/m, kinetics of lateral pressure variation shows a weak but reproducible decrease of π, which can be explained by a condensation of molecules at the lipid interface induced by DNA interaction with the monolayer [30]. Brewster angle microscopy reflectance and ellipsometry measurements demonstrate an increase of the layer thickness. PM-IRRAS spectra of both the BGTC/DOPE (3/2) monolayer at 20 mN/m alone or after DNA injection are presented in Figure 17 in the 1800–900 cm^{-1} region. The main bands are the ν(C=O) of DOPE ester groups around 1737 cm^{-1} and the $\nu_{as,s}$ (PO_2^-) around 1220 and 1080 cm^{-1}, respectively [31, 32]. The BGTC contributions in the 1700–600 cm^{-1} region cannot be seen due to the strong negative band around 1660 cm^{-1} due to the dispersion of the refractive index of water in this spectral range [15, 33]. When DNA was injected under the lipid monolayer, the PM-IRRAS spectrum (Fig. 17) shows a significant

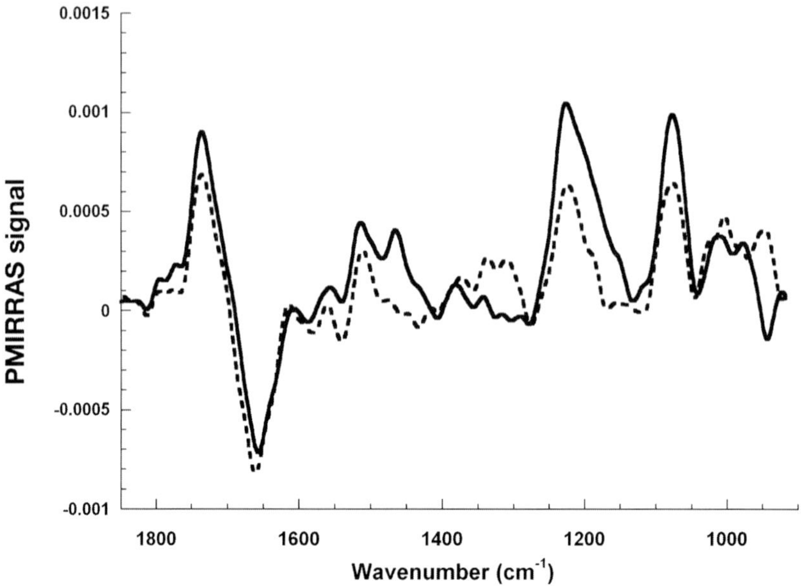

FIG. 17 PM-IRRAS spectra of DNA interaction with a BGTC/DOPE (3/2) monolayer at 20mN/m: (- - -) BGTC/DOPE (3/2), 20 mN/m; (—) + DNA, R(+/–) = 5.

increase of the $\nu_{as,s}$ (PO_2^-) around 1220 and 1080 cm^{-1}. Since we can assume that DNA preferentially interacts with BGTC (no interaction was observed with a pure DOPE monolayer) and also that phosphate groups of DOPE will not be perturbed by DNA interaction, we can conclude that the increase of the $\nu_{as,s}$(PO_2^-) is due to accumulation of DNA phosphate groups under the monolayer and then to the interaction of DNA strands with the monolayer. One can also note that, whatever the charge ratio, the intensity ratio of DNA bands $I(PO_2^-{}_{as})/I(PO_2^-{}_{s}) > 1$. In the isotropic conformation of DNA in buffer, the intensity ratio of the asymmetric PO_2^- at 1220 cm^{-1} versus symmetric PO_2^- at 1080 cm^{-1}, $I(PO_2^-{}_{as})/I(PO_2^-{}_{s}) < 1$. This ratio $I(PO_2^-{}_{as})/I(PO_2^-{}_{s})$ will be used as a marker of the DNA orientation. Indeed, transition moments of ν_{as}(PO_2^-) and ν_{s}(PO_2^-) point in perpendicular directions in the DNA double helix (Scheme 2). Thus, the intensity ratio of these two bands is sensitive to this group orientation and allows us to estimate the orientation of DNA. In the present case, the intensity ratio indicates an anisotropic orientation of the DNA under the lipid monolayer; this favors the asymmetric PO_2^- vibration moment over the symmetric one and means that the double helix of DNA is mainly parallel to the interface plane [30].

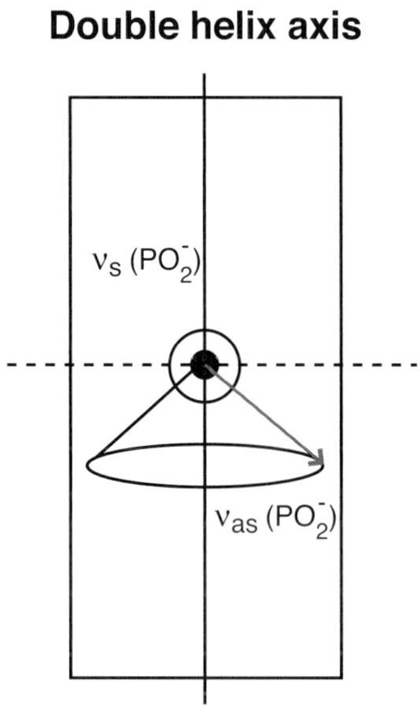

SCHEME 2 Projections of the PO_2^- asymmetric and symmetric transition moments in the double helix of DNA.

Formation of an asymmetric lipid bilayer stabilized by DNA by compression beyond the collapse

As described previously, compression beyond the collapse of phospholipid monolayers can lead to the formation of stable multilayers at the air–water interface. The behavior of the mixed adsorbed DNA-BGTC/DOPE (3/2) monolayer was investigated beyond the collapse (around $\pi \approx 46$ mN/m, which corresponds to a molecular area of 50 $Å^2$/mol). The compression was then carried on to divide the molecular area by a factor of 2. Brewster angle microscopy pictures display an important and homogeneous increase of reflectance and are characteristic of an organized collapse from a bi- to a tridimensional system.

To go further in the identification of these thick layers, PM-IRRAS spectra were recorded. In the 1800–900 cm^{-1} region, they display an increase in the intensity of all bands compared to the spectrum of a pure compact BGTC/DOPE monolayer at 40 mN/m (Fig. 18A). Since all the bands can contain contributions of both lipid and DNA, we investigate in parallel the 3000–2800 cm^{-1} spectral region of the asymmetric ($\nu_{as}(CH_2)$) and symmetric ($\nu_s CH_2$) stretching bands located at 2920 and 2850 cm^{-1}, respectively, mainly characteristic of the lipids chains and free of DNA contribution (Fig. 18B). Their intensity is multiplied by a factor of about 2 compared to those measured with the pure lipid monolayer.

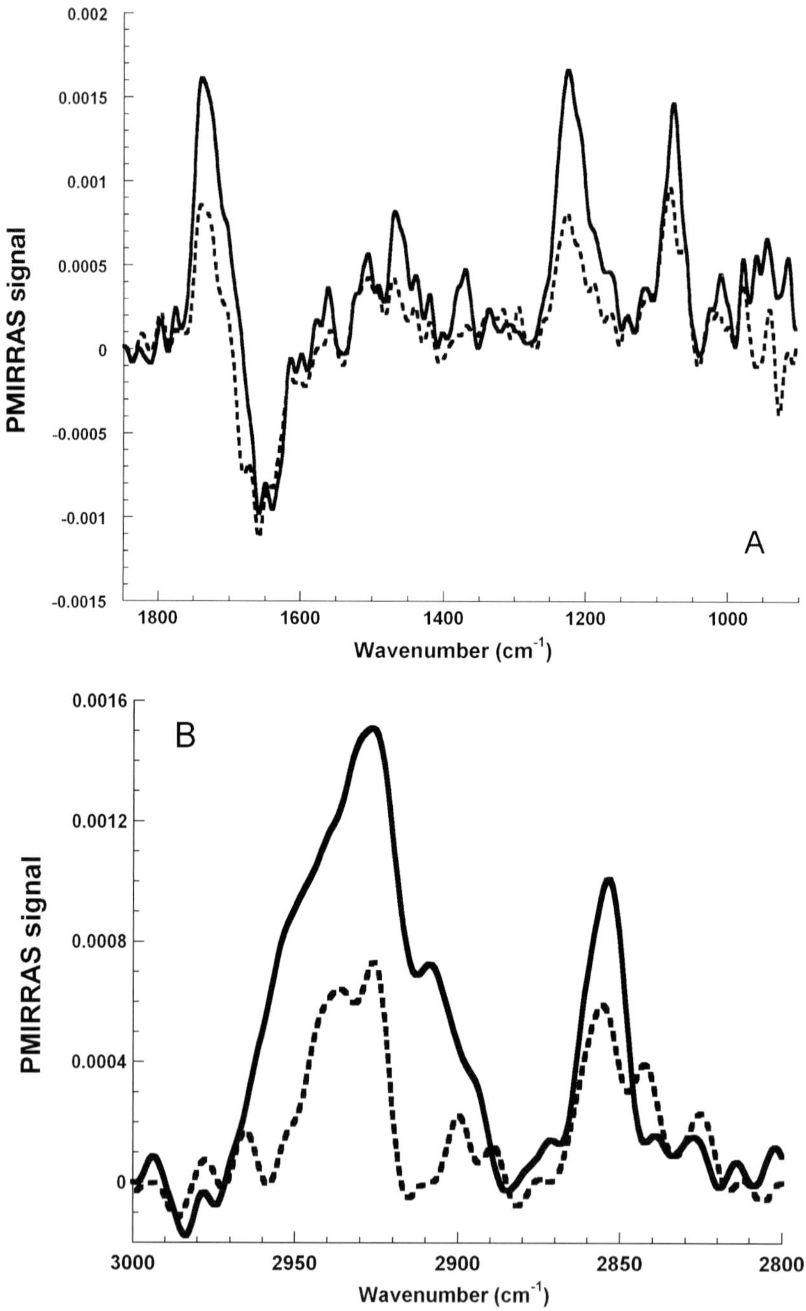

FIGURE 18 PM-IRRAS spectra of a DNA-BGTC/DOPE (3/2) layer after compression beyond the collapse. (A) 1850–900 cm^{-1} region and (B) CH region, - - - BGTC/DOPE (3/2), 40 mN/m, before collapse, — + DNA R(±) = 5 beyond the collapse.

This value proves the formation of a lipid bilayer at the surface. The important broadening of the bands and their shift toward higher frequencies correspond to the perturbation of the organization of the lipid acyl chains. In particular, it means that the number of acyl chains having a gauche conformational defect has increased. In parallel, it was shown that no bilayer is formed during the compression beyond the collapse of a pure BGTC/DOPE (3/2) monolayer which collapses with the formation of aggregates (data not shown) and that it is the presence of DNA adsorbed under the layer which induces the bilayer formation. In the 1800–900 cm^{-1} region, the intensity increase noticed for all bands shows both the DNA interaction and the bilayer formation. When DNA is in default, all the main band intensities are multiplied by a factor almost equal to 2, but since the band intensities are sensitive both to the surface concentration of the molecules and to their orientation, it is quite difficult to decorrelate both phenomena, which can occur simultaneously with antagonist effects on band intensities. In parallel, a shift of the $\nu_{as,s}(PO_2^-)$ toward lower frequencies is observed compared to the case of pure BGTC/DOPE lipid monolayer at 40 mN/m showing both the DNA phosphate contribution due to its interaction with the bilayer and perhaps also a reorientation of the DOPE phosphate group due to the bilayer formation.

To explain such a behavior, the hypothesis of the formation of an asymmetric lipid bilayer on the top of the DNA layer was formulated. Indeed, it may be suggested that the BGTC molecules remain in the layer facing the aqueous phase due to their strong interactions with the DNA adsorbed underneath, while DOPE (which does not directly interact with DNA in the subphase) flip–flops beyond compression to form a top layer.

CONCLUSIONS

The main disadvantage of PM-IRRAS with respect to classical reflection absorption spectroscopy is the complexity of both the optics and electronics of the setup. Considering the optical part, PM-IRRAS requires a spectrometer with a simultaneous double-channel acquisition. Secondly, the entrance polarizer and PEM have to be perfectly aligned to avoid multireflections, which would spoil the polarization modulation. In addition, the infrared beam has to be well centered at the entrance of the modulator to have a homogeneous modulation of the entire beam. Concerning the electronics, PM-IRRAS needs a high quality lock-in amplifier having a very low time constant (less than 0.3 ms). Obviously, it is from the complexity of the setup that PM-IRRAS takes its advantages.

The principal advantage of the technique fundamentally originates from the use of a double-beam spectrometer. It allows us to rid the spectra of all the isotropic absorptions of the environment in real time and diminishes the recording time. PM-IRRAS is therefore useful to study reactions with relatively slow kinetics (a few minutes) in biological systems, such as protein adsorption at the air/lipid monolayer interface for example. If needed, it is possible to get information on systems with much quicker kinetics (a few seconds) [22]. In this study, modification of the secondary structure of the photosystem II protein just after its spreading on a water surface has been revealed.

Another advantage of PM-IRRAS is the relative simplicity and the efficiency of its surface selection rule. Coupled with spectral simulations, only one spectrum is required to determine the orientation of molecular groups. As a comparison, conventional IRRAS requires at least six spectra (with three different angles of incidence and both p and s polarizations) to get orientational information [34]. This implies to be able to maintain the system under study in a stable state during at least 2 h.

Finally, PM-IRRAS is more efficient than IRRAS to detect vibrational modes which transition moments are perpendicular to the dielectric substrate. As an illustration, the band progression due to the methylene wagging modes has been observed for the first time in infrared spectra of a Langmuir monolayer thanks to PM-IRRAS [35]. This method is also very useful to determine the orientation of peptides and proteins. In particular, α-helices perpendicular to the interface give rise to a very characteristic spectral signature of Amide I and Amide II bands (see e.g., the study on bacteriorhodopsin reported in [36]).

REFERENCES

[1] Greenler RG. *J. Chem. Phys.* 1966; 44:310–315.

[2] Greenler RG. *J. Chem. Phys.* 1969; 50:1963–1968.

[3] Blaudez D, Buffeteau T, Cornut J-C, Desbat B, Escafre N, Pézolet M, Turlet J-M. *Appl. Spectrosc.* 1993;47:869–874.

[4] Saccani J, Buffeteau T, Desbat B, Blaudez D. *Appl. Spectrosc.* 2003;57:1260–1265.

[5] Balkansky, M. *Optical Properties of Solids.* (Abèlès, Ed.). Amsterdam: North Holland, **1972**, Chap. 8.

[6] Griffiths PT, de Haseth JA. *Fourier Transform Infrared Spectrometry.* New York: Wiley; 1986 295.

[7] Yamamoto K, Ishida H. *Appl. Spectrosc.* 1994;48:775–787.

[8] Yamamoto K, Ishida H. *Vibr. Spectrosc.* 1994;8:1–36.

[9] Hasegawa T, Nishijo J, Umemura J, Theiss W. *Anal. Scienc.* 2001;17:i697–i700.

[10] Cornut I, Desbat B, Turlet J-M, Dufourcq J. *Biophys. J.* 1996;70:305–312.

[11] Le Calvez E, Blaudez D, Buffeteau T, Desbat B. *Langmuir.* 2001;17:670–674.

[12] Bellet-Amalric E, Blaudez D, Desbat B, Graner F, Gauthier F, Renault A. *Biochim. Biophys. Act.* 2000;1467:131–143.

[13] Boncheva M, Vogel H. *Biophys. J.* 1997;73:1056–1072.

[14] Saccani J, Castano S, Beaurain F, Laguerre M, Desbat B. *Langmuir.* 2004;20:9190–9197.

[15] Blaudez D, Turlet J-M, Dufourcq J, Bard D, Buffeteau T, Desbat B. *J. Chem. Soc. Faraday Trans.* 1996;92:525–530.

[16] Buffeteau T, Le Calvez E, Castano S, Desbat B, Blaudez D, Dufourcq J. *J. Phys. Chem.* 2000;104:4537–4544.

[17] Fezoua-Boubegtiten Z, Desbat B, Brisson A, Lecomte S. *Biochim. Biophys. Acta, Biomembr.* 2010;1798:1204–1211.

[18] Castano S, Desbat B, Laguerre M, Dufourcq J. *Biochim. Biophys. Act.* 1999;1416:176–194.
[19] Castano S, Desbat B, Dufourcq J. *Biochim. Biophys. Act.* 2000;1463:65–80.
[20] Ulrich WP, Vogel H. *Biophys. J.* 1999;76:1639–1647.
[21] Lavoie H, Blaudez D, Vaknin D, Desbat B, Ocko BM, Salesse C. *Biophys. J.* 2002;83:3558–3569.
[22] Gallant J, Desbat B, Vaknin D, Salesse C. *Biophys. J.* 1998;75:2888–2899.
[23] Lavoie H, Desbat B, Vaknin D, Salesse C. *Biochemistry.* 2002;41:13424–13434.
[24] Shanmuk S, Dluhy RA. *Vibr. Spectrosc.* 2004;36:167–177.
[25] Vigneron JP, Oudrhiri N, Fauquet M, Vergely L, Bradley JC, Basseville M, Lehn P, Lehn JM. *Proc. Natl. Acad. Sci. USA.* 1996;93:9682–9686.
[26] Oudrhiri N, Vigneron JP, Peuchmaur M, Leclerc T, Lehn JM, Lehn P. *Proc. Natl. Acad. Sci. USA.* 1997;94:1651–1656.
[27] Pitard B, Oudrhiri N, Lambert O, Vivien E, Masson C, Wetzer B, Hauchecorne M, Scherman D, Rigaud JL, Vigneron JP, Lehn JM, Lehn P. *J. Gene Med.* 2001;3:478–487.
[28] Luton D, Oudrhiri N, de Lagausie P, Aissaoui A, Hauchecorne M, Julia S, Oury JF, Aigrain Y, Peuchmaur M, Vigneron JP, Lehn JM, Lehn P. *J. Gene Med.* 2004;6:328–336.
[29] Densmore CL, Giddings TH, Waldrep JC, Kinsey BM, Knight V. *J. Gene Med.* 1999;1:251–264.
[30] Castano S, Delord B, Février A, Lehn JM, Lehn P, Desbat B. *Langmuir.* 2008;24:9598–9606.
[31] Furer VL. *Appl. Spectrosc.* 1990;53:860–863.
[32] Mantsch HH, McElhaney RN. *Chem. Phys. Lipids.* 1991;57:213–226.
[33] Dluhy RA. *J. Phys. Chem.* 1986;90:1373–1379.
[34] Gericke A, Flach CR, Mendelsohn R. *Biophys. J.* 1997;73:492–499.
[35] Pelletier I, Bourque H, Buffeteau T, Blaudez D, Desbat B, Pézolet M. *J. Phys. Chem.* . 2002;106:1968–1976.
[36] Blaudez D, Boucher F, Buffeteau T, Desbat B, Grandbois M, Salesse C. *Appl. Spectrosc.* 1999;53:1299–1304.

Chapter 3

Infrared spectroscopy for characterization of biomolecular interfaces

Robert J. Hamers, Xiaoyu Wang, Ryan Franking, Rose Ruther and Courtney Stavis
Department of Chemistry, University of Wisconsin-Madison, 1101 University Avenue Madison, WI 53706, USA

Infrared spectroscopy can provide unique insights into the chemistry of surfaces for biological studies [1–4]. Because the conformations of biological molecules impact their vibrational frequencies [1], infrared spectroscopy has long been used to characterize surfaces for use in biological studies. While many studies have used infrared spectroscopy to characterize the adsorption of proteins and other biological molecules to surfaces [5–7], there is increasing interest in the intentional functionalization of surfaces with molecular/biomolecular layers that will provide selectivity, binding only to specific targets in solution [6,8]. Applications include biosensing, layers to control inflammatory response in biomedical implants, and the use of surface chemistries to control the growth and proliferation of cells.

Biointerface Characterization by Advanced IR Spectroscopy. DOI: 10.1016/B978-0-444-53558-0.00003-5

While fundamental studies can often be carried out on specialized substrates, emerging applications often require the use of a wider variety of substrates in both bulk and thin-film form. Particularly when these materials are insulators or semiconductors, there is a need to develop an understanding of the factors that control sensitivity and signal-to-noise with different materials. One of the biggest challenges to the use of infrared spectroscopy in biological research is the relatively low sensitivity and the fact that surfaces of practical interest many have strong absorption features that obscure the useful information. As a consequence, the successful use of infrared for biological research demands paying close attention to the detailed optical properties of the relevant interfaces and understanding how to optimize the sensitivity. In a practical sense, optimizing conditions for infrared are largely controlled by the reproducibility of the reference sample that is typically used to calculate the effective absorbance.

The ability of infrared spectroscopy to identify the types and number of different chemical functional groups present at a surface in a nondestructive manner has contributed to its widespread application in chemistry and biology [7,9–12,13,14]. While many studies have used infrared spectroscopy to characterize the adsorption of proteins and other biological molecules to reflective materials such as gold [14–16], the use of infrared to characterize adsorption onto dielectric materials is more complex and requires careful optimization in order to achieve maximum sensitivity for biomolecular layers [9,17].

While FTIR spectroscopy can be performed in different geometries, here we focus on the single-bounce external reflection mode. This approach has somewhat lower sensitivity than alternatives such as the attenuated total internal reflection (ATR) geometry [5]. However, the single-bounce geometry is the most versatile method because it can be used on both transparent and absorbing substrates and does not require any special sample fabrication. These are major advantages when attempting to use FTIR to characterize technologically relevant surfaces such as silicon and glass [17].

A major difference between dielectric materials and metals is that the dielectric properties of these materials result in reduced reflectivity, which in turn means that there is less light reaching the detector of the FTIR system. In general, the signal-to-noise ratio of the FTIR system depends on the specific detectivity (D^*) of the detector (a measure of the detector's signal-to-noise ratio) and the range of input values resolved by the systems' analog-to-digital converter [18]. At low intensities, the instrument's ability to discriminate very small changes in intensity between sample and reference samples is diminished. Some dielectric materials (such as glass) also have reflectivities that can vary by several orders of magnitude as a function of wavelength

For biological analysis of proteins, the amide region, particularly the so-called amide I band, is often used to identify the conformation of fibrinogen and other proteins in liquids [19–24] and on surfaces [11–12,14]. Conformational analysis is based on the fact that the amide I band (a C=O vibration) shifts in frequency because of hydrogen-bonding interactions with nearby amine groups

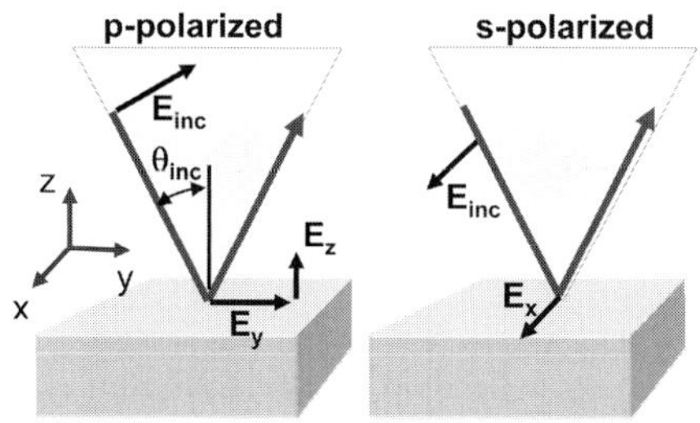

FIGURE 1 **Schematic illustration of optical geometry for single-bounce external-reflection FTIR and definitions of relevant parameters.**

in a manner that is characteristic of the local conformation of the protein. Changes in bonding between β-sheets, helices, localized turns, and other conformations produce small changes in shape and frequency shifts of $\sim$10–20 cm^{-1} in this peak [11,12,25–27]. Hence, there is a great desire to understand the factors that control the intensity, symmetry, and frequency of infrared spectra of proteins on diamond and other surfaces. In this chapter, we will review the basic optics of the single-bounce external-reflection geometry for analysis of molecular and biomolecular layers and illustrate the results for molecular and biomolecular layers of relevance.

Figure 1 depicts the geometry that will be used, along with the definitions of relevant directions of the electric field.

GENERAL CONSIDERATIONS FOR SEMICONDUCTOR SURFACES

For biomolecules directly linked to semiconductors and other dielectric materials, the factors controlling sensitivity of various detection methods are readily described using the Fresnel equations, which describe the reflectivity of the sample [13]. Several factors emerge from the simple reflectivity model. First, that the changes in normalized reflectivity ($\Delta R/R$) and/or the absorbance (A) can be positive or negative, depending on the conditions of the experiment. According to the Kramers–Kronig laws [28], a change in absorption $k(\tilde{\nu})$ must be accompanied by a change in the index of refraction $n(\tilde{\nu})$. At surfaces both effects can be important and of comparable importance. Notably, dispersion effects can affect both the shapes and frequencies of the peaks [10,17,29]. This can be very important in analysis of surface species such as proteins, as the frequencies of the vibrational modes are often used to interpret molecular conformations. In general, the absorption component can be described in terms of a symmetric Lorentzian or Gaussian lineshape (depending on the factors controlling the linewidth), while the index of refraction change is highly asymmetric.

The index of refraction change associated with an absorption band lying between frequencies ν_1 and ν_2 can be approximated as follows [29]:

$$n(\tilde{\nu}_0) = n(\infty) + \frac{1}{\pi}\int_{\nu_1}^{\nu_2} \frac{k(\tilde{\nu})d\nu}{(\tilde{\nu} - \tilde{\nu}_0)}$$

Figure 2 depicts changes in $n(\tilde{\nu})$ and $k(\tilde{\nu})$ associated with a typical infrared vibrational mode, simulated here using parameters derived experimentally for the ~1730 cm^{-1} C=O vibrational mode in thin films of poly(methyl methacrylate), which has been used as a model system [17,29]. When absorption features typically have $k(\tilde{\nu})$ with a symmetric Lorentzian or Gaussian peak shape, the index of refraction has an asymmetric shape that combines a derivative-like shape with a static change in the limiting value of n from ~1.4 at low frequencies to ~1.35 at high frequencies. While the specific values and frequencies depend on the nature of the particular vibrational mode, the general features shown here are expected for any vibrational absorption. The distinctive shapes, particularly the asymmetry associated with $n(\tilde{\nu})$ can be readily distinguished in the FTIR spectra, allowing a visual assessment of the extent to which $n(\tilde{\nu})$ and $k(\tilde{\nu})$ each contribute to the overall spectra.

To understand the factors that control sensitivity and line shape, it is useful to consider separately the effect of the index of refraction $n(\tilde{\nu})$ and the actual absorbance $k(\tilde{\nu})$. To do so, we use the Fresnel reflectivity calculations to predict the size and shape of measured absorption features as a function of angle. Here, we use a modification of the method of Hansen [30], as reported by Tolstoy [13]. Briefly, the complex index of refraction $\tilde{n}_j = n_j + ik_j$ and complex permittivity of each layer are calculated as $\tilde{\varepsilon}_j = n_j^2 - k_j^2 + 2in_jk_j$.

The cosines of the complex angles of refraction are then given by Snell's law as $\cos\varphi_j = \left[1 - \left(\frac{n_j \sin\varphi_1}{\tilde{n}_j}\right)^2\right]^{1/2}$ and the complex indices of refraction are $\xi_j = \tilde{n}_j \cos\varphi_j$.

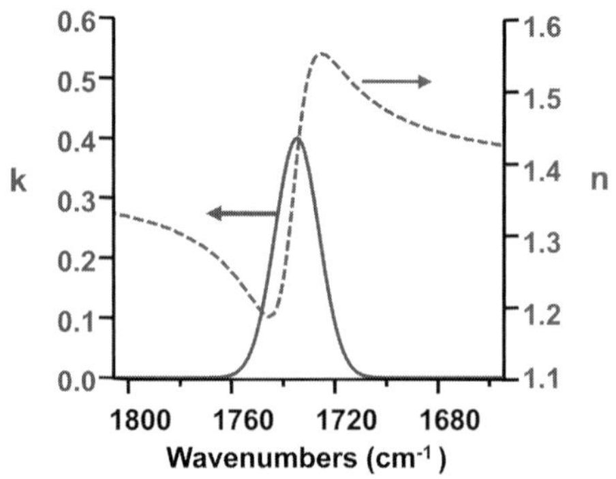

FIGURE 2 **Typical variation in $n(\tilde{\nu})$ and $k(\tilde{\nu})$ associated with a C=O stretching vibration of a molecular layer on a surface, after. [29].**

For three or more layers, a phase shift term is included for the center layers (of noninfinite thickness) as $\beta_i = \frac{2\pi d_j \xi_j}{\lambda}$, representing the phase shift of light of wavelength λ passing through the jth material of thickness d_j. Each interface between two materials is represented as a (2 × 2) matrix:

$$M_j = \begin{vmatrix} \cos\beta_j & \frac{-i\sin\beta}{q_j} \\ -iq_j\sin\beta_j & \cos\beta \end{vmatrix}$$

for p-polarized light and

$$\begin{vmatrix} \cos\beta_j & \frac{-i\sin\beta_j}{p_j} \\ -ip_j\sin\beta_j & \cos\beta_j \end{vmatrix}$$

for s-polarized light, where

$$p_j = \sqrt{\tilde{\varepsilon}}\cos(\varphi_j)$$

and

$$q_j = \frac{\cos\varphi_j}{\sqrt{\tilde{\varepsilon}}}.$$

The overall optical response of an N-layer system is then calculated as the matrix product $M = M_2 M_3 \ldots = \begin{bmatrix} m_{11} & m_{12} \\ m_{21} & m_{22} \end{bmatrix}$.

The complex reflection coefficients for p- and s-polarized light are then:

$$r_p = \frac{(m_{11} + m_{12}p_N)p_1 - (m_{21} + m_{22}p_N)}{(m_{11} + m_{12}p_N)p_1 - (m_{21} + m_{22}p_N)}$$

and

$$r_s = \frac{(m_{11} + m_{12}q_N)q_1 - (m_{21} + m_{22}q_N)}{(m_{11} + m_{12}q_N)q_1 - (m_{21} + m_{22}q_N)}$$

and the total reflectance $R_p = r_p^* r_p$ and $R_s = r_s^* r_s$.

Figure 3 illustrates some of the more important aspects of single-bounce FTIR spectra. Panels 3(a) and 3(b) depict the intrinsic behavior of light reflecting from a dielectric material, while 3(c) and 3(d) show the effect of introducing a weak absorber. Figure 3(a) shows the electric field ratio, which is the ratio of the electric field at the surface to that of the incident light. Because the underlying silicon is able to partially polarize on the time scale of the incident electromagnetic field oscillations, it creates an oscillating image dipole that enhances the incident field component perpendicular to the surface plane and reduces the field component parallel to the plane. With s-polarized light, the electric field ratio is always less than 1, because the image dipole at the surface partially cancels that of the incident beam E_o. P-polarized light exhibits a

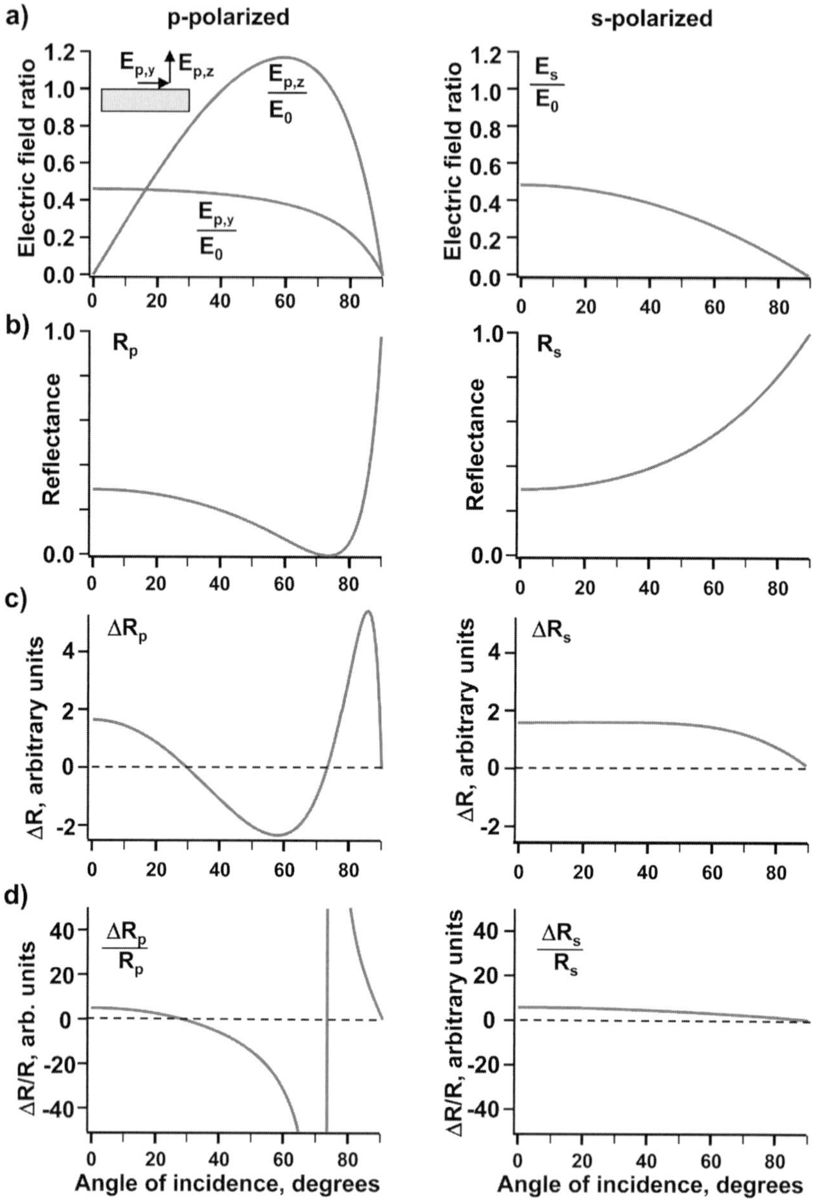

FIGURE 3 Optical parameters calculated from Fresnel equations for an organic monolayer on silicon, using p-polarized (right) and s-polarized (left) radiation.

complex dependence that can be understood from the fact that the incident electric field E_o can be decomposed into two orthogonal components lying parallel to and perpendicular to the sample surface. The component of the incident electric field lying parallel to the surface, $E_{p,y}$, is similar to that of s-polarized light and is partially canceled by the image dipole of the substrate; the net field parallel to the surface decreases monotonically as the angle of incidence θ_{inc} increases. However, the component of the incidence electric field lying perpendicular to the sample surface (labeled $E_{p,z}$) is enhanced by the surface image dipoles such that $\frac{E_{pz}}{E_0}$ can be greater than 1; for silicon it reaches a maximum near $\theta_{\text{inc}} = 60^{\text{o}}$ degrees and then decreases toward zero. Panel 3(b) shows the absolute reflectance of the sample for p-polarized (R_{p}) and s-polarized (R_{s}) light. Here, it becomes apparent that over most of the accessible range of incident angles the reflectivity using p-polarized light is smaller than that using s-polarized light. Of particular significance is that for dielectric materials the reflectivity drops to zero at Brewster's angle; for silicon this occurs at an angle of $\sim 73^{\text{o}}$.

Infrared spectra of molecular and/or biomolecular films are typically presented in terms of either the normalized change in reflectivity $\frac{\Delta R}{R_0} = \frac{R_{\text{sample}} - R_0}{R_0}$ or absorbance $A = -log_{10}\left(\frac{R_{\text{sample}}}{R_0}\right)$, thereby measuring the reflectivity of the sample of interest (R_{sample}) with respect to that of a reference sample (R_O) as a standard. For small absorbances $\ln\left(\frac{R_{\text{sample}}}{R_0}\right) = \ln\left(\frac{R_0 - \Delta R}{R_0}\right) = \ln\left(1 - \frac{\Delta R}{R_0}\right) \approx \frac{\Delta R}{R_0}$. Taking into account the difference between base 10 and base e logarithms, $A \approx \frac{-1}{2.303}\left(\frac{\Delta R}{R}\right)$.

Figure 3(c) shows the change in reflectance (ΔR) when a weak absorber is introduced while Figure 3(d) shows the normalized change in reflectivity $\Delta R/R$. For s-polarized light, the presence of an absorber *increases* the reflectivity ($\Delta R > 0$) of all angles; this corresponds to *negative* absorbances for all angles of incidence. The sensitivity ($\Delta R/R$) predicted using s-polarized light decreases monotonically as the angle of incidence is increased because the surface electric field (Fig. 3(a)) decreases and because the overall reflectivity increases.

For p-polarized light the situation is more complex. At small angles of incidence, the presence of a molecular layer is predicted to increase the reflectivity (corresponding to a *negative* absorbance). At intermediate angles (~28–73°), the reflectivity decreases (positive absorbance), while at even steeper angles ($\theta_{\text{inc}} > \theta_{\text{Brewster}}$) the reflectivity again increases (negative absorbance). The behavior for ($0 < \theta_{\text{inc}} < \theta_{\text{Brewster}}$) can be simply understood as a competition between the in-plane and out-of-plane components of the electric field depicted in Figure 3(a): interaction of an isotropic molecular absorber with the electric field perpendicular to the surface ($E_{p,z}$) decreases the reflectance ($\Delta R < 0$), but interaction with $E_{p,y}$ increases the reflectance ($\Delta R > 0$). At smallangles of incidence $E_{p,y} \gg E_{p,z}$ and the net effect is $\Delta R > 0$ just as with s-polarized light, while at larger angles the effect of the vertical component

$E_{p,z}$ becomes dominant. A particularly notable feature using p-polarized light is that the absorbance has a discontinuity at Brewster's angle, where the total reflectivity R_p goes to zero. This discontinuity leads to two angles of maximum sensitivity, one on each side of Brewster's angle; for angles immediately adjacent to Brewster's angle, ΔR can be relatively large while the total reflectivity (R_p) remains comparatively small.

The above considerations suggest that the overall sensitivity $\Delta R/R$ for single-bounce external reflection measurements is the highest using p-polarized light near Brewster's angle. However, the signal-to-noise ratio of the FTIR detectors and the analog-to-digital converters are usually optimized for higher intensity levels [18]. Consequently, there will be a system-dependent optimum angle at which sufficient light strikes the detector while also taking advantage of the higher $\Delta R/R$ near Brewster's angle. An infrared system using a typical extended source (e.g., Globar or tungsten source) will not have a perfectly collimated source, and typical geometric parameters yield an angular divergence with a full angle on the order of several degrees. Thus, maximum sensitivity in practice is generally achieved using an angle approximately 5° higher or lower than Brewster's angle.

To test the validity of these calculations, Figure 4 shows experimental data from an organic molecular layer on a silicon(1 1 1) surface. This particular molecular layer was prepared using a common photochemical grafting procedure in which an organic alkene will graft to H-terminated Si(1 1 1) when illuminated with ultraviolet light [31–35]. In the data shown here, the grafted molecule also had a second functional group consisting of a trifluoroacetic acid-protected amine group that was then deprotected to yield a surface functionalized with primary amine groups [35,36]. This is a convenient pathway toward a wide range of functionalized surfaces, since there are a number of excellent coupling reactions that can link DNA, peptides, and other biomolecules to amine-terminated surfaces [36–38].

In the C–H region, the s-polarized spectrum shows downward peaks (negative absorbances) associated with the symmetric and asymmetric C–H stretching vibrations of the alkyl chain. A highly asymmetric feature is observed at ~1600 cm^{-1}, and very weak, downward (negative absorption) features near 1450 cm^{-1}. Spectra measured at other angles of incidence with p-polarized light exhibited similar features and are not shown here. The p-polarized spectra show a very rich behavior. The C–H stretching modes are negative at $\theta_i = 30°$, disappear at 35°, and reappear as positive peaks at 40–65°; at higher angles yet the peaks flip sign once again. Similarly, the peaks for the C–H deformation modes at 1370 and 1465 cm^{-1} are small (but positive) for $\theta_i \sim 30°$, and then they increase and reach a maximum around 60°. At higher angles, the peak shape becomes complex and then inverts, becoming negative again for angles of 70–75°. This behavior almost perfectly reproduces the behavior predicted in Figure 3. However, there are some small discrepancies. In particular, the N–H deformation mode is negative (rather than positive) at intermediate angles, and

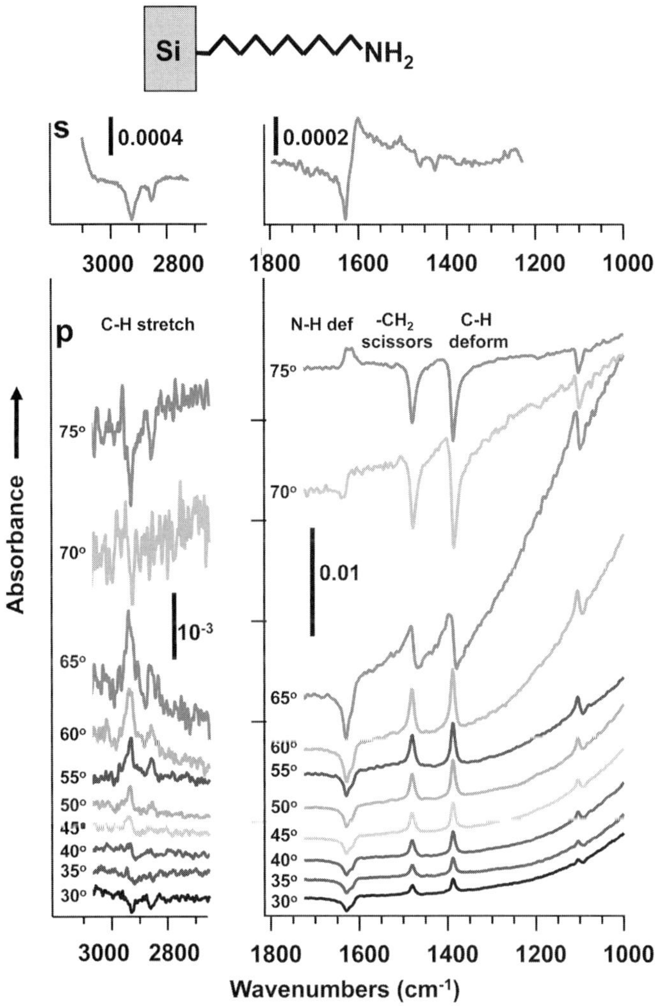

FIGURE 4 **Experimental measurements of infrared absorption from amine-bearing organic layer on Si(1 1 1), as described in the text.**

the C–H deformation modes and scissors modes do not become negative at the very lowest angles of incidence, even though the C–H modes do. These deviations most likely arise from the fact that the predictions in Figure 3 were based on an isotropic layer, while a molecular layer at a surface has a preferred orientation that can be reflected in the anisotropy of molecular transition dipoles. At moderate angles of incidence (from zero up to Brewster's angle), the interaction of the incident electric field with transition dipoles oriented parallel to the surface increase the total reflectivity (giving rise to negative absorbances) while the interaction with dipoles oriented perpendicular to the surface decrease the reflectivity and give rise to positive absorbances. For layer in which the vibrational transition dipole is randomly oriented (i.e., an isotropic layer), the total response is the vectorial combination of the horizontal and vertical components and gives the overall result shown in the bottom panel of Figure 3. For isotropic samples, the introduction of an organic layer always increases the reflectivity (negative absorbances) when measured with s-polarized light. For p-polarized light at small angles of incidence, the in-plane component of the electric field is large and the out-of-plane component is large. As the angle of incidence increases, there will be an intermediate angle of $\sim 30^{\circ}$ at which these two contributions cancel, giving rise to a net zero absorption. As the angle of incidence is increased further, the vertical component of the electric field dominates and isotropic layers give rise to positive absorbances. Finally as the angle cross through Brewster's angle, the reflectivity changes sign once again. In the case of the experimental data in Figure 4, the experimental data indicate that the N–H deformation mode preferentially lies parallel to the surface plane, while the C–H deformation and scissors modes have some preferential orientation perpendicular to the surface. In principle a more detailed analysis of these angle-dependent IR spectra can provide a way of determining structural features such as the molecular tilt angle [17].

Absorption versus dispersion

Because the absorption features of molecular monolayers are in general weak (typically absorbances of 10^{-3} or small) most analyses of sensitivity considerations in FTIR have focused primarily on conditions for optimizing the overall sensitivity as measured by the absorbance or by $\Delta R/R_0$ [17]. However, a second factor that must be taken into consideration is the influence of *dispersion* on the quality of the spectra [29]. According to the Kramers–Kronig relationship, a change in absorption $k(\tilde{v})$ is always accompanied by a change in the index of refraction $(\tilde{v})$. The effect of the variation in $n(\tilde{v})$ is to introduce asymmetry, or dispersion, into the measured line shapes. Dispersion can be important because it can lead to complicated line shapes, shifts the peak locations, and make it difficult to extract information about peak intensities from the spectra [29]. As one example, in Figure 4 the s-polarized vibrational mode near 1600 cm^{-1} has a strongly dispersive character.

The influence of the wavelength-dependent changes in n can also be estimated by noting that the change in the index of refraction is proportional to the change in the index of refraction and for a single peak can be approximated by $n(\upsilon_0) = n_\infty + \frac{1}{\pi}\int_{\upsilon_1}^{\upsilon_2} \frac{k(\upsilon)d\nu}{(\upsilon - \upsilon_0)}$. As depicted in Figure 2, a typical absorption peak with a maximum $k = 0.4$ will be accompanied by a wavelength-dependent change in the index of refraction of $\sim \pm 0.2$ in traversing an absorption peak.

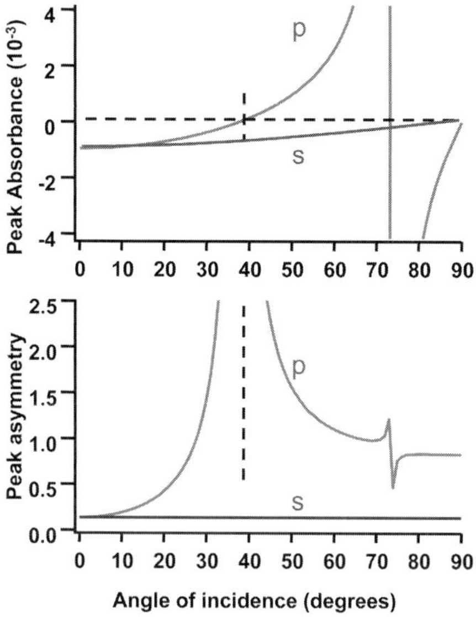

FIGURE 5 Peak absorbance (a) and Peak asymmetry (b) predicted for a typical organic layer on silicon. At ~ 40°, the influence of k vanish, leaving perfectly dispersive peaks.

Using these as typical values of what might be commonly observed in infrared spectra of other molecular/biomolecular layers, we used the Fresnel calculations to calculate the change in absorbance that would be produced by each.

Figure 5(a) shows the change in absorbance as a function of angle (similar to Fig. 3(d) but now plotted as absorbance instead of $\Delta R/R$), and Figure 5(b) shows the dispersion-induced asymmetry in peak shape; both calculations were performed using parameters typical of a C=O vibration of a molecular layer on silicon. For purposes of illustration, we operationally define the peak asymmetry as the predicted change in absorbance due to a change of $n = 0.2$ (from 1.4 to 1.2) divided by that induced by the change in absorbance from $k = 0$ to $k = 0.4$, when measured at the same polarization and angle of incidence: $\textit{Peak asymmetry} = \frac{|A_{n=1.4,k=0} - A_{n=1.2,k=0}|}{|A_{n=1.4,k=0.2} - A_{n=1.4,k=0}|}$. The chosen values approximate the changes in n and k that are encountered as the frequency is traversed across a typical absorption feature. The plot in Figure 5(b) shows that s-polarized spectra are generally more symmetric than p-polarized spectra. For p-polarized spectra at an angle of incidence of ~40° (for silicon), the effect of k vanishes (in Fig. 3(c), $\Delta R = 0$), leaving peaks that are almost perfectly dispersive in shape; consequently, the peak symmetry, as defined above, becomes infinite. At higher angles, the peaks become more symmetric, with a small discontinuity generated near Brewster's angle.

Biological molecules at surfaces: nonspecific binding of fibrinogen on H-terminated silicon surfaces

The nonspecific binding of proteins such as fibrinogen at surfaces is important in a range of technological and biomedical applications. Consequently, there have been numerous studies investigating the binding of fibrinogen to surfaces [12,27,39,40]. Figure 6 shows FTIR spectra of silicon that were H-terminated by an initial immersion in HF solution and then etched in NH_4F solution to render the surfaces clean and hydrophobic [41–44]. These samples were then immersed in a solution of ~0.2 mg/ml human fibrinogen in 0.1 M $NaHCO_3$ (pH 8.3) for 20 min at room temperature. Rinsing followed, including an initial 15 min soak in a wash-off buffer (pH 7.4) consisting of 0.3 M NaCl, 20 mM Na_2PO_4, and 2 mM EDTA (commonly known as 2X SSPE) with 1% added

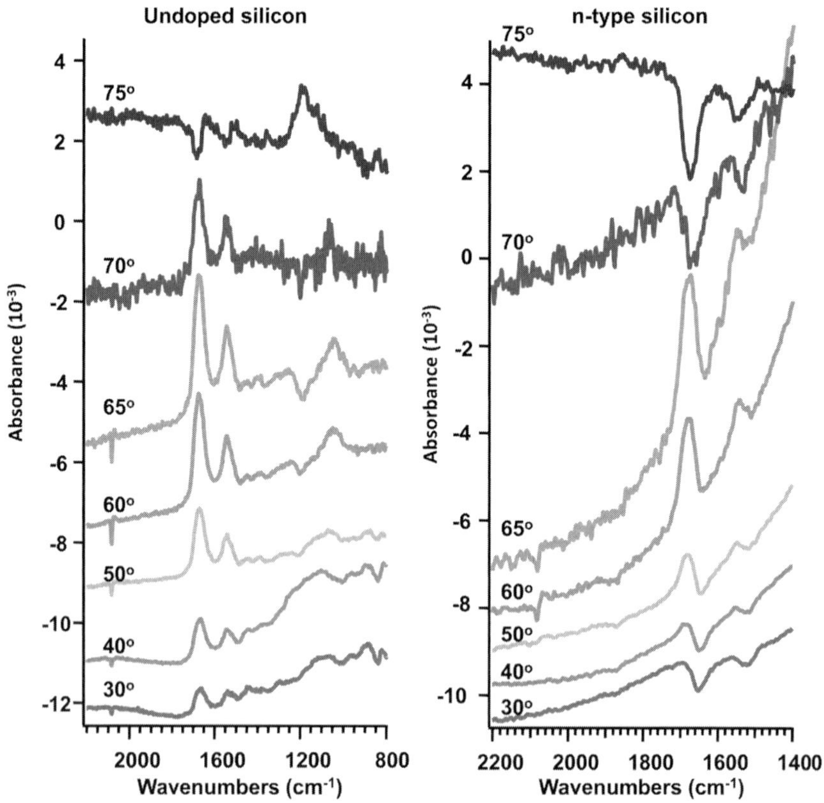

FIGURE 6 FTIR spectra from nonspecifically bonded layer of fibrinogen on undoped silicon and an doped (n-type) silicon wafer.

Triton X-100 surfactant, a second 5 min soak in 2X SSPE, and two 5 min soaks in deionized H_2O. The samples were dried in a stream of purified dry nitrogen and measured immediately.

Figure 6 shows data for two samples, one of which was intrinsic (undoped) silicon, and another that was intentionally doped to enhance its conductivity; the resulting free carriers enhance the reflectivity of the sample and also make it opaque, so that the backside of the sample is not probed and no interference effects are created in the sample. The undoped sample shows that the absorption features increase in size as the angle of incidence is increased from 30° to its largest value at 65°, approaching Brewster's angle. However, because at this angle the intensity of reflected light is very low, a closer inspection of the noise in the spectrum shows that the best signal-to-noise ratio is actually obtained at angles of ~ 60°. At small angles of ~30°, the peaks are smaller and more dispersive in shape.

Figure 6 also shows a comparison with an n-type doped sample. While the overall reflectivity is increased by doping, the presence of free carriers also induces a much stronger dispersion into the complex index of refraction of the silicon substrate. The overall behavior is qualitatively similar to that observed with the undoped sample, although the dispersion effects overall are more pronounced; this is most clearly observed at angles of 40–50°, where the undoped sample shows relatively symmetric peaks while the doped sample shows significantly more asymmetry to the peaks. As expected from the Fresnel modeling in Figure 3, at the smallest angles of 30° the peaks are inverted or dispersive; at 40° almost perfectly dispersive, and at angles of ~60–65° the peaks are the strongest, before inverting again at higher angles.

Comparison of the doped versus, undoped samples shows that the potential benefits of having more light reaching the detector (a consequence of doping) are largely offset by the more strongly wavelength-dependent background from the bulk silicon, particularly at low frequencies where the response to the free carriers is most pronounced. The steeper background on the doped samples in Figure 6 can be through a polynomial fit; however, this does not eliminate the apparent shift. Figure 7 shows spectra from two doped and two undoped sample; in both cases a fifth-order polynomial was fit to the featureless regions outside the amide bands and used to create an appropriate smooth background. The resulting baseline-corrected spectra are shown in Figure 7. For the two samples measured on doped silicon, there is still a pronounced shift, with the spectrum measured at 30° showing a negative absorption feature with its extremum at 1653 cm^{-1}, while the positive peak measured at 65° shows a maximum at 1678 cm^{-1}. The measurements on undoped silicon are at 1672 and 1666 cm^{-1}. The increased shift on the doped samples is attributed to the fact that doping the sample introduces additional dispersion via the substrate and can be reproduced in Fresnel calculations by including dispersion of the underlying Si substrate.

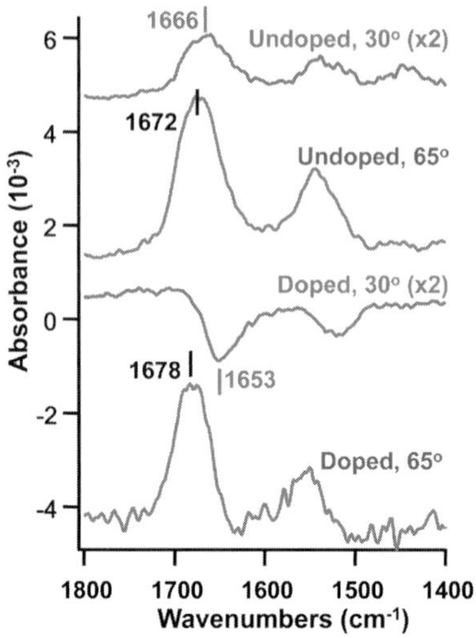

FIGURE 7 **Expanded via of amide region from Figure 6 for angles of incidence of 30° and 65°, showing shifts in apparent peak frequencies. Spectra for the doped samples have been expanded two times to facilitate comparison. Note that on the doped sample, the shifts in the apparent peak frequencies are greater than those on the undoped sample.**

Protein binding on a diamond thin film

The adsorption of biological molecules onto thin films poses additional complexities. For example, a number of studies have shown that diamond thin films can provide greatly enhanced chemical stability, leading to biological interfaces that can be stored for extended periods of time and/or subjected to relatively harsh chemical treatments with little or no degradation [36,45–48]. Diamond is also of interest as a coating material for prosthetic implants and other devices in which the hardness of diamond is of interest. In the case of thin film, the reflectivity of the sample is a strong function of both angle and wavelength. Interference effects within the thin film cause a strongly wavelength-dependent reflectivity as illustrated in Figure 8 for a ~810 nm thick diamond film on an undoped Si substrate [49]. In the case of the diamond thin films, the surface electric field is affected by the combined effects of reflection at the top surface and the reflection from the diamond–silicon interface. Depending on the relative phase of these two, the effect can be to enhance the electric field or to cancel it. The total effect can be accurately modeled using the Fresnel approach described above, but now including four layers consisting of (1) air, (2) the molecular layer, (3) the diamond thin film, and (4) the underlying silicon substrate.

Figure 9 shows the results of such a Fresnel model. For a film that is thin as compared with the wavelength of infrared light, the behavior is qualitatively

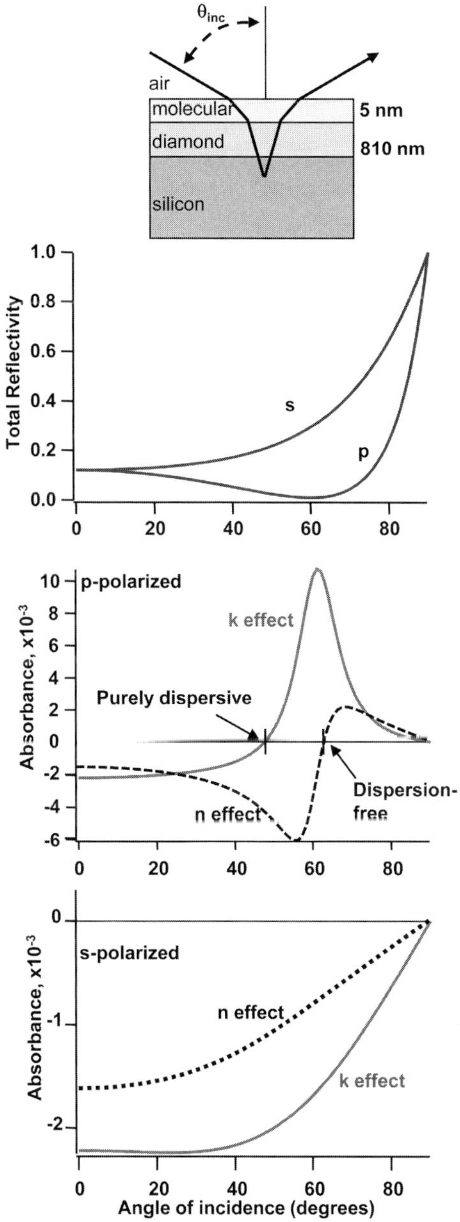

FIGURE 8 **Fresnel calculations for 810-nm thick diamond thin film on a silicon substrate. Fresnel reflectivity calculations for 810-nm diamond thin film on silicon substrate. The influence of changes in n (''n effect'') and of k (''k effect'') is shown for p-polarized and s-polarized light. Note that the reflectivity curve for p-polarized light shows a single maximum that closely corresponds to the angle at which dispersion effects are minimized.**

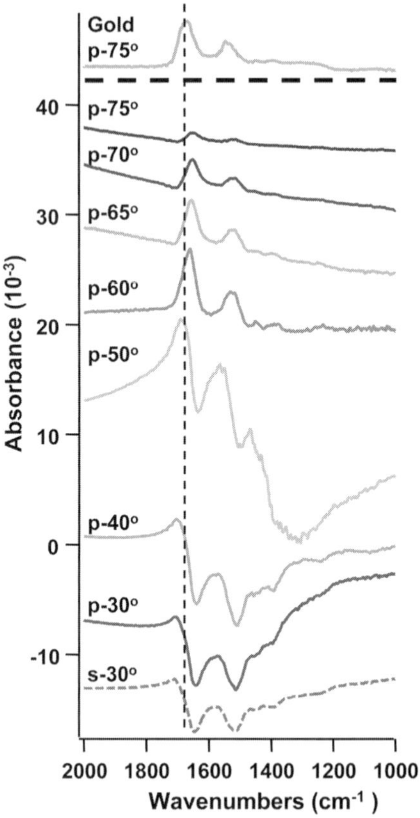

FIGURE 9 Experimental measurements of fibrinogen nonspecifically adsorbed onto H-terminated diamond thin film measured at two different angles. A single s-polarized spectrum is shown (bottom). For comparison, the spectrum of a thin film of fibrinogen adsorbed on gold is also shown. The narrow vertical line is shown to highlight the shifts in apparent peak frequency that occur.

similar to what is observed for a bare silicon surface, although with some important differences. At any given wavelength, the reflectivity has a minimum value, but the reflectivity never drops to zero and occurs at a slightly smaller angle (61°) than Brewster's angle for a semi-infinite sample. This minimum is often referred to as a "pseudo-Brewster's angle." S-polarized light is again predicted to yield weak, negative absorbances. P-polarized light is predicted to yield negative absorbance at shallow angles of incidence, passing through zero near $\theta_{inc} = 50°$, and exhibiting a pronounced peak near $\theta_{inc} = 61°$, slightly smaller than Brewster's angle of 67° predicted for diamond, $n = 2.37$. A large enhancement in sensitivity is again observed near Brewster's angle, but in this case the peak is broader, somewhat shifted, and perhaps most importantly, does not change sign at the minimum in reflectivity. Thus, unlike the semi-infinite bulk sample, in the case of the diamond thin film, it is possible to perform measurements at or very close to the reflectivity minimum.

Perhaps even more interesting is to examine the influence of dispersion. In order to explore how variations in $n(\tilde{\upsilon})$ and $n(\tilde{\upsilon})$ affect the infrared spectra, we used the Fresnel equations to calculate the reflectance as a function of angle $R(\theta)$ for two different values of n (n effect) and two different values of k (k effect) to simulate the changes in absorbance with ($R(\theta)$) and without ($R_0(\theta)$) a organic layer. In a manner similar to silicon, we calculated the apparent absorption due to a change in k from $k = 0$ to $k = 0.4$ at constant $n = 1.4$ (k effect) and also the apparent absorption due to a change in $n = 1.4$ to $n = 1.6$ at constant $k = 0.4$ (n effect). The resulting reflectance spectra were then represented as absorbances using $A(\theta) = -log_{10}\left(\frac{R(\theta)}{R_0\theta}\right)$, thereby simulating how the changes in n and k associated with a molecular layer are expected to affect the observed absorption spectra measured at different angles.

In a manner similar to the previous discussion of silicon, in the case of the diamond thin film when using p-polarized light the apparent absorption is predicted to change sign at an angle of incidence of ~48°, measurements at this angle are predicted to yield purely dispersive peaks. However, it is also notable that there is also an angle where the effect of dispersion (n effect) vanishes; moreover, this angle is almost identical to the angle where the effect of absorption (k effect) is maximized, leading to maximum sensitivity and nearly dispersion-free peaks.

Figure 10 shows experimental data for fibrinogen nonspecifically bound on the H-terminated surface of an 810-nm thick diamond thin film on undoped silicon [49]. As anticipated, using s polarized light (here shown only for $\theta_{inc} = 30°$) and using shallow angles of p-polarized light, the peaks are negative and exhibit rather pronounced dispersive effects, visible from the clear asymmetry in the overall peaks and the background. Near $\theta_{inc} = 50°$ (where the Fresnel model predicted perfectly dispersive behavior) the peak shape becomes highly distorted, clearly dominated by the dispersive effects of the peaks. At higher angles of incidence, the intensities become clear and at angles near $\theta_{inc} = 60–65°$ the peak intensities have strong intensity and are nearly symmetric in shape (Fig. 10). At higher angles of incidence, the peaks decrease in amplitude. Thus, the experimental data clearly confirm that in the thin-film geometry, there is an optimal range of angles θ_{inc}= 65–65° that simultaneously optimizes both amplitude and symmetry of the peaks.

Influence of substrate material on FTIR spectra

One of the biggest advantages over the single-bounce external reflection method is that it is able to measure spectra of molecular and biomolecular layers on surfaces that do not transmit light in the range of interest. Of these, glass is a

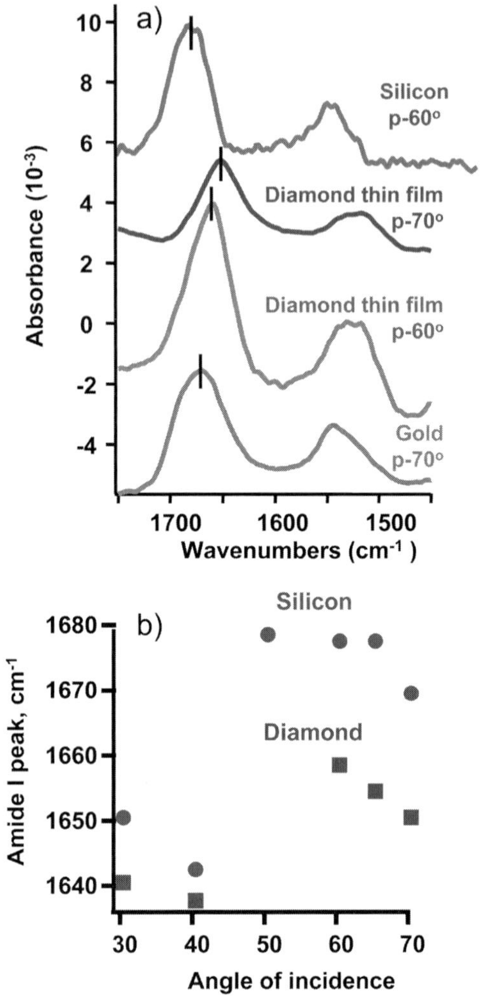

FIGURE 10 **(a) Comparison of the amide bands for fibrinogen adsorbed onto surfaces of bulk silicon, an 810-nm diamond thin film on silicon, and a gold thin film. (b) Peak frequencies for the amide I band on silicon and diamond fibrinogen measured at different angles of incidence.**

particularly important example because of its widespread use in a variety of biomolecular arrays such as gene chips and protein chips. Germanium is a favorite material because its very high index of refraction (4.01 at 1600 cm^{-1}) and nearly complete transparency in the infrared allows it to be used in internal reflection geometry even in contact with liquid solutions [12]. For example, Lu and Park [12] modified the surfaces of a germanium ATR element and used deconvolution methods to analyze how the changes in surface chemistry impacted the protein conformation. The adsorption of fibronectin onto Ge surfaces with different surface functional groups was investigated [50]. Silicon has also been widely used as a substrate for infrared studies because of its importance in microelectronics and because it also has a moderately high index of refraction in the mid-infrared (3.4 at 1600 cm^{-1}) and because its native oxide is relatively stable in water, enabling a wider range of surface chemistries compared with germanium.

Because the overall reflectivity (material 1) and the substrate (material 2) varies roughly like

$$\frac{n_2 - n_1}{n_2 + n_1} \approx \frac{n_2 - 1}{n_2 + 1},$$

the amount of infrared radiation impinging on the detector is highest for germanium and silicon and much less for materials such as diamond ($n = 2.35$) and glass. Glass is a particularly challenging material because it is opaque through most of the infrared [51] and has a low index of refraction, so that only small fraction of the incident light is reflected; consequently, one often runs into sensitivity considerations due to the detector noise. In addition, for examining bands at moderately low frequencies (below $\sim$ 1800 cm^{-1}), glass can have n and k values that change rapidly with frequency, leading again to dispersive effects. Figure 11(a) shows the absolute reflectivity of a common glass used in making microscope slides, also commonly used in gene chip technologies. The absolute reflectivity is low throughout most of the infrared region, but is dominated by a strong discontinuity in the reflectance at approximately 1275 cm^{-1}, changing by more than a factor of 10^3 in a span of less than 50 cm^{-1}. This peak arises primarily from the longitudinal optical phonon band (ν_{LO}) [13], with additional contributions in the 1100–1300 cm^{-1} region from functional groups (such as BO_4^- anions and/or PO_4^- anions) that are often present in glasses. Figure 11(b) shows selected spectra obtained of a PMMA thin film on a glass substrate, in order to characterize the magnitude and symmetry of the peaks. Figure 11(c) shows a prediction from Fresnel theory for a substrate with $n \sim 1.2$ and $k \sim 0.2$, a value estimated from the experimental reflectivity curves and in reasonable agreement with prior estimates [13]. First, the low dielectric constant effective reduces the magnitude of the image dipole induced in the material, thereby reducing the impact of perpendicular dipoles and increasing the importance of the parallel components. Consequently, the absorbance at small angles (below Brewster's angle) is always negative. Secondly, the nonzero value of k has the effect of broadening the otherwise sharp discontinuity predicted in the absorbance, making the absorption features less strongly dependent on the incident angle than would be the case for a lossless dielectric.

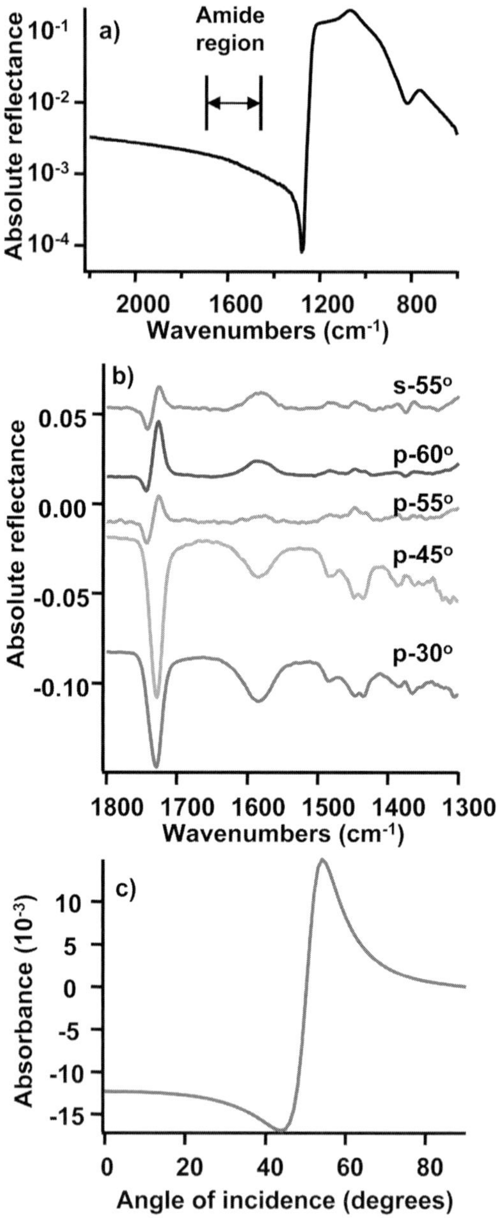

FIGURE 11 (a) Absolute reflectivity of glass surface measured with p-polarized light at $\theta_{inc} = 60^{\circ}$. (b) Infrared absorption of thin films of PMMA at different angles of incidence of p-polarized and s-polarized light, showing maximum in sensitivity for p-polarized light at angles of 30–40°. (c) Predicted absorbance versus angle of incidence for molecular layer on glass surface assuming that the molecular layer has a peak of k = 0.4 and a constant n = 1.4. The effect of nonzero k for the glass substrate is to broaden the otherwise-sharp discontinuity near Brewster's angle, yielding a wider range of angles for which significant sensitivity enhancement is obtained.

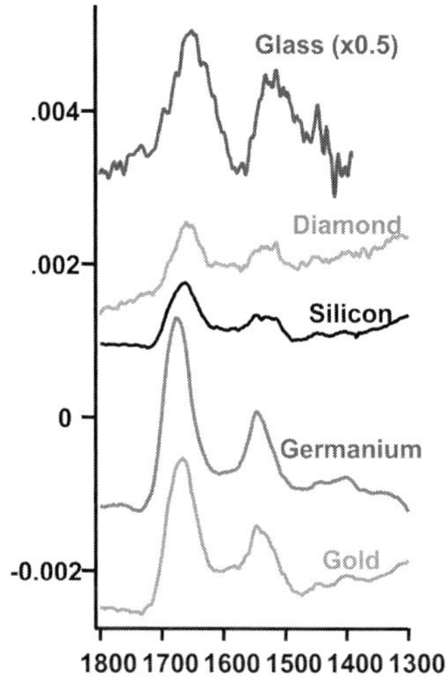

FIGURE 12 **Comparison of spectra of nonspecifically bonded fibrinogen on five different surfaces. Note that the absolute absorbances cannot be directly compared because the absolute coverages on each surface are unknown. However, the signal-to-noise is clearly worst on glass and best on germanium and gold.**

Figure 12 compares spectra of fibrinogen obtained on glass with different substrates. The absolute intensities on the different substrates cannot be easily compared because there is no independent knowledge that the amount of fibrinogen adsorbed on the different materials is the same. The most notable differences are in the signal-to-noise ratio of the measurements. Fibrinogen and other molecular layers yield a comparatively large absorbance n(note that the vertical scale has been reduced by two times for the glass sample compared with the others) but the signal-to-noise ratio is comparatively poor because the total amount of light reaching the detector is small. Diamond (here a bulk sample with $n = 2.35$ and $k = 0$) similarly has a moderately poor S/N ratio due to its low absorbance. Silicon, germanium, and gold with their higher dielectric constants yield the highest S/N ratio. A second difference is in the frequencies of the amide I band. As noted earlier, the peak frequencies can be shifted due to dispersion effects. Dispersion-induced shifts are most pronounced on materials such as gold, germanium, and silicon. Allara and coworkers showed that dispersion effects on metallic substrates always shifted the absorption peaks to higher frequency [29]. Our results show that this general trend holds for other substrates: silicon and diamond show the lowest peak frequencies for the amide I band, while germanium and gold show the highest.

Conformational analysis of proteins on surfaces by FTIR

Frequencies of amide I bands have been correlated with specific conformations such as α-helical, β-sheet, and random coil structures. For example, β-sheets give rise to a modes typically in the range of 1620–1640 cm^{-1}, and vibrational modes at 1650–1658 cm^{-1} are typically associated with the α-helix although there is some overlap with random structures [5]. One of the most important potential uses of FTIR spectroscopy in biological research is to investigate changes in conformations of proteins at surfaces.

To test the ability of FTIR to analyze conformational changes of proteins at surfaces, we compared spectra of antibodies to *E.coli* (K12 strain) that were covalently linked to diamond thin films. Figure 13 outlines the basic process used to covalently link antibodies to diamond thin films. First, a molecule bearing a protected primary amine group was grafted to the H-terminated diamond surface using photochemical grafting procedures described previously [36, 52]. The molecule used in these studies was trifluoroacetamide-protected aminodecene, referred to as TFAAD. After deprotection in HCl, this yields a surface terminated with primary amine groups; this same procedure was used to produce the amine-terminated silicon surfaces depicted in Figure 4. After grafting to the diamond thin film and deprotecting to yield primary amine groups, glutaraldehyde was used as a bifunctional linker to bind free amine groups of the antibody to the amine-modified diamond sample [46].

Figure 13 shows the resulting FTIR spectra for several *E. coli*-modified diamond thin films measured at $\theta_{inc} = 63^{o}$ using p-polarized light. The bottom spectrum shows a fresh sample immediately after preparation, clearly showing features from the amide I (~1650 cm^{-1}) and amide II (1500–1550 cm^{-1}) bands. The amide I peak shows a main peak at 1643 cm^{-1} and a shoulder near ~1672 cm^{-1}. To test whether we could identify changes in the amide I peak associated with changes in the secondary structure, the fresh sample was intentionally denatured by immersion in water at 70°C for 20 min, followed by rinsing to remove the dissolved salts. The resulting denatured spectrum shows a clear change, particularly a decrease in the high-frequency peak at 1643 cm^{-1} and an increased peak at 1662 cm^{-1}. We also investigated the effect of storing the sample at 37 °C in 0.1 molar phosphate-buffered saline (PBS) solution. The resulting stored spectrum again shows a sharp peak at 1643 cm^{-1}, although some broadening is observed. The data suggest that storing the sample induces some subtle changes in the shape of the amide I band between stored and fresh samples, but the differences are less pronounced than those observed on the denatured sample. Further studies will be needed to clearly identify the specific conformational changes that accompany these changes in the FTIR spectrum. Nevertheless, the data do establish that it is possible to directly observe covalently grafted Ecolab antibodies and to observe changes in the FTIR spectrum that reflect changes in the antibody conformation.

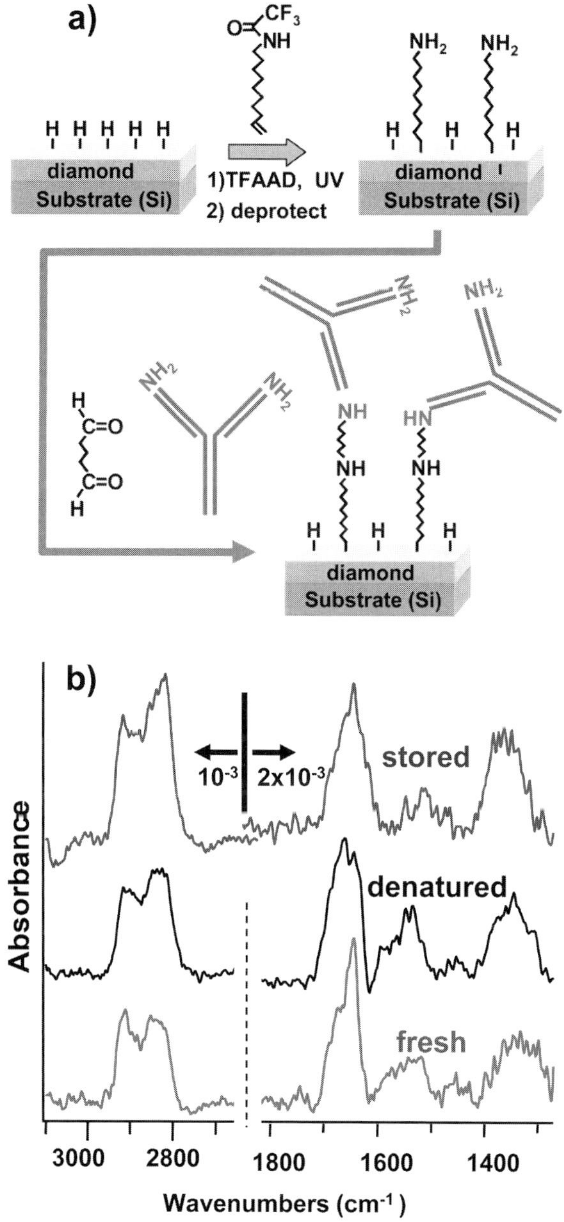

FIGURE 13 Infrared spectra of E. coli antibodies covalently immobilized onto H-terminated diamond thin film sample using p-polarized light, $\theta_{inc} = 63^o$. Spectra are shown for a freshly prepared sample, a sample after denaturating in 70°C water for 20 min, and for a sample stored for 3 days in 37°C PBS buffer solution.

CONCLUSIONS AND OUTLOOK

FTIR spectroscopy is a powerful tool for characterizing molecular and biomolecular layers on surfaces. The single-bounce external reflection geometry is extremely versatile because it can be applied to both transparent and absorbing materials. However, optimum conditions are highly sample dependent. The examples discussed here show how the combination of Fresnel modeling with experimental measurements of model systems such as PMMA thin films yields important fundamental insights into the infrared spectra, including (1) how to optimize experimental conditions for maximum sensitivity, (2) how to understand the intensities and shapes of the peaks, and (3) the fundamental origins of material-dependent and measurement-dependent frequency shifts of spectral features such as the amide bands of proteins at surfaces.

ACKNOWLEDGMENTS

This work was supported in part by the National Science Foundation grants CHE-0911543, DMR-0706559, and DMR-0520527. RJH thanks Dr. John Carlisle of Advanced Diamond Technology and Dr. James E. Butler for supplying diamond thin films used in these studies.

REFERENCES

[1] Surewicz WK, Mantsch HH, Chapman D. Determination of protein secondary structure by Fourier-transform infrared-spectroscopy – a critical-assessment. *Biochemistry* 1993;32(2), 389–394.

[2] Marsh D, Muller M, Schmitt FJ. Orientation of the infrared transition moments for an alpha-helix. *Biophys. J.* 2000;78(5), 2499–2510.

[3] Kim J, Somorjai GA. Molecular packing of lysozyme, fibrinogen, and bovine serum albumin on hydrophilic and hydrophobic surfaces studied by infrared-visible sum frequency generation and fluorescence microscopy. *J. Am. Chem. Soc.* 2003;125(10), 3150–3158.

[4] Roach P, Farrar D, Perry CC. Surface tailoring for controlled protein adsorption: effect of topography at the nanometer scale and chemistry. *J. Am. Chem. Soc.* 2006;128(12), 3939–3945.

[5] Chittur KK. FTIR/ATR for protein adsorption to biomaterial surfaces. *Biomaterials* 1998;19(4–5), 357–369.

[6] Branch DW, Wheeler BC, Brewer GJ, Leckband DE. Long-term stability of grafted polyethylene glycol surfaces for use with microstamped substrates in neuronal cell culture. *Biomaterials* 2001;22(10), 1035–1047.

[7] Steiner G, Tunc S, Maitz M, Salzer R. Conformational changes during protein adsorption. FT-IR spectroscopic imaging of adsorbed fibrinogen layers. *Anal. Chem.* 2007;79(4), 1311–1316.

[8] Alcantar NA, Aydil ES, Israelachvili JN. Polyethylene glycol-coated biocompatible surfaces. *J. Biomed. Mater. Res.* 2000;51(3), 343–351.

[9] Remes Z, Kromka A, Kozak H, Vanecek M, Haenen K, Wenmackers S. The infrared optical absorption spectra of the functionalized nanocrystalline diamond surface. *Diam. Relat. Mat.* 2009;18(5–8), 772–775.

[10] Chabal YJ. Surface infrared-spectroscopy. *Surf. Sci. Rep.* 1988;8(5–7), 211–357.

[11] Soderquist ME, Walton AG. Structural-changes in proteins adsorbed on polymer surfaces. *J. Colloid Interface Sci.* 1980;75(2), 386–397.

[12] Lu DR, Park K. Effect of surface hydrophobicity on the conformational-changes of adsorbed fibrinogen. *J. Colloid Interface Sci.* 1991;144(1), 271–281.
[13] Tolstoy VP, Chernyshova IV, Skryshevsky VA. *Handbook of Infrared Spectroscopy of Ultrathin Films.* Hoboken, NJ: John Wiley and Sons, Inc; 2003.
[14] Roach P, Farrar D, Perry CC. Interpretation of protein adsorption: surface-induced conformational changes. *J. Am. Chem. Soc.* 2005;127(22), 8168–8173.
[15] Onodera K, Hirano-Iwata A, Miyamoto KI, Kimura Y, Kataoka M, Shinohara Y, Niwano M. Label-free detection of protein–protein interactions at the gas/water interface through surface infrared spectroscopy: discrimination between specific and nonspecific interactions by using secondary structure analysis. *Langmuir* 2007;23(24), 12287–12292.
[16] Terrettaz S, Ulrich WP, Vogel H, Hong Q, Dover LG, Lakey JH. Stable self-assembly of a protein engineering scaffold on gold surfaces. *Protein Sci.* 2002;11(8), 1917–1925.
[17] Brunner H, Mayer U, Hoffmann H. External reflection infrared spectroscopy of anisotropic adsorbate layers on dielectric substrates. *Appl. Spectrosc.* 1997;51(2), 209–217.
[18] Griffiths PR, de Haseth JA. *Fourier Transform Infrared Spectrometry.* Hoboken, NJ: John Wiley; 2007.
[19] Surewicz WK, Mantsch HH. New insight into protein secondary structure from resolution-enhanced infrared-spectra. *Biochimica Et Biophysica Acta* 1988;952(2), 115–130.
[20] Surewicz WK, Olesen PR. On the thermal-stability of alpha-crystallin—a new insight from infrared-spectroscopy. *Biochemistry* 1995;34(30), 9655–9660.
[21] Yang WJ, Griffiths PR, Byler DM, Susi H. Protein conformation by infrared-spectroscopy—resolution enhancement by Fourier self-deconvolution. *Appl. Spectrosc.* 1985;39(2), 282–287.
[22] Susi H, Byler DM. Protein-structure by Fourier-transform infrared-spectroscopy–2nd derivative spectra. *Biochem. Biophys. Res. Commun.* 1983;115(1), 391–397.
[23] Byler DM, Susi H. Examination of the secondary structure of proteins by deconvolved FTIR spectra. *Biopolymers* 1986;25(3), 469–487.
[24] Susi H, Byler DM. Fourier deconvolution of the amide I Raman band of proteins as related to conformation. *Appl. Spectrosc.* 1988;42(5), 819–826.
[25] Dong A, Huang P, Caughey WS. Protein secondary structures in water from 2nd-derivative amide I infrared spectra. *Biochemistry* 1990;29(13), 3303–3308.
[26] Schwinte P, Voegel JC, Picart C, Haikel Y, Schaaf P, Szalontai B. Stabilizing effects of various polyelectrolyte multilayer films on the structure of adsorbed/embedded fibrinogen molecules: an ATR-FTIR study. *J. Phys. Chem. B* 2001;105(47), 11906–11916.
[27] Tunc S, Maitz MF, Steiner G, Vazquez L, Pham MT, Salzer R. *In situ* conformational analysis of fibrinogen adsorbed on Si surfaces. *Colloids Surf B Biointerfaces* 2005;42(3–4), 219–225.
[28] Yamamoto K, Masui A, Ishida H. Kramers-Kronig analysis of infrared reflection spectra with perpendicular polarization. *Appl. Opt.* 1994;33(27), 6285–6293.
[29] Allara DL, Baca A, Pryde CA. Distortions of band shapes in external reflection infrared spectra of thin polymer films on metal substrates. *Macromolecules* 1978;11:1215–1220.
[30] Hansen WN. Electric fields produced by propagation of plane coherent electromagnetic radiation in a stratified medium. *J. Opt. Soc. Am.* 1968;58(3), 380–388.
[31] Strother T, Cai W, Zhao XS, Hamers RJ, Smith LM. Synthesis and characterization of DNA-modified silicon (111) surfaces. *J. Am. Chem. Soc* 2000;122(6), 1205–1209.
[32] Stewart MP, Buriak JM. Photopatterned hydrosilylation on porous silicon. *Angew. Chem. -Int. Edit.* 1998;37(23), 3257–3260.
[33] Linford MR, Chidsey C. ED. Alkyl monolayers covalently bonded to silicon surfaces. *J. Am. Chem. Soc.* 1993;115(26), 12631–12632.
[34] Cicero RL, Linford MR, Chidsey C. ED. Photoreactivity of unsaturated compounds with hydrogen-terminated silicon(111). *Langmuir* 2000;16(13), 5688–5695.

[35] Wang XY, Ruther RE, Streifer JA, Hamers RJ. UV-induced grafting of alkenes to silicon surfaces: photoemission versus excitons. *J. Am. Chem. Soc.* 2010;132(12), 4048.
[36] Yang WS, Auciello O, Butler JE, Cai W, Carlisle JA, Gerbi JE, Gruen DM, Knickerbocker T, Lasseter TL, Russell JN Jr, Smith LM, Hamers RJ. DNA-modified nanocrystalline diamond thin-films as stable, biologically active substrates. *Nat. Mater.* 2002;1(4), 253–257.
[37] Yang WS, Butler JE, Russell JN, Hamers RJ. Direct electrical detection of antigen–antibody binding on diamond and silicon substrates using electrical impedance spectroscopy. *Analyst* 2007;132(4), 296–306.
[38] Cai W, Peck JR, van der Weide DW, Hamers RJ. Direct electrical detection of hybridization at DNA-modified silicon surfaces. *Biosens. Bioelectron.* 2004;19(9), 1013–1019.
[39] Jackson M, Mantsch HH. The use and misuse of FTIR spectroscopy in the determination of protein-structure. *Crit. Rev. Biochem. Molec. Biol.* 1995;30(2), 95–120.
[40] Liedberg B, Ivarsson B, Lundstrom I. Fourier-transform infrared reflection absorption-spectroscopy (FT-IRAS) of fibrinogen adsorbed on metal and metal-oxide surfaces. *J. Biochem. Biophys. Methods* 1984;9(3), 233–243.
[41] Dumas P, Chabal YJ, Jakob P. Morphology of hydrogen-terminated Si(1 1 1) and Si(1 0 0) sufaces upon etching in HF and buffered-HF solutions. *Surf. Sci.* 1992;269:867.
[42] Chabal YJ. Infrared spectroscopy of semiconductor sufaces: H-terminated silicon surfaces. *J. Mol. Struct.* 1993;292:65.
[43] Jakov P, Chabal YJ. Chemical etching of vicinal si(111): dependence of the surface structure and the hydrogen termination on the pH of the etching solutions. *J. Chem. Phys.* 1991;95(4), 2897.
[44] Garcia SP, Bao H, Hines MA. Understanding the ph dependence of silicon etching: the importance of dissolved oxygen in buffered HF etchants. *Surf. Sci.* 2003;541:252.
[45] Lu MC, Knickerbocker T, Cai W, Yang WS, Hamers RJ, Smith LM. Invasive cleavage reactions on DNA-modified diamond surfaces. *Biopolymers* 2004;73(5), 606–613.
[46] Stavis C, Clare TL, Butler JE, Radadia AD, Carr R, Zeng H, King WP, Carlisle JA, Aksimentiev A, Bashir R, Hamers RJ. Surface functionalization of thin-film diamond for highly stable and selective biological interfaces. *Proc. Natl. Acad. Sci..*, 2010.
[47] Lasseter TL, Clare BH, Abbott NL, Hamers RJ. Covalently modified silicon and diamond surfaces: resistance to nonspecific protein adsorption and optimization for biosensing. *J. Am. Chem. Soc.* 2004;126(33), 10220–10221.
[48] Clare TL, Clare BH, Nichols BM, Abbott NL, Hamers RJ. Functional monolayers for improved resistance to protein adsorption: oligo(ethylene glycol)-modified silicon and diamond surfaces. *Langmuir* 2005;21(14), 6344–6355.
[49] Hamers, R. J.; Stavis, C.; Pokhrel, A.; Franking, R.; Ruther, R. E.; Wang, X.; Cooperrider, M. C.; Zheng, H.; Carlisle, J. A.; Butler, J. E. Characterization of molecular and biomolecular layers on diamond thin films by infrared reflection–absorption spectroscopy. *Diam. Relat. Mater.*, accepted, in press.
[50] Cheng SS, Chittur KK, Sukenik CN, Culp LA, Lewandowska K. The conformation of fibronectin on self-assembled monolayers with different surface-composition – an FTIR/ATR study. *J. Colloid Interface Sci.* 1994;162(1), 135–143.
[51] Iimura K, Nakajima Y, Kato T. A study on structures and formation mechanisms of self-assembled monolayers of *n*-alkyltrichlorosilanes using infrared spectroscopy and atomic force microscopy. *Thin Solid Films* 2000;379(1–2), 230–239.
[52] Sun B, Colavita PE, Kim H, Lockett M, Marcus MS, Smith LM, Hamers RJ. Covalent photochemical functionalization of amorphous carbon thin films for integrated real-time biosensing. *Langmuir* 2006;22(23), 9598–9605.

Chapter 4

Infrared analysis of biomolecule attachment to functionalized silicon surfaces

Norman A. Lapin,[1] Oliver Seitz[2] and Yves J. Chabal[2]

[1] *BioSamPle Solutions, LLC, 116 Research Drive, Bethlehem, PA 18015*

[2] *University of Texas at Dallas, Department of Materials Science and Engineering Laboratory for Surface and Nanostructure Modification, 800 W. Campbell Road, RL 10 Richardson, TX 75080 (USA)*

INTRODUCTION

The purpose of this chapter is to assess the utility of infrared (IR) spectroscopy (in some cases, supported by complementary techniques) in studying biomolecule attachment to inorganic surfaces used in a variety of biosensing applications. To illustrate the attachment of protein and other biomolecules to such

Biointerface Characterization by Advanced IR Spectroscopy. DOI: 10.1016/B978-0-444-53558-0.00004-7

surfaces, the model primarily discussed here is the biotin–streptavidin (SA) coupling system. This model system is selected for its robustness and broad use in biosensing and biotechnology [1,2], a popular example of which is Invitrogen (formerly Dynal) Dynabeads® that are SA coated for the subsequent attachment of biotinylated DNA probes that can hybridize to specific DNA and RNA target sequences. For comparison, systems of biomolecule attachment different from the biotin–SA system are also discussed, as are several relevant substrates, linkers, and attachment schemes for biomolecule surface coupling.

The biotin–(strept)avidin interaction (where (strept) denotes the analog protein SA, produced by the bacterium Streptomyces avidinii) has several robust characteristics and unique features that make it ideal as a general linkage system in a variety of applications as well as a model system for the study of biorecognition.

The noncovalent lock-and-key interaction of avidin or SA with biotin is characterized by an affinity constant of 10^{13}–10^{15} M^{-1} (L/mol), the highest known to exist in nature. The robustness and high affinity interaction of biotin and SA are very useful in spectroscopic studies, as setup conditions may differ from the ideal biological environment necessary to maintain protein integrity [3]. Weak interactions between protein and surface can result in loss of protein or protein denaturation. In contrast, the biotin–SA bond can only be broken under extreme conditions (e.g., pH 4, high temperature and salt concentration) and even stabilizes the SA protein structure (increasing the protein denaturation temperature from 75°C to 112°C) [3]. The specificity of SA–biotin binding ensures that binding is directed only to the target of interest. Hence, biotinylated molecules can be specifically bound to (strept)avidin-linked molecules or (strept)avidin-coated surfaces for applications in molecular recognition or as molecular markers.

Biotin is a small molecule (244 Da) that, when introduced into biologically active macromolecules, does not affect in most cases their biological activity (e.g., enzymatic catalysis or antibody binding). Biotin can also derivatize small molecules (e.g., mononucleotides or thyroid or steroid hormones) that subsequently retain their native activities as enzyme substrates or binding targets of specific antibodies. (Strept)avidin is a tetrameric protein whose each four lobes are ~15 kDa in size. Each lobe contains a cleft (or space) into which biotin has a conformational fit, stabilized primarily by hydrogen bonds at its ureido and other moieties [2].

For sensing, solid inorganic substrates need to be considered. A variety of substrates have been investigated for biomolecule attachment including metal oxides such as titanium oxide [4] and aluminum oxide [5], metals such as gold [6–8], III–V semiconductors such as GaAs [9], as well as silicon and silicon oxide [10–13]. Metals are usually coupled with thiol groups, while metal-oxides react in general with silanes or phosphonates [6, 10, 11, 14]

There is much interest in the chemical and biofunctionalization of silicon and silicon oxide surfaces. Silicon is the material most extensively used for devices in electronic and bioelectronic applications. Silicon dioxide as glass (amorphous silicon oxide) is the most common substrate used in biosensing applications due to its transparency to visible light – allowing fluorescence tagging and quantification of biomolecules. However, glass is largely opaque to the IR region of light and thus hinders direct spectroscopic investigation, while silicon is highly transparent in this region. Since the native oxide of silicon (or its oxide after standard cleans) is both ultrathin (1 – 100 nm) and chemically similar to glass, oxide-terminated silicon is an ideal substrate for the IR study of biomolecule attachment to silicon oxide [15, 16].

A common method to control biomolecule attachment to these surfaces is through small linking molecules, some of which form self-assembled monolayers (SAMs) [15, 17]. These small linkers contain a moiety that reacts with the substrate and a functional group at the other end that can react with the next molecular layer. Generally, there are three major functional groups that can be observed with IR: amines, aldehydes, and carboxylic acids. Primary amines enable attachment to the surface of proteins or other targets via ester or aldehyde groups on the target. Aldehyde groups react with amines. While amine groups are probably the most commonly used head group for linker attachment to targets, their weak IR absorption makes them difficult to track in quantitative analysis. Such difference in intensity is well emphasized in Figure 1. The two surfaces are coated with equivalent silanes molecules, which differ only by their head groups. As deduced by the amount of hydrocarbon due to the same chain length (11 carbons in this case), their coverage are similar. However, the intensity of

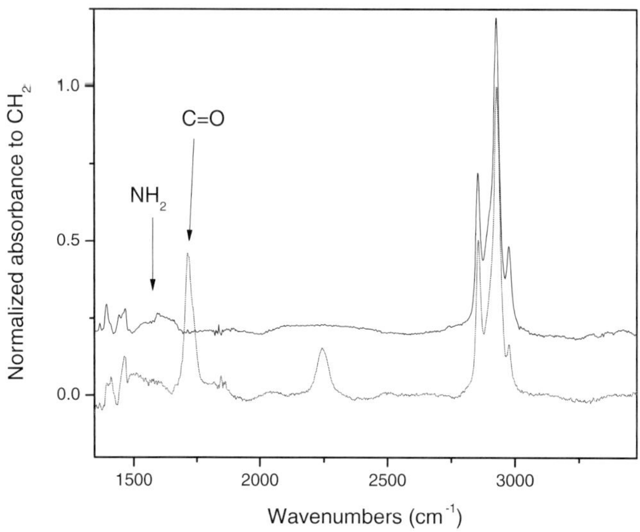

FIGURE 1 Absorption spectra (normalized to the CH_2 asymmetric stretching mode) of amine and aldehyde terminated surfaces with silane molecules containing 11 carbons in their chains. For an equivalent amount of molecules on the surface, the intensity of the amine mode is far weaker than the intensity of the aldehyde group.

the amine group vibration mode is far weaker than the intensity of the carbonyl group, rendering the aldehyde head group more suitable for quantitative analysis.

Carboxylic acid groups are generally converted to an activated ester via EDC/NHS chemistry to link to amines; or are converted to amines via EDC/NHS completed with a diamine molecule. Carboxylic acids also have a very strong IR signature and each reaction step can be followed very clearly with IR spectroscopy to determine reaction efficiency [12].

This chapter largely focuses on the coupling of the biotin–SA system to silicon oxide via silane. In this attachment scheme, shown in Figure 2, aminopropyltriethoxysilane (APTES) is covalently bound to silicon oxide via Si–O–Si bonds to form the aminopropylsiloxane (APS) surface. NHS-derivatized biotin then reacts with the primary amines of APS to form an amide bond linking biotin to the surface. SA is subsequently adsorbed to the biotinylated surface via the steric fit of a surface-bound biotin molecule into one or two of four available clefts on the SA protein. This lock-and-key fit is stabilized by hydrogen bonds between biotin and protein residues (amino acid moieties) internal to the cleft. One to three solution-facing SA clefts are then available at the surface for binding of a variety of biomolecules including other proteins, antibodies, DNA, or small molecules that can themselves be biotinylated for specific binding to the SA surface.

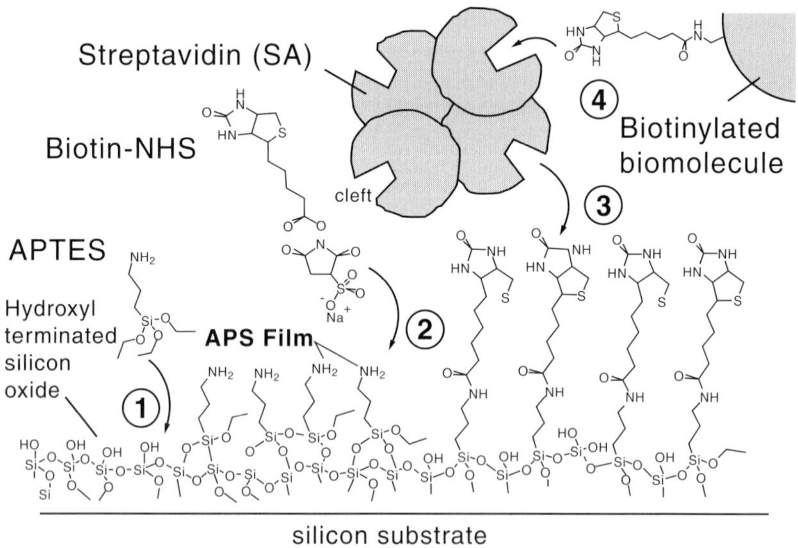

FIGURE 2 Scheme representing APTES molecules that are covalently bound to silicon oxide forming a 3-aminopropylsiloxane (APS) film with Si–O–Si bonds to the surface (1); biotin–NHS is covalently bound to the amine-terminated surface forming amide bonds (2); SA is biospecifically attached to biotin via the SA cleft (3); then biotinylated biomolecules can be biospecifically attached to SA (4). Reproduced from Ref. [15] with permission.

Studying such a complex system, made of successive layers, can be done using various techniques which generally mainly focus on the last layer, that is, the target biomolecules. Moreover, in most cases, the recognition of the biomolecules is obtained because of the attachment of specific marker to the protein. As most common examples can be mentioned labeled protein such as SA – Cy3 or SA – FITC which can be detected using fluorescence technique, or SA – Metal (gold, iron...) which can be used for XPS investigation. In all cases the biomolecule needs a modification of its pristine, which is not required for recognition when using IR spectroscopy. Some other advantages of IR spectroscopy for biomolecular studies are mentioned in the following.

IR radiation excites molecular vibrations of molecules upon its absorption by molecular bonds, and this absorption is used for bond identification. The overwhelming majority of absorbers are organic and hence IR is very useful for biomolecular identification.

IR spectroscopy provides specific advantages in analyzing biofunctionalized surfaces. Its first-order utility is in verification of the presence of specific compounds and chemical bonds in an experimental sample. In a biosurface context, this includes characterization of initial chemical functionalization of the surface, identification of small molecules such as biotin and other ligands, and identification of protein signature (amide bands of the protein backbone) verifying protein presence, and even quantitative analysis when IR absorbance areas are calibrated to known quantities of protein or labeled protein [11,18]. With prior knowledge of expected surface reactions, identification of the attachment type (covalent, ionic, and electrostatic) is possible. IR can be used to identify and verify specific chemical bonds formed and broken upon covalent attachment of molecules to the surface, whereas molecules that retain their original moieties are merely physisorbed to the surface. In the absence of calibration, semiquantitative assessment of covalent bonding efficiency can be assessed by comparison of predefined band areas before and after chemical events.

IR spectroscopy can enable especially high sensitivity for detection of biochemical species at surfaces. Liu *et al.* noted a detection sensitivity of 0.1 pmol SA/cm^2 at amine-terminated gold substrates [11]. Such sensitivities are vital to application in the development of protein and DNA microarrays, whose aim is to minimize the spot size (currently ~100 μm diameter) while maximizing the sensitivity of biomolecules on each spot for high throughput multiplex of thousands of proteins at once [19].

By identifying characteristic features of each chemical layer and biolayer, it is possible to track the behavior of individual layers through a succession of chemical and biochemical attachment steps. Growth, degradation, and structural changes in each layer can be monitored before and after each process step and the effect of structure and integrity of base layers on that of subsequently attached layers can be investigated. When correlated to the biosensor function (beyond the scope of this chapter), it may be possible to correlate device quality and functionality to biolayer features and integrity.

Silicon surface functionalization

Linker attachment on silicon-based substrates can be divided into two types: attachment to silicon oxide surfaces and attachment to oxide-free surfaces. The quality of the initial chemical functionalization of silicon or silicon oxide is key to the subsequent success of biomolecule attachment on these surfaces. The use of IR spectroscopy makes it possible to characterize this initial chemical layer in detail. In addition to detailing the different linker attachments in the two mentioned configurations, this section will focus on the type of information that can be learned about these surfaces using IR spectroscopy.

Functionalization of silicon oxide surfaces

The oxide surface is the most commonly used for functionalization of silicon substrates. The silicon oxide interface can be thermally or chemically grown, or natively present due to natural oxidation of silicon in ambient conditions. At such surfaces, silane molecules are most widely used. While it is advantageous to use these molecules based on the extensive literature for attachment processes, ease of synthesis and ready availability with the desired functional groups, there are drawbacks that will be discussed in the following. We present next the advantages of using IR spectroscopy for characterization of the silanization process, and then some limitations are highlighted.

An example of APS IR absorption spectra is presented in Figure 3, where optimization of APS layers was investigated by Pasternack *et al.* [20]. In this particular case, IR spectroscopy was used to determine the influence of temperature during the adsorption process on the resulting layer. More specifically, Figure 3 shows the spectra characteristic of Si surfaces functionalized with APTES molecules when the chemical reaction with SiO_2 is complete (after 48 h of incubation) and performed at room temperature (Fig. 1(a)), 50 °C (Fig. 1(b)), and 70 °C (Fig. 1(c)). The most prominent features for all spectra are located between 950 and 1250 cm^{-1} (herein referred to as the SiO–X region), where Si–O–Si and Si–O–C modes have been reported. This broad feature is dominated by two bands centered at ~1035 and ~1150 cm^{-1}, which are assigned to SiO_2 transverse optical (TO) and longitudinal optical (LO) stretching modes, respectively, and a mode at 1107 cm^{-1} attributed to Si–O–C. CH_x stretching modes are present between 2850 and 2990 cm^{-1} for all incubation temperatures. More specifically, at room temperature, the CH_2 asymmetric and symmetric stretching modes are observed at 2932 and 2864 cm^{-1}, respectively, confirming the presence of the propyl chain of the APTES molecule. Also present is the CH_3 asymmetric mode at 2975 cm^{-1} seen most for the surface prepared at room temperature, while the symmetric CH_3 stretching mode is not always fully apparent. The observation of the CH_3 mode suggests that some ethoxy groups have not been completely hydrolyzed as previously suggested and, consequently, that the APTES molecules are not fully bonded

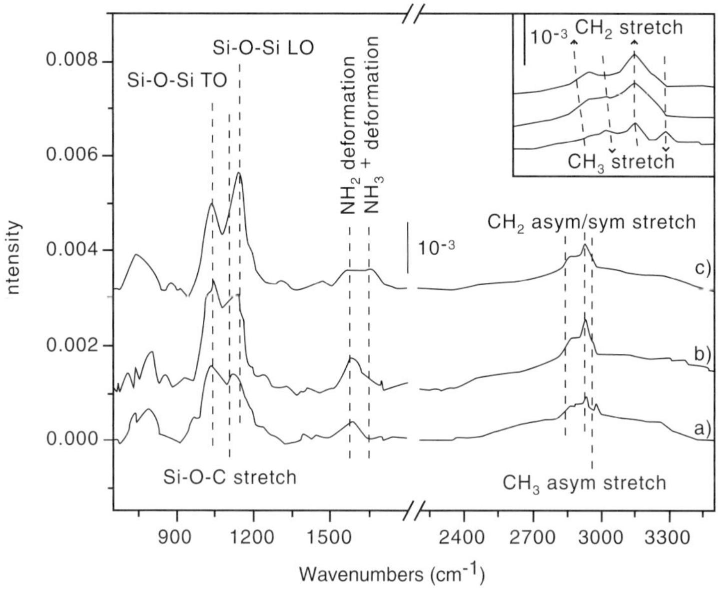

FIGURE 3 Transmission IR spectra for the APS film prepared at (a) room temperature, (b) 50 °C in situ, and (c) 70 °C in situ displayed on a common scale. The spectra are referenced to the SiO_2 surface. The inset shows the CH_x region (2750–3050 cm^{-1}) with the same ordering (a–c). Reproduced from Ref.[20] with permission.

either with the SiO_2 surface and/or with a neighboring APTES molecule (horizontal polymerization). Interestingly, the area of the CH_3 stretching mode drops significantly (by a factor of 70%) when the sample is prepared in solution at 70 °C.

The remaining spectral features include an absorbance feature at 1570 cm^{-1}, corresponding to the NH_2 scissor vibration and confirming the presence of the NH_2 terminal group of APTES molecules. In addition to the NH_2 scissor mode, another feature at 1610 cm^{-1}, corresponding to the asymmetric $–NH_3^+$ deformation mode, is sometimes present (e.g., Fig. 1(c),). The weak feature at 1490 cm^{-1} is also assigned to the symmetric $–NH_3^+$ deformation mode. The latter increases in intensity with temperature; as evidenced by the behavior of the broad absorption feature on the low frequency side of the $–NH_2$ bending mode at 1610 cm^{-1}. The presence of these modes suggests that when samples are exposed to air after synthesis, water molecules are weakly bonded to the NH_2 groups, thereby allowing for the protonation of the amine. The formation of $–NH_3^+$ termination (as indicated by the growth of the mode at 1610 cm^{-1}) at higher incubation temperatures implies increasing stability of the ionized form of the amino group with increasing temperature, possibly influencing the mechanism though which the APS film may interact with its surrounding environment or subsequent chemical treatment.

In depth *in situ* IR absorption spectroscopy investigation of this system has elucidated parameters important to the self-assembly of APTES on chemically grown silicon oxide (SiO_2) to form the APS film. Preannealing to 70 °C produces significant improvements to the quality of the film: the APS film is denser, and the Si–O–Si bonds between the molecules and the SiO_2 surface are more structured and ordered with only a limited number of remaining unreacted ethoxy groups. In contrast, postannealing the functionalized SiO_2 samples after room temperature reaction with APTES (i.e., *ex situ annealing*) does not lead to any spectral change, suggesting that postannealing growth has no strong effect on the horizontal polymerization as suggested earlier. The combined IR and ellipsometry data show that the higher the solution temperature, the denser and thinner the APS layer is for a given immersion time.

In another IR spectroscopy study, the presence and intensity of the LO phonon mode induced upon silanization was shown to be correlated to the formation of an intimate Si–O–Si bond at the interface that extends the underlying oxide matrix [21]. The characteristic LO absorption frequency, slightly blue-shifted from the LO mode associated with the initial underlying oxide thickness, makes it possible to distinguish truly interfacial Si–O–Si bonds from other Si–O–Si bonds formed within the APTES layer (or other examples of silane layers), such as those formed during polymerization. IR spectroscopy is the only technique that can provide such precise information, and was shown to be essential to study both synthesis and stability of thin organic layers on oxidized silicon [21].

IR spectroscopy is therefore ideal to identify issues associated with APS layers, and therefore provides guidance for improvement in the formation of the film. Indeed, the growth of such layers is still subject to many problems, as discussed in the following section.

Activation methods for optimal results

Silane attachment on SiO_2 surfaces requires the presence of hydroxyl groups for attachment of the molecules. These hydroxyl groups can be generated by different processes, the most common being piranha or oxygen plasma treatments. While these treatments are usually harmless when blank surfaces are used (silicon samples or glass substrates), they can be problematic for patterned devices because of their influence on surrounding surface functionalization. Piranha solution (sulfuric acidhydrogen peroxide; 3/7; v/v) is usually used to clean the surface for all residual hydrocarbons and metallic ions. In patterned devices, this treatment can be harmful to any metallic areas such as electrical contacts.

Similarly, oxygen plasma treatment is a convenient dry treatment for surface cleaning and hydroxyl group formation. However, this treatment generates interface roughness and creates electrical defects/traps in the oxide layer, caused by high-energy ions that can negatively affect the performance of the associated electrical devices. While electrical measurements are sensitive to interface damage, IR spectroscopy has identified and quantified the structural damage resulting from oxygen plasma treatment as shown in Figure 4.

Figure 4 shows the absorption spectra of a freshly cleaned SiO_2/Si substrate before (a) and after (b) oxygen plasma treatment (RIE plasma, 300 mTorr, 35 W for 20 s on each side). Spectrum (c) is the difference absorption spectrum emphasizing the changes due to the plasma treatment. There is a clear perturbation of the two components (LO and TO phonon modes) of the silicon oxide characterized by a loss of intensity. The loss of intensity of the higher frequency component of the LO mode indicates that the thickness of the oxide matrix has been reduced, that is, the surface Si–O–Si bonds have been reacted.

The degree of damage can be quantified using an appropriate calibration. Previous IR absorption studies of silicon samples (FZ, double sides polished) have shown that samples with 10 Å silicon oxide (on each side determined by spectroscopic ellipsometry) gives an integrated area of 0.33 cm^{-1} for a measurement performed in transmission (~70° incidence). The integrated area is typically measured between 900 and 1300 cm^{-1} to fully include the LO and TO modes of the SiO_2. In Figure 4, the integrated area of the SiO_2 modes recorded

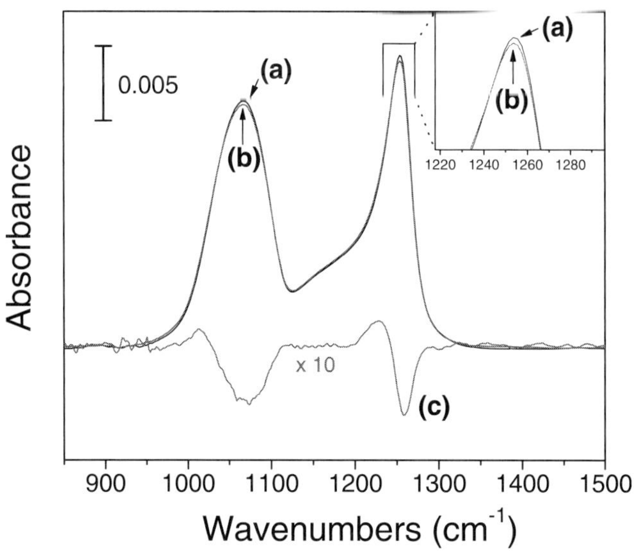

FIGURE 4 **Absorption spectra of freshly cleaned SiO_2/Si substrate before (a) and after (b) O_2 plasma treatment. Spectrum (c) is the difference absorption emphasizing the changes due to the plasma treatment. A removal/disturbance of the SiO_2 absorption bands occurs during the treatment. Reproduced from Ref. [16] with permission.**

before and after oxygen plasma treatment, decreases from 2.86 cm to 2.84 cm^{-1}. These areas correspond to thicknesses before and after treatment of 8.6 and 8.1 nm, respectively. The oxygen plasma treatment leads to the removal/damage of the top 5 Å of the SiO_2 layer. Since one monolayer of SiO_2 has been defined in the literature to span a range of thicknesses (from 0.16 to 0.47 nm) [22], the observe damage/removal is therefore confined to the two first monolayers.

APS layer formation is highly dependent on atmospheric conditions

Observed sensitivity of molecular layer formation to preprocess humidity suggests that initial environmental conditions play a significant role in both growth and stability of the layer. Since all APTES adsorptions to silicon wafers are carried out inside a moisture-free glovebox, it is the humidity prior to entry to this glovebox that was measured. Ambient humidity was measured as atmospheric moisture in g water/m^3 air prior to APTES adsorption on silicon oxide surfaces. Figure 5(a) shows the behavior of APS layer growth with increasing atmospheric moisture (as measured by the local weather station at the time of APTES adsorption). At low moisture content, Si–O–Si band area (indicative of APS thickness) is low and repeatable. At high atmospheric moisture (>6 g water/m^3 air) the thickness of the APS layer varies widely with increasing variability up to approximately fivefold at 18 g water/m^3 air. It is noted that

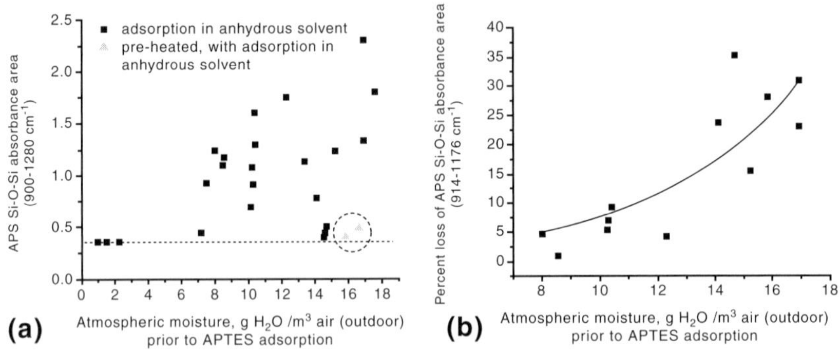

FIGURE 5 (a) Amount of APS adsorbed to the silicon oxide surface (Si–O–Si region, ~900 cm^{-1} – 1280 cm^{-1} band area) versus presilanization atmospheric moisture content (outdoor humidity reading with indoor experiment). Black squares indicate APTES adsorption in an anhydrous solvent. Additional samples were preheated to 100–200°C, flushed with dry nitrogen gas and then sealed from atmospheric moisture prior to exposure to APTES in the anhydrous solvent (circled triangles). (b) APS Si–O–Si band area percentage loss (taken as a measure of surface instability) after the APS surface was exposed to biotin–NHS in deionized water for 60 min and subsequently sonicated in deionized water for 10–15 min versus presilanization atmospheric moisture content (outdoor humidity reading with indoor experiment). Least squares fit, R^2) 0.69. Reproduced from Ref. [15] with permission.

the surfaces were prepared indoors while the moisture content readings are outdoor. Nonetheless, a relationship exists suggesting sensitivity of the APS growth to initial moisture conditions. It has been reported previously that the thicker the water layer at silicon oxide surfaces, the more the growth of silane at the surface [20, 23, 24]. As a negative control, surfaces were heated immediately prior to the entry of silicon wafers into the glovebox for APTES adsorption (triangles in Fig. 5(a)), on a day when environmental moisture readings were high. APS growth was observed to be significantly decreased, returning near to values at low atmospheric moisture. These results corroborate the evidence that surface water content has a significant effect on APS growth. Based on these observations, the greater the humidity, the thicker the layer of water molecules at the silicon oxide surface is likely to be.

In the high atmospheric moisture range prior to APTES adsorption (>6 g water/m^3 air), as moisture increases, there is a greater percentage loss of the Si–O–Si band complex (Fig. 5 (b)). This trend suggests that stability of the APS layer decreases with greater numbers of water molecules in the environment.

APS layer formation is highly uncontrolled and unstable in aqueous media

Silane surface modification for biological application has mostly been associated with the use of short molecules such as APTES or C4-ald because of their extensive literature. However, recently it was shown that such choices are not optimal when stability and reproducibility of the outcome are required [16].

The ability to control the adsorption process (reproducibility of the outcome) as well as the stability of the SAM in a buffer solution was investigated for short amine alkyl chain length using spectroscopic ellipsometry, which is more useful than IR spectroscopy for amine-terminated molecules for reasons mentioned earlier. In this work, the thicknesses obtained as a function of immersion times are compared before and after 2 h immersion in the buffer solution.

Figure 6 shows the bar graph of the resulting APTES layer thickness as a function of treatment time. The black bars represent the (mono)-layer thicknesses obtained just after the grafting process is completed. Each data point taken after a given treatment time is obtained on a freshly prepared sample. The adjacent red bars indicate the thickness obtained for the same sample after a subsequent 2 h immersion in phosphate buffered saline (PBS). Right after grafting, the thickness values range from 0.4 to 1.2 $\pm$ 0.1 nm. Importantly, note that the results described here represent a single set of experiments. If another experiment were performed, the data would vary. This is because the nature and thickness of the layer obtained by attaching APTES is not reproducible. Therefore, Figure 6 highlights the important point that the adsorption is often incomplete or polymerization takes place (initiated by an uncontrolled amount of water adsorbed on the surface). In this case, if some polymerization occurs, the range of thicknesses obtained for APTES layers is still reasonable since it

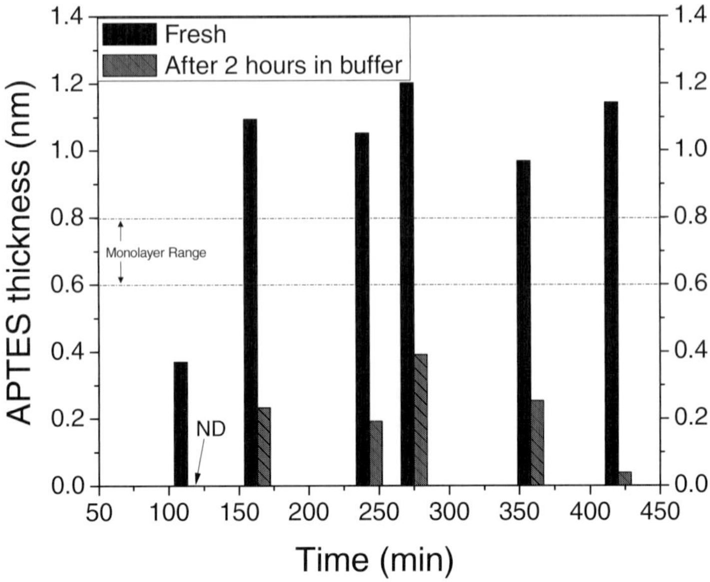

FIGURE 6 Film thicknesses determined by spectroscopic ellipsometry of (mono)layers made from APTES molecules as a function of adsorption time. Black columns are measurements right after the surface is freshly prepared; red/patterned columns are of the same sample after a given adsorption time, immersed in a 0.1 × PBS buffer solution for 2 h. All thicknesses have an error of ±0.1 nm. For all the different (mono)layers, a drastic loss of molecules on the surface is observed. ND = none detected. Reproduced from Ref. [16] with permission.

does not greatly exceed the monolayer range indicated in the graph (7 ± 1 Å). These values are somewhat lower than what is more commonly obtained for APTES adsorption (~2–5 nm), which again points to the variations that can occur for APTES due to variations in humidity, range of temperature (in the laboratory and the solution), and the manipulator (sample drying time, sample transfer time, etc.).

After immersion in the buffer solution, the thickness decreases substantially, no matter what the starting thickness is, indicating a loss of material. At least 60% of the APTES is removed from the surface.

To highlight these observations, an IR study was performed with a layer involving an aldehyde head group, which was shown to yield films that are similar to APTES layers [16].

The FTIR absorption spectra shown in Figure 7 correspond to (a) a freshly Si–SiO_2-C4-ald prepared surface, and (b) the similarly coated surface treated for 2 h in the buffer solution. Both surfaces are referenced to the initial Si–SiO_2 surface. The spectrum 7(c) shows the difference spectrum, highlighting the chemical modification of the surface, such as the loss of aldehyde groups. The

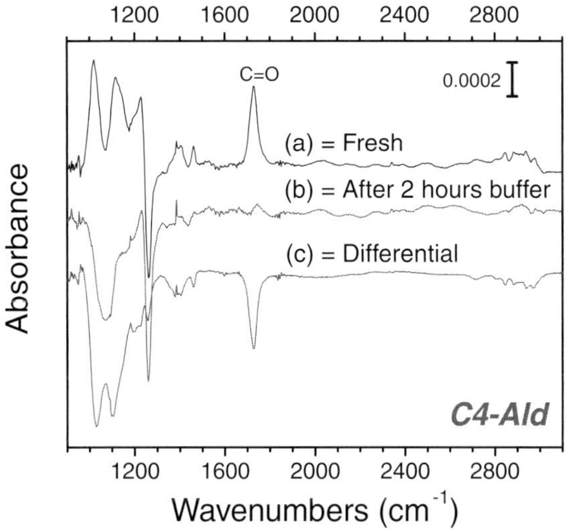

FIGURE 7 Absorption spectra of (a) freshly assembled C4-ald molecules on SiO_2/Si substrate, and (b) after assembling and left 2 h in a buffer solution. Both spectra are referenced to a clean SiO_2/Si substrate before plasma treatment. Spectrum (c) is the differential absorption emphasizing the changes due to buffer treatment. After buffer treatment, most of the molecules have been removed from the surface. Reproduced from Ref. [16] with permission.

initial surface, obtained after immersion in C4-ald solution in toluene overnight, is characterized by absorption in the 2900 cm^{-1} (CH_x) region originating from the chain hydrocarbons, and by the strong and distinct signature of the C4-ald molecules at 1728 cm^{-1} corresponding to the C=O stretch bond from the aldehyde group. The integrated area under this C=O stretch bond is 0.0203 cm^{-1}. Other modes associated with the attached molecules are also visible: a peak centered at 1462 cm^{-1} corresponding to the CH_2 bending mode of the chain, and a double structure (1384 and 1400 cm^{-1}) related to the aldehyde C–H modes. At lower frequencies (950–1270 cm^{-1} where Si–O–Si and Si–O–C modes are usually reported) a precise assignment of the observed peaks is difficult because of superposition of positive and negative bands. The two positive bands at ~1020 and ~1116 cm^{-1} are part of a broad band associated to the silane attachment (SiO–X where X = Si or C). Indeed, this broad band is composed of TO and LO phonons, respectively, arising from an increase in oxide thickness (Si–O–Si matrix), and from the Si–O–C stretch mode due to the unhydrolyzed silane molecules. The presence of negative bands (arising from spectral features in the reference spectrum) associated with the plasma treatment (an issue that was discussed previously) hinders the thorough interpretation of this region, as was done for the APS layer formation. The two negative peaks are centered at ~1070 and 1260 cm^{-1} and correspond to the perturbation of the TO and the LO modes of the initial 6.7 nm thick SiO_2 layer, respectively.

The stability of the self-assembled layer was measured by recording the IR spectrum before and after a 2 h immersion in the buffer solution. Figure 7(b) shows these absorbance spectra referenced to the initial Si–SiO_2 surface (spectra a and b), and the difference spectrum (c = b–a). The most striking feature is the nearly total disappearance of the C=O stretching mode at 1728 cm^{-1}. The integrated area measured is reduced to ~0.0021 cm^{-1}, namely ~10% of the integrated area of the freshly coated sample. There is also a loss (~65%) of the hydrocarbon modes in the 2900 cm^{-1} region, which is consistent with the C = O loss although less pronounced because of hydrocarbons contamination usually present in this range. At lower frequencies, only the two negative features are observed at 1068 and 1262 cm^{-1}, due to a loss of SiO_2 during the plasma treatment as previously discussed. There is no gain (positive contribution) in this spectral range (900–1200 cm^{-1}), that is, no extra Si–O–Si or Si–O–C modes corresponding to the presence of the siloxane on the surface can be detected.

Spectroscopic ellipsometry measurements for this system had only shown that the layer thickness decreased upon treatment. The IR absorption measurements make it possible to understand that the instability of the silane molecules is caused by a weak attachment between the SiO_2 surface and the Si–O from the molecules. Indeed if this attachment was strong enough, a positive Si–O signature would still be present in the spectra even after disappearance of the C = O signature.

IR measurements have also guided an optimization of the silanization process. Indeed, it was recently shown that replacing the most commonly used short chain molecules (APTES or C4-ald) used in silanization with their longer counterparts could considerably improve the quality and reproducibility of these SAMs [16].

The FTIR absorption spectra obtained on the Si–SiO_2–*C11-ald* samples are shown in Figure 8. For this SAM, ellipsometry has shown that, after 8 h, a full monolayer was reliably grafted, while shorter immersion time could lead to submonolayer films. In contrast to their shorter counterpart, APTES, no polymerization occurred. The spectra 8(a) and 8(b) show the absorbance of the freshly prepared SAM and after 4 h in the buffer solution, respectively. Both are referenced to the initial Si–SiO_2 surface before plasma treatment. Spectrum 8(c) is the differential absorbance spectra (c = b–a) emphasizing the effect of the buffer solution on the SAM. After SAM adsorption, there is a clear signature of the longer hydrocarbon chains in the 2900 cm^{-1} region. The bands at 2927 and 2856 cm^{-1} are assigned to the asymmetric and symmetric methylene stretch modes, respectively. Their frequencies are usually correlated with the degree of chain packing. For dense packing of alkyl chain molecules, the asymmetric stretch for such chains (~ 12 carbons) had been reported at 2920 cm^{-1} for a methyl head group and at 2926 cm^{-1} for a carboxylic head group. The measured value of 2927 cm^{-1} is higher than the value for an optimum quality SAM with the same head groups (i.e., improvement is possible), but it is sufficient to provide a substantial stability as shown below. The peak at 2978 cm^{-1} is

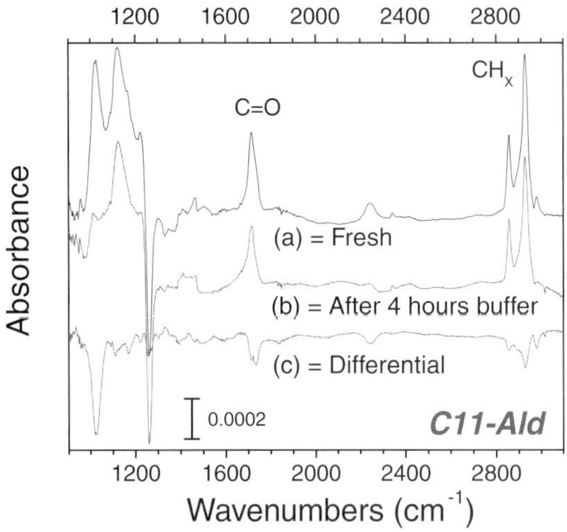

FIGURE 8 Absorption spectra of (a) freshly assembled C11-ald molecules on the SiO_2/Si substrate, and (b) after assembling and left 4 h in a buffer solution. Both spectra are referenced to a clean SiO_2/Si substrate before plasma treatment. Spectrum (c) is the differential absorption emphasizing the changes due to the buffer treatment. A very little amount of molecules have left the surface after the buffer treatment. Reproduced from Ref. [16] with permission.

assigned to the asymmetric methyl stretch mode, which shows that not all the ethoxysilyl groups have been hydrolyzed during the reaction involved in attaching the SAM layer to the SiO_2 surface. The peak corresponding to the C = O stretch bond from the aldehyde group is clearly identified at 1714 cm^{-1}. It is positioned at a lower wavenumber than the peak observed on the Si–SiO_2–*C4-ald* sample (1728 cm^{-1}), indicating a better order and interaction between the carbonyl groups. The CH_2 bending mode (1463 cm^{-1}) and the double structure related to the aldehyde C–H modes (1410 and 1390 cm^{-1}) are observed as in the case of the short chain molecules. The positive peaks corresponding to the additional Si–O–Si and Si–O–C from the silanization process are observed in the 950–1250 cm^{-1} region. The loss of intensity is also visible at 1050 and 1260 cm^{-1} due to the plasma treatment. In addition to these expected vibrational features, an additional peak is always observed at 2250 cm^{-1} and is attributed to the formation of Si–H with oxygen back-bonds.

After a 4 h immersion in the buffer solution, the SAM layer is hardly affected. Spectrum 8(b) confirms that most of the molecules (92%) are still on the surface, as quantified by measuring the integrated area under the C=O stretch band at 1714 cm^{-1} (0.0150 cm^{-1}). In the hydrocarbon stretching region, the peak related to the CH_3 mode is not detected anymore. This removal during the buffer treatment is more clearly seen in the differential spectra 8(c) where a loss of CH_2 (2927 and 2856 cm^{-1}) and CH_3 (2978 cm^{-1}) is observed (overall

~18% loss compared to the initial amount of hydrocarbons). Since the buffer solution is mainly water based, this CH_x loss is mostly explained as a process of hydrolization of the ethoxysilyl group with removal of the $–O–CH_2–CH_3$ group. A part of the decrease can also be attributed to physisorbed chain removal, which is consistent with the loss of C=O stretch intensity.

The examples given above for the functionalization of oxidized silicon surfaces with silanes molecules highlight the importance of IR spectroscopy to characterize and therefore guide the process. It also emphasizes that functionalized oxide surfaces are not as stable as necessary for long-term sensing devices. An alternative approach involves functionalization of oxide-free silicon surfaces (typically H-terminated surfaces), as described in the next section.

Functionalization of oxide-free silicon surfaces

There has been some reluctance to consider oxide-free silicon surfaces as viable templates for field effect transistor-based sensors because of the electrical quality of the SiO_2/Si interface and the belief that a thin thermal oxide layer is therefore an absolute requirement to minimize interface states. However, progress made in functionalization of H-terminated Si surfaces, such as the advantages of reproducibility and control compared to SiO_2 grafting, has kindled a genuine interest in oxide-free surfaces for biosensor applications [13]. Again, IR spectroscopy has played a central role in the development of wet processing methods.

There are two known strategies to obtain a reactive Si surface via wet chemistry, namely (i) hydrogen-terminated Si and (ii) halogen-terminated Si. To obtain a H-terminated Si surface two etching processes are employed. Treatment of Si(100) wafers with 1–2% HF (aqueous) produces atomically flat mono, di, and trihydride terminated Si surfaces. Si(111) wafers treated with 40% NH_4F (aqueous) result in monohydride-terminated Si surfaces (Si–H). The H-terminated Si surfaces are stable in air for few tens of minutes only and get oxidized if exposed for longer time; however, this time is enough to prevent (sub)oxidation and process to the next functionalization step.

To obtain halogen terminated Si surfaces, the most common process consists of preparing chloro-terminated Si surfaces by treating the H-terminated Si(111) with PCl_5 at 80–100°C using benzoyl peroxide as a radical initiator in chlorobenzene [25–27].

Once the reactive Si surfaces are obtained, the organic SAM can be formed on the surface by appropriate chemistry. An example is via hydrosilylation reaction. The alkyl monolayers are prepared on the Si–H surface by inserting a 1-alkene molecule, with the desired group (protected or unprotected) at the other end of the molecule. The reaction is thermally induced but can also be promoted by UV irradiation. Other pathways are also used such as electrochemical reaction using diazonium molecules, or reactions with organomagnesium, organolithium commonly describe as Grignard reaction. Less common but also possible, alcohols compounds are used to create a SAM where a single oxygen is present at the interface between the organic layer and the silicon substrate. The different SAM functionalizations on oxide-free Si surfaces have been recently discussed by Aureau and Chabal in a review where Si preparations and functionalization techniques are described in detail [28].

As an example, we present here the surface functionalization of a Si(111) surface using hydrosilylation reactions, involving the reaction of C=C bonds with Si–H. In order to enable further attachment of biological species, a carboxylic group is required. Since amine or aldehyde groups are not commercially available for such alkene molecules, the carboxylic group becomes a preferred choice. Moreover, carboxylic acid terminated SAM on Si–H surface can be achieved, using two different pathways. The first one involves molecules where the carboxyl group is present, that is, nonprotected. The second involves molecules that contain a protected carboxyl group, with methyl or ethyl groups, that is, requiring further chemistry to deprotect [12].

While both preparations can lead to good quality layers, free of interfacial oxide, higher quality SAMs can be achieved using protected molecules [12]. The quality of such carboxylic-terminated SAMs has been investigated using IR spectroscopy and is presented in Figure 9 for SAM layers prepared via one-step and two-step processes. Both spectra are referenced to the hydrogenated surfaces. In both cases, there is no evidence for the formation of interfacial SiO_2 in the range 900–1200 cm^{-1}, although there is reaction of a fraction of Si–H bonds at the surface (negative band at 2082 cm^{-1}) The attachment of the molecule is best evidenced by the presence of the carbonyl group at ~1715 cm^{-1}. The ultimate assessment of the SAM quality comes from the degree of packing

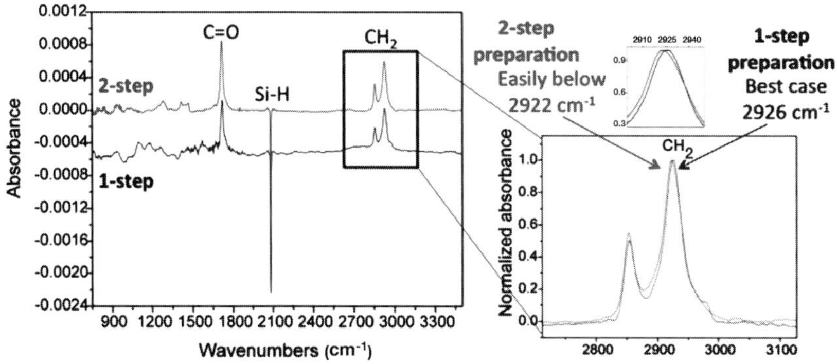

FIGURE 9 **Absorption spectra of carboxylic acid terminated alkyl chains via the one- and two-step processes. Both spectra are referenced to a freshly prepared Si -H surface (Right). Enlargement of the hydrocarbon region emphasizing the differences between the CH_2 asymmetric stretching modes obtained via the two processes (Left). The two-step process leads to a denser film compared to the one-step one.**

density. A sensitive measure of the SAM packing is the values of the asymmetric CH_2 stretch modes. For highly packed films, a stronger interaction among neighboring alkyl chains leads to a lower stretch frequency. For these two systems, with the two-step process mode, the alkyl chain frequency appears at 2921 cm^{-1}, which is a lower frequency than obtained by the one-step process (2926 cm^{-1}). This difference in quality can account for interfacial growth of oxide when any further step is performed on the one-step process SAM, such as attachment of protein, that could be unacceptable in some cases when a high quality interface is required (e.g., for subsequent electrical measurements). Again, IR spectroscopy has played an important role in the optimization of the silicon/SAM interface.

Now that methods for functionalizing both oxidized and oxide-free silicon surfaces have been presented, we turn to the next processing step, namely the attachment of biotin to the linker SAMs.

Biotinylation

This section focuses on understanding and interpreting spectra of the biotinylation of planar surfaces. Biotinylation of amine-terminated surfaces by the biotin derivative, biotin *N*-hydroxysuccinimide ester (biotin–NHS) is in common use and is studied here as a model system. Comparative spectra can indicate whether the biotin has chemically or physically adsorbed to the surface and identify the specific chemical bonds formed. Furthermore, spectral features can be tracked as a means of observing the effects of various surface processing steps including whether there has been degradation of the biotinylated surface as well as the effects of sonication and solvents on surface organization and quality.

The appearance of various bands in the IR spectrum of a sample on a planar surface directly indicates the presence of specific chemical bonds or species at the surface. These bands may point to the presence of specific chemical and biological compounds at the surface such as haptens (e.g., biotin) and proteins (e.g., SA) or to the formation of new bonds due to reaction of these compounds with the surface. Conversely, the disappearance of bands (relative to previous spectra) may suggest breaking of bonds and removal of compounds from the surface.

Furthermore, the area under each band is a semiquantitative indicator of the relative amount of bonds formed or material present (compared to other species at the surface and/or previous amount of the specific bond at the surface).

Since some bands are too close to others to be resolved (and hence areas measured accurately), band complexes can be deconvolved either manually or automatically with software programs such as Origin. The distribution of vibrational states of molecules at the surface can often be well approximated by a Gaussian or Lortenzian distribution. Lortenzian character could be indicative of forced resonance of molecules due to their local environment at the surface, whereas Gaussian behavior suggests more random motions of molecules and

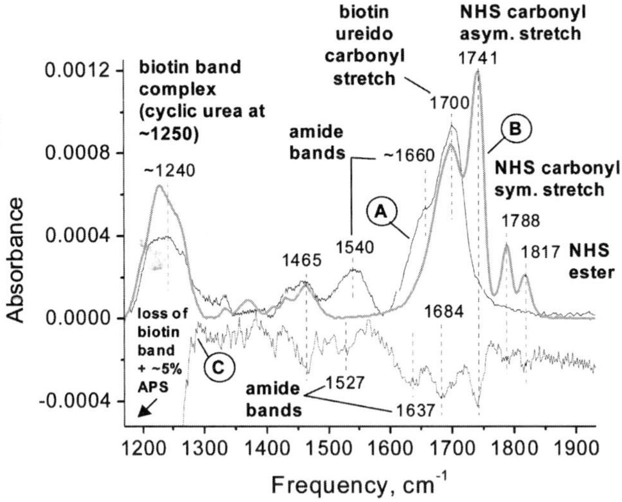

FIGURE 10 Representative FTIR spectra of: (A) an APS surface exposed to biotin–NHS (for covalent attachment) and referenced to the APS surface prior to exposure; (B) a silicon oxide surface exposed to biotin–NHS (for physisorption) and referenced to the preexposed silicon oxide surface; (C) a sonicated biotin–NHS treated APS surface referenced to the same biotinylated surface before sonication (amplified five times). Bands in spectrum C are negative due to decrease in amount of molecules after sonication. Reproduced from Ref. [15] with permission.

hence freedom from influence of the local environment. In a number of cases, one fit is clearly better than the other, suggesting that the behavior of the molecules at the surface varies with environmental conditions.

Figure 10 shows spectra of the biotinylation of an amine-terminated surface. Spectrum B shows the features of physisorbed biotin–NHS with all moieties intact. Spectrum A shows features of biotin chemisorbed to an amine-terminated surface, where the absence of reacted NHS bands and the appearance of amide bands indicate chemical attachment to primary amines at the surface. Spectrum C shows the loss of both amide and NHS bands after sonication of the surface of spectrum A, suggesting that a small amount of physisorbed biotin–NHS that remained at the surface was removed and that physisorbed biotin–APS complexes were also removed, since sonication cannot break the amide bonds.

Within spectrum A, bands of both the biotin ureido (part of the biotin head group) and the amide I moiety overlap. While it is clear that there are two bands present in this complex (the peak of the ureido band at 1700 cm^{-1} and the amide I band as a shoulder at 1660 cm^{-1}), their individual areas are not accessible.

Upon deconvolution of these bands, a Lorentzian fits the ureido well in Figure 11 but in other cases a Gaussian fit was best. Such variation in fits suggests variation in the molecular environment of each biotinylated surface formed. At some surfaces where a Gaussian fit applies, biotin ureido moieties may be less densely packed and thus more free of the influence of neighboring

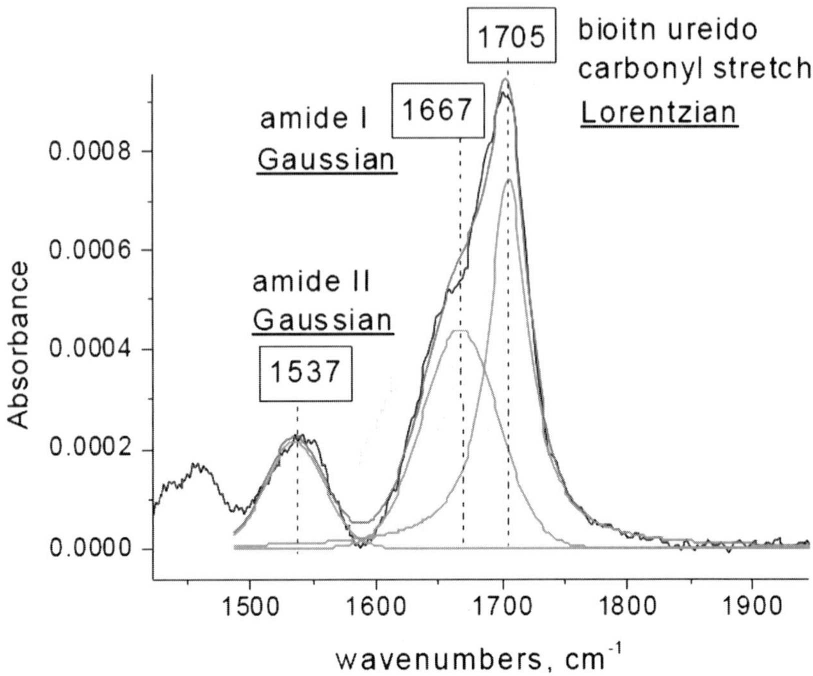

FIGURE 11 Deconvolution of biotinylated surface referenced to the APS surface. Biotin ureido carbonyl stretch and amide bands are deconvolved using Gaussian and Lorentzian peak fitting.

molecules. At surfaces where a Lorentzian fits the ureido band best, biotin molecules may be more densely packed where they can influence each other's motion and resonance more directly.

Alternatively, the amide II band (proportional in area to amide I) could be used to quantify the chemical reaction of biotin with the surface.

BEHAVIOR OF BIOTINYLATED SURFACES IN DIFFERENT ENVIRONMENTS

IR features of the biotinylated surface are observed to vary with different adsorption conditions and even follow a pattern that may reflect changes in the molecular organization of the surface and surface stability. When correlated with the device function, such changes in IR features may be of value in evaluating surface quality. The behavior of IR features of biotinylated surfaces and surfaces exposed to water (e.g., during protein adsorption) vary depending on the solvent used for biotin adsorption.

Figures 12 and 13 show upon adsorption of biotin–NHS to the APS surface, corresponding changes in the biotin ureido band position near 1700 cm^{-1} versus changes in band area at 1250 cm^{-1}, also attributed to the biotin ureido [15]. For

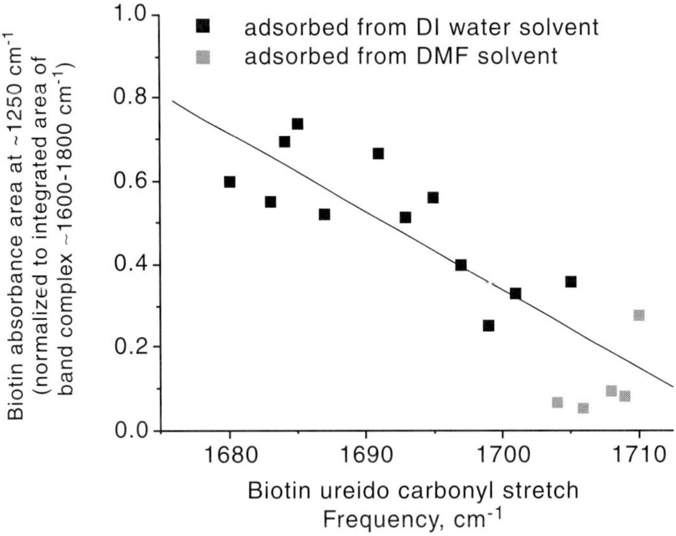

FIGURE 12 Change in biotin ~1250 cm^{-1} band area versus biotin ureido carbonyl frequency. The ~1250 cm^{-1} band area normalized to the 1600–1800 cm^{-1} band complex area versus biotin ureido carbonyl stretch frequency (1600–1800 cm^{-1}). The band area at ~1250 cm^{-1} decreases as ureido carbonyl stretch frequency increases R^2 = 0.71.

anhydrous adsorption conditions (with anhydrous dimethylformamide (DMF) as the solvent), the ureido band is consistently between 1704 and 1710 cm^{-1}. This frequency has higher precision (within 6 cm^{-1}) from experiment to experiment than for aqueous adsorption conditions where the ureido band is lower in frequency and spans a greater range (1680–1705 cm^{-1}). The 1250 cm^{-1} band is also observed to decrease or even disappear when the ureido frequency is near or greater than 1700 cm^{-1} (Fig. 12).

These observations may suggest that biotin adsorption in the aqueous solvent, where rapid degradation of the surface by water likely occurs, creates surfaces that are more disorganized, while anhydrous solutions that do not disrupt the Si–O–Si bonds between APS and the silicon oxide surface allow more organized and stable biotinylated surfaces to form. The repeatability of the ureido band position near 1710 cm^{-1} in this case might therefore reflect greater order of biotin at the surface. We hypothesize that biotinylated surfaces characterized by ureido carbonyl stretch frequencies blue-shifted past 1700 cm^{-1} may have greater molecular organization and stability than those characterized by lower ureido frequencies. This assertion may then imply that biotinylated surfaces formed in aqueous solvent whose ureido band position is greater than 1700 cm^{-1} are more well organized or similar in structure to biotinylated surfaces formed in anhydrous solvent than surfaces with ureido bands at lower

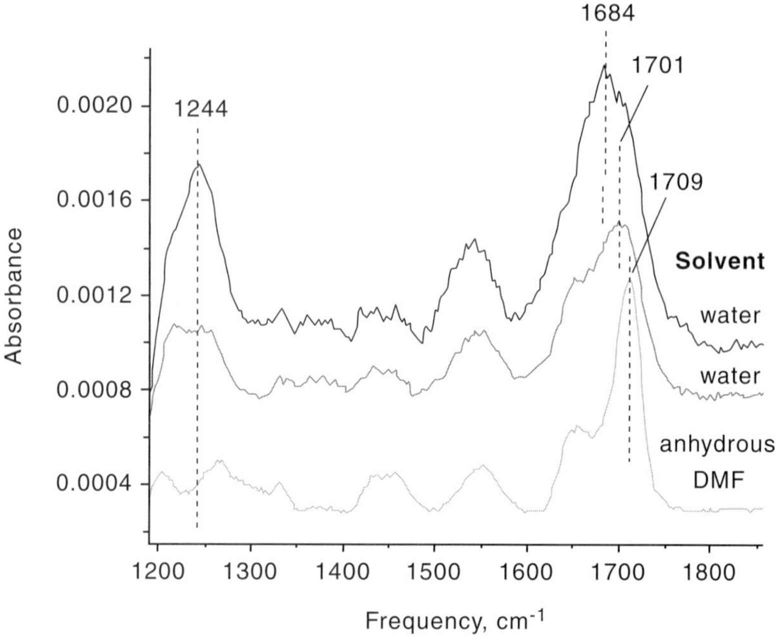

FIGURE 13 Selected spectra of biotinylated surfaces corresponding to data of Figure 12 demonstrating that the area of the biotin band at ~1250 cm^{-1} decreases with increase in the frequency of the ureido carbonyl stretch (1680–1700 cm^{-1}).

frequencies. This observation highlights an effect that is opposite to what is usually observed, namely that increased order is observed to cause a blue-shift toward 1700 cm^{-1} instead of increased organization resulting in red-shifted frequencies.

Figure 14 shows two pairs of spectra (aqueous, (a), and anhydrous, (b)) where the upper spectrum of each pair is that of biotinylation and the lower spectrum is that of loss of IR features after water rinse of the surface. In spectral pair (a), for biotin adsorption in aqueous medium, most of the ureido band is at lower frequency and this band is what is lost upon the exposure of the biotinylated surface to water. In spectral pair (b), for biotin adsorption in anhydrous medium, in the upper spectrum, the peak of the ureido band is near 1710 cm^{-1}, yet most of the band loss is still near 1680 cm^{-1}. In all, loss of IR bands after water rinse are more pronounced near 1680 cm^{-1} for water-solvent biotinylated surfaces than anhydrous-solvent surfaces and overall band loss is greater for water-solvent surfaces as well. Figure 16 shows that after exposure of the biotinylated surface to water, the band area at 1250 cm^{-1} decreases as well.

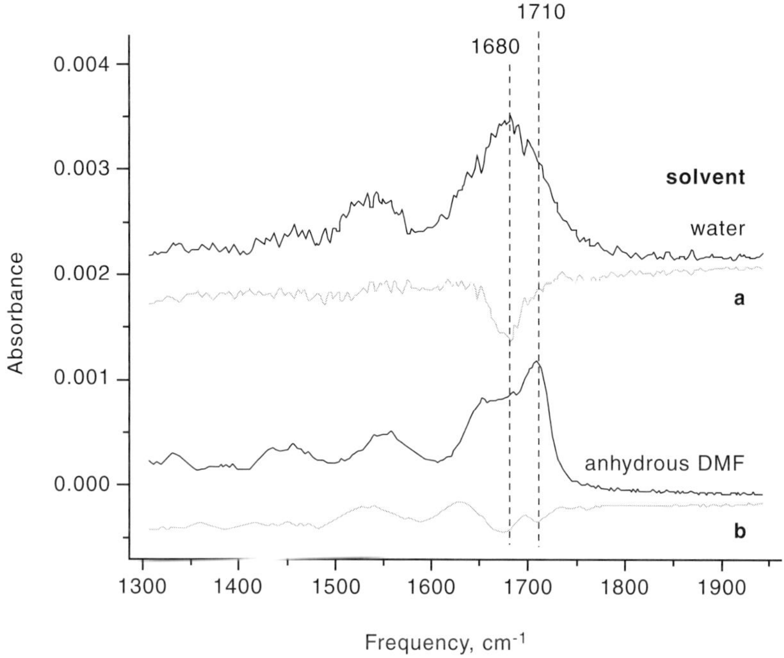

FIGURE 14 **Spectra of biotinylated surfaces before and after the rinse step. Spectra of biotinylated surfaces referenced to respective APS-terminated surfaces (black spectra) and corresponding spectra of the rinsed biotinylated surface referenced to the biotinylated surface before rinsing (gray spectra). Biotin–NHS dissolved in DI water (a) and in anhydrous DMF (b) prior to adsorption to the APS surface.**

The observed IR band loss after exposure of the biotinylated surface to aqueous solution occurs only at frequencies associated with the biotin ureido and not amide bands or other moieties such as siloxanes. Hence, the molecule is still attached to the surface. Since aqueous rinsing is not known to break the chemical bonds of biotin, the molecules are most likely still present at the surface. A possible explanation for the observed band loss might be shielding or other changes in the biotin ureido carbonyl dipole moment due to changes in the local environment of the biotin molecules at the surface. Although the physical significance of such changes in the biotin ureido is not clear, observations suggest that they are more pronounced at biotinylated surfaces with lower frequency ureido bands. That these low frequency ureido bands change in response to environmental conditions (i.e., exposure to water) more easily than their higher frequency counterparts further corroborates the data suggesting that the molecules these bands identify are less organized and less stable than their higher frequency counterparts.

Protein attachment

Upon exposure of the biotinylated surface to avidin or SA, IR spectroscopy is useful for verifying the presence of the protein at the surface. When calibrated to spectra of (strept)avidin coated surfaces of known mass, the amount of protein at the surface can be quantified as well [18]. In the absence of a calibration curve, comparison of band areas can give an approximation of the relative amount of protein gained or lost from the surface.

Figure 15a shows adsorption of SA to a biotinylated surface. Amide I and II bands, characteristic of the backbone of all proteins, are observed at 1640 cm^{-1} and 1530 cm^{-1} respectively, confirming the presence of SA at the surface. Comparison of spectra in Figure 15(a) and (b) show that SA attachment to the biotinylated surface is due only to specific adsorption of SA to biotin, since adsorption does not occur when SA is preblocked with biotin (i.e., SA clefts are presaturated with biotin, leaving none free to bind to surface bound biotin, Fig. 15(b)).

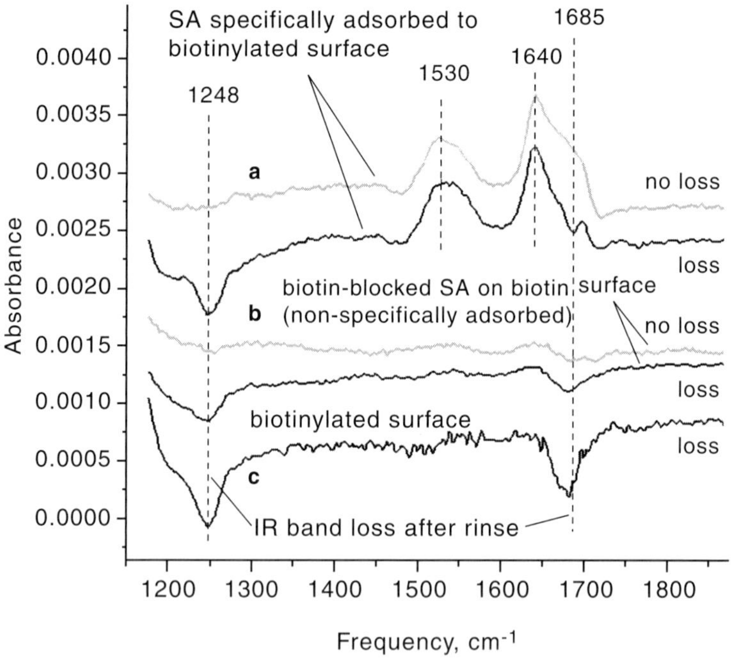

FIGURE 15 Representative FTIR spectra of (a) biotinylated surfaces exposed to SA and referenced to preexposed biotinylated surfaces; (b) biotinylated surfaces exposed to biotin-blocked SA and referenced to preexposed biotinylated surfaces; (c) a biotinylated surface rinsed with phosphate buffered saline (PBS) and PBS with Tween20 (PBST) referenced to the same surface prior to rinse. "Loss" and "no loss" indicate the presence or absence of band loss in above spectra. All samples were rinsed per methods section protocol. All biotinylated reference surfaces were sonicated prior to spectra collection. Reproduced from Ref. [15] with permission.

Behavior of the biotinylated surface upon protein adsorption

In some cases, the biotin ureido band area decreases after SA adsorption to the biotinylated surface (Fig. 15(a), "loss" at 1248 and 1685 cm^{-1}) or when the biotinylated surface is washed with the aqueous solution in the absence of SA (Figs. 15(c) and 16). Based on changes in the biotinylated surface after exposure to water, the observed band loss is likely caused by the aqueous solution that the protein is present in during adsorption. On some biotinylated surfaces, however, this loss is not observed after protein adsorption. Such fluctuation in this band loss phenomena likely related to the same variations in the biotinylated surface is observed in Figures 12–14. The biotinylated surfaces in Figure 15 were prepared in aqueous solvent, yet based on the distribution of IR characteristics in Figure 12, some of these surfaces may behave similarly to surfaces prepared in anhydrous solution, which exhibit less band loss near 1680 cm^{-1} after exposure to water.

Biotinylation and protein attachment to oxide-free silicon surfaces

In this section, we describe briefly the biotinylation and SA attachment on a surface free of oxide. This surface has been less commonly used, but is

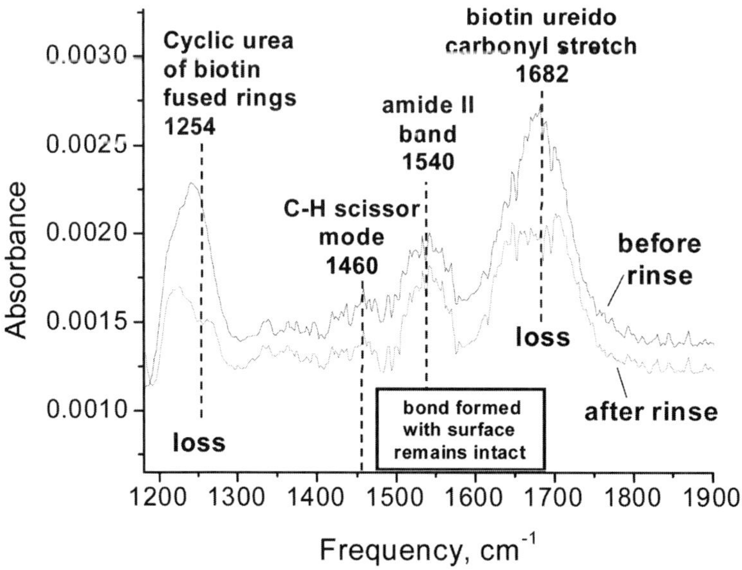

FIGURE 16 FTIR spectra of (top) an APS surface exposed to biotin–NHS, referenced to the APS surface, and (bottom) the same surface after rinsing with PBST/PBS/DI water, referenced to the APS surface. While spectral decrease in the biotin ureido is observed at 1254 and 1682 cm^{-1}, amide and aliphatic scissor mode band areas did not decrease. Reproduced from Ref. [15] with permission.

becoming increasingly more attractive because the process is better controlled. Moreover, functionalization of these surfaces can be done very successfully, without altering the features or quality of the interface (important in the case of taking electrical measurements at the surface).

To prepare an oxide-free surface for biotin and SA attachment, several steps are required to be taken into account.

a. The head groups of the molecules at the starting surface must be activated via reagents such as EDC.
b. Biotin is required to have the appropriate derivatization to react with the head groups at the surface. In this case, the appropriate biotin derivatization is the hydrazide group.

Figure 17 shows the absorption spectra of a Si (100) surface after ethyl undecylenate SAM attachment followed by the attachment of biotin-hydrazide, and then SA attachment. Except for the spectrum of the monolayer, which is referenced to the hydrogenated silicon surface, all spectra are differential, such that each spectrum is referenced to the spectrum after the previous processing step.

After functionalization of the hydrogen-terminated silicon surface with the COOeth-terminated SAM, the surface shows no presence of interfacial oxide in the 900–1200 cm^{-1} region. Moreover, the presence of the carbonyl group and

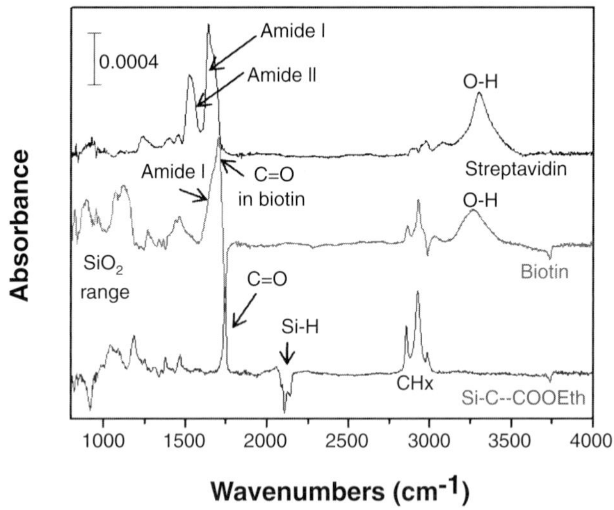

FIGURE 17 **FTIR absorption spectra of a hydrogen-terminated silicon surface functionalized with ethyl undecylenate (Si–C–COOEth), a biotinylated surface, and a streptavidin coated surface. Except the first spectrum referenced to the hydrogen-terminated silicon surface, all spectra are differential, that is, referenced to spectrum of the previous step.**

the CH_x groups at the 1740 and 2900 cm^{-1} regions, respectively, attests to the success of the functionalization.

After activation of the surface to create carboxylic acid groups, and subsequent treatment with EDC and biotin-hydrazide for 90 min in the aqueous solution, no growth of oxide is observed. The features in the region 900–1200 cm^{-1} are related to the biotin attachment. Moreover, the attachment of the biotin on the surface is clearly observed via the negative band at 1740 cm^{-1}, related to the reaction of the COOEth group with the Hydrazide groups to form amide bonds. The peaks observed in the 1500–1750 cm^{-1} are equivalent to the one observed and described in the case of biotinylation of the APS surface.

After biotinylation, SA is the attached to the surface. After 2 h in PBS (aqueous buffer), SA attachment is clear, due to the observation of the amide I and II bands. Note also that no siloxane bands are observed in this case and hence no suboxidation of the interface has occurred.

Compared with the degradation in the aqueous solution of silane-based biotinylated surfaces, this effect is profound and may have important implications for applications that require a high degree of surface stability.

While IR spectroscopy can provide a great deal of information about the attachment chemistry and degradation of surfaces, IR spectroscopy cannot easily be used to differentiate between different proteins. An example is given in the attachment of bovine serum albumin (BSA) on the Si–C surface (Fig. 18).

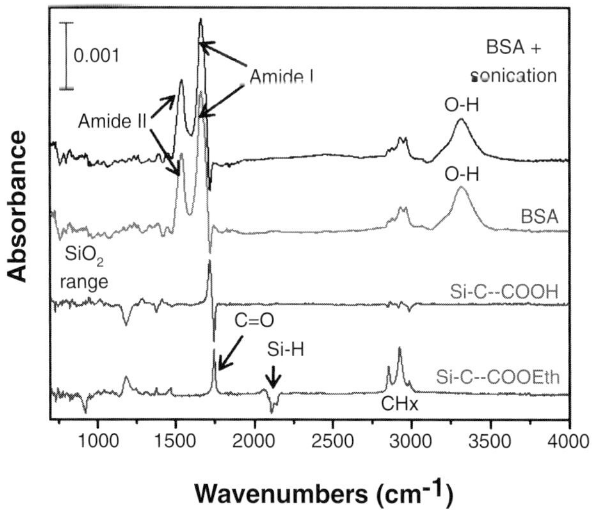

FIGURE 18 FTIR absorption spectra of the Si–H surface functionalized with ethyl undecylenate (Si–C–COOEth), after activation to the carboxylic acid group, after BSA attachment (BSA), and after the treatment of the BSA coated sample in a sonication bath. Except the first spectrum that is referenced to the Si–H surface, all spectra are differential, meaning that they are referenced to the previous step.

Both SA and BSA are identified by their amide I and II bands due to peptide chains present in all proteins.

CONCLUSIONS

IR absorption investigations of biotin–SA attachment at silicon-based surfaces have demonstrated several key advantages of using FTIR to overcome limitations in biointerface analysis. FTIR, uniquely suited for such analyses, can track specific compounds through successive process steps, directly monitor the efficiency of reactions with the surface and uncover nanoscale and molecular level changes with high sensitivity that provide new insights into surface quality and the root causes of its degradation.

In such IR investigations, each interfacial layer in the construction of biotin–SA coupled surfaces has been separately and continuously monitored: from oxide and oxide-free surface preparation to amine functionalization to biotinylation through protein attachment. While ellipsometry has been used to verify layer formation at the nanoscale, it can only track changes in the uppermost layer and does not differentiate between chemical compounds. XPS can identify elemental composition of layers, but is not layer specific, since organic building blocks carbon, nitrogen, and oxygen repeat in each layer. Tracking base layers such as APS through complete surface processing has shed light on multidimensional issues to its stability. It is likely that atmospheric humidity determines the thickness of the water layer on the silicon surface prior to the establishment of a controlled anhydrous environment that causes the profound observed effects on APS layer growth and stability.

The unique IR signatures of the APS layer has permitted its monitoring through the subsequent process of biotinylation in the aqueous solution in which its degradation was clearly observed. Even more sensitive examples of layer differentiation can be found at the SAM/silicon interfaces. In the preparation of carboxylic acid terminated SAMs on oxide-free silicon, IR spectroscopy was instrumental in uncovering the formation of interfacial oxide between the silicon and SAM and in revealing that a one-step process is inferior to the two-step process. Likewise, on oxide surfaces, due to differences in Si–O–Si bonding structure between silicon oxide and terminal groups of APTES, APS layer stability was directly observable due to decrease in APS Si–O–Si band intensity [20]. At the same APS surface, the reaction efficiency was clearly quantified by the band absorption (integrated area) of methylene vibrations present only in unreacted APTES ethoxy groups. Moreover, the observation of modification of the 5 Å thick surface region of plasma-treated silicon oxide is a poignant example of the utility of high sensitivity IR spectroscopy in uncovering potential root causes of molecular attachment inefficiency and electrical device failure.

For the biotinylation step, the ability to specifically track chemical moieties has revealed the presence of unreacted biotin–NHS at the postbiotinylated surface and its removal by sonication (along with biotin–APS chains likely desorbed by water attack on APS-to-surface Si–O–Si bonds).

Protein attachment, investigated after biotinylation was again observed in parallel to biotin modification by the aqueous solution. While it is difficult to determine the causes of such modification or even what its physical meaning is, the recognition of two independent simultaneous events would not have been possible without the molecular signature differentiation possible only with IR characterization.

This final biointerface process step brings to light certain limitations of IR analysis. The structure of the complex environment in which SA attaches to a biotinylated surface is not entirely resolvable by IR spectroscopy. The observed disappearance of the biotin head group related IR bands (cyclic urea at 1250 cm^{-1} and ureido carbonyl stretch at 1700 cm^{-1}) is puzzling, particularly because it is not accompanied by the loss of any other bands (such as amide bands) that might suggest desorption of the entire layer. It is highly unlikely that the covalent bonds connecting the biotin head group to the rest of the molecule could be broken by water (the way Si–O–Si bonds can be). Hence, electronic screening had to be invoked, presumably caused by the rearrangement of the surface molecules that can screen all or part of the IR absorption by biotin moieties. However, the structural implications of these changes are not clear.

In a related issue, since IR absorption itself is a function of fluctuation in the dipole moment of bonds, changes in molecular dipole moments due to changes in local environment can affect band intensity, leading to inaccuracies in quantification based on the band area. While the strength of IR spectroscopy lies in its sensitivity and specificity in the interrogation of small molecules and moieties, this same feature limits its scope in the investigation of large biomolecules such as proteins. Because all proteins (including antibodies, (strept)avidin, fibrinogen, etc.) share the common molecular structure of the peptide chain, their IR signatures are nearly indistinguishable. The amide I and II bands at 1530 and 1640 cm^{-1} appear due to IR absorption of peptide bonds and obscure most other molecular differences between proteins, although secondary structure of peptide chains (e.g., their large-scale conformations as alpha helices and beta sheets) is discernable by deconvolution of the amide I band. However, tertiary structure, or protein folding, is often the critical determinant to protein function and hence is the subject of other biomolecular techniques.

Although noncovalent interactions and the internal structure and function of biomacromolecules is largely outside the scope of IR measurement, the utility of IR spectroscopy in biointerfacial investigations has not yet been fully tapped. In many investigations of biomacromolecules at surfaces, the techniques employed (such as fluorescence microscopy and surface plasmon resonance) cannot directly determine covalent attachment of molecules to surfaces and such investigations often do not include complementary analysis to determine the extent to which such attachment has taken place. IR microscopy enables analysis in two dimensions, and is therefore amenable to the investigation of microarrays [29]. As such, it could be used more extensively to determine microspot quality in terms of covalent attachment of ligands, DNA, or proteins to microarray surfaces [30]. Many biotechnology applications also employ biofunctionalized microbeads (such as Dynabeads® mentioned in the introduction) which could be interrogated by IR with samples in pellet form or even *in situ* aqueous solution in a liquid container through attenuated total reflection geometry at a container wall. There is hence a potential for expansion of IR usage in biotechnology and biosensor quality control.

ACKNOWLEDGMENTS

This work is supported in part by the Texas Higher Education Coordinating Board (Norman Hackerman Adv. Res. Program), and the National Science Foundation (Grant CHE-0911197).

REFERENCES

[1] Diamandis EP, Christopoulos TK. *Clin. Chem.* 1991;37:625–636.
[2] Wilchek M, Bayer EA, Livnah O. *Immunol. Lett.* 2006;103:27–32.
[3] Gonzalez M, Bagatolli LA, Echabe I, Arrondo J. LR, Argarana CE, Cantor CR, Fidelio GD. *J. Biol. Chem.* 1997;272:11288–11294.
[4] Xiao SJ, Textor M, Spencer ND, Sigrist H. *Langmuir* 1998;14:5507–5516.
[5] Saprigin AV, Thomas CW, Dulcey CS, Patterson CH, Spector MS. *Surf. Interface Anal.* 2005;37:24–32.
[6] Brogan KL, Schoenfisch MH. *Langmuir* 2005;21:3054–3060.
[7] Lahav M, Vaskevich A, Rubinstein I. *Langmuir* 2004;20:7365–7367.
[8] Spinke J, Liley M, Schmitt FJ, Guder HJ, Angermaier L, Knoll W. *J. Chem. Phys.* 1993;99:7012–7019.
[9] Ding X, Moumanis K, Dubowski JJ, Frost EH, Escher E. *Appl. Phys. A: Mater. Sci. Amp; Process.* 2006;83:357–360.
[10] Dupont-Filliard A, Billon M, Livache T, Guillerez S. *Anal. Chim. Acta* 2004;515:271–277.
[11] Liu Z, Amiridis MD. *Surf. Sci.* 2005;596:117–125.
[12] Seitz O, Dai M, Aguirre-Tostado FS, Wallace RM, Chabal YJ. *J. Am. Chem. Soc.* 2009;131:18159–18167.
[13] Seitz O, Fernandes P G, Mahmud G A, Wen H-C, Stiegler HJ, Chapman R A, Vogel EM, Chabal YJ, *Langmuir* 2011;27:7337–7340.
[14] Hanson EL, Schwartz J, Nickel B, Koch N, Danisman MF. *J. Am. Chem. Soc.* 2003;125:16074–16080.
[15] Lapin NA, Chabal YJ. *J. Phys. Chem. B* 2009;113:8776–8783.
[16] Seitz O, Fernandes PG, Tian R, Karnik N, Wen H-C, Stiegler HJ, Chapman RA, Vogel EM, Chabal YJ. J. Mater. Chem., 2011, 21, 4384.
[17] Su X, Wu Y-J, Robelek R, Knoll W. *Langmuir* 2004;21:348–353.
[18] Pradier CM, Salmain M, Liu Z, Methivier C. *Surf. Interface Anal.* 2002;34:67–71.
[19] Haab BB, Dunham MJ, Brown PO. *Genome Biol.* 2001;2(RESEARCH0004), 1–13.
[20] Pasternack RM, Amy SR, Chabal YJ. *Langmuir* 2008;24:12963–12971.

[21] Tian R, Seitz O, Li M, Hu W, Chabal YJ, Gao J. *Langmuir* 2010;26:4563–4566.
[22] Kuznetsova TF, Rat'ko AI, Eremenko SI. *Russ. J. Appl. Chem.* 2003;76:198–202.
[23] Kallury K. MR, Macdonald PM, Thompson M. *Langmuir* 1994;10:492–499.
[24] Tripp CP, Hair ML. *Langmuir* 1995;11:1215–1219.
[25] Rivillon Amy S, Michalak DJ, Chabal YJ, Wielunski L, Hurley PT, Lewis NS. *The J. Phys. Chem. C* 2007;111:13053–13061.
[26] Rivillon S, Brewer RT, Chabal YJ. Appl. Phys. Lett., 2005, 87, 173118.
[27] Rivillon S, Chabal YJ, Webb LJ, Michalak DJ, Lewis NS, Halls MD, Raghavachari K. *J. Vac. Sci. Technol. A* 2005;23:1100–1106.
[28] Aureau, D.; Chabal, Y. J. *Functionalization of Semiconductor Surfaces*. Formation of Organic Monolayers through wet chemistry, (F. Tao, S.L. Bernasek, Eds.) VCH-Wiley:Heidelberg, 2011.
[29] Mossoba MM, Al-Khaldi SF, Kirkwood J, Fry FS, Sedman J, Ismail AA. *Vib. Spectro.* 2005;38:229–235.
[30] Soultani-Vigneron S, et al. *J. Chromatogr. B* 2005;822:304–310.

Chapter 5

Attenuated total reflection infrared (ATR-IR) spectroscopy, modulation excitation spectroscopy (MES), and vibrational circular dichroism (VCD)

Thomas Bürgi

Département de Chimie Physique, Université de Genève, 30 Quai Ernest-Ansermet, 1211 Genève 4, Switzerland

GENERAL INTRODUCTION

In this chapter, we describe the potential and limitations of attenuated total reflection infrared (ATR-IR) spectroscopy, modulation excitation spectroscopy (MES), and vibrational circular dichroism (VCD) for the investigation of

Biointerface Characterization by Advanced IR Spectroscopy. DOI: 10.1016/B978-0-444-53558-0.00005-9

bio-interfaces. The three techniques are completely independent of each other although ATR-IR and MES can be combined, as will be shown below. We will therefore discuss the three techniques independently in sub-chapters. ATR-IR is based on the total internal reflection of a light beam at an interface formed between two media. The resulting field at the interface is used for spectroscopy. The evanescent nature of this field allows one to probe the layer close to the interface formed between the internal reflection element (IRE) and the sample. Therefore, the field probes only a small part of the bulk fluid phase, such that molecules at the interface can be studied even in the presence of strongly absorbing solvents such as water. This makes the technique particularly interesting for applications in biology, where water is *the* solvent. The electric field can furthermore be polarized, which can be used for orientation measurements. MES is based on the stimulation of the sample under investigation by a periodic change of an external parameter such as temperature, pressure, pH, or concentration, just to mention a few. The periodic system response to this stimulation (excitation) can be followed using a spectroscopic technique like ATR-IR by recording time-resolved spectra. The latter are then transformed into phase-resolved spectra using a phase-sensitive detection (PSD) scheme. MES leads to a drastic suppression of noise, which is very welcome when studying molecules at interfaces by spectroscopic techniques. Furthermore, MES helps one to disentangle complex overlapping spectra arising from several species.

VCD is the differential absorption of left- and right-circularly polarized light. VCD spectra selectively spot chirality, that is, if the sample is not chiral or racemic no signals are observed. VCD spectra are sensitive to the conformation of a molecule and also the secondary structure of proteins and peptides. VCD of adsorbates on flat surfaces is very difficult for several reasons (sensitivity, polarization effects) and has not yet been reported. However, VCD of molecules adsorbed on metal nanoparticles has been demonstrated, which is a promising perspective.

ATR-IR spectroscopy

Introduction and historical background

The phenomenon of internal reflection was already known by Newton [1]. The principles of ATR spectroscopy or internal reflection spectroscopy were elaborated 50 years ago [2–4]. In 1967, Harrick [3] showed that the principle of total internal reflection can be used for spectroscopy. When hitting an interface between two adjacent media a light beam is reflected. We call a reflection internal if the light beam approaches the interface from the optically denser medium, that is, from the medium with the higher refractive index (Fig. 1).

At an angle of incidence above the critical angle θ_c (measured with respect to the surface normal) all the intensity is internally reflected. This so-called total reflection is used, for example to guide light within optical fibers over large distances with hardly any loss. If the rare medium (refractive index n_2) is absorbing, the evanescent electric field will be attenuated due to absorption

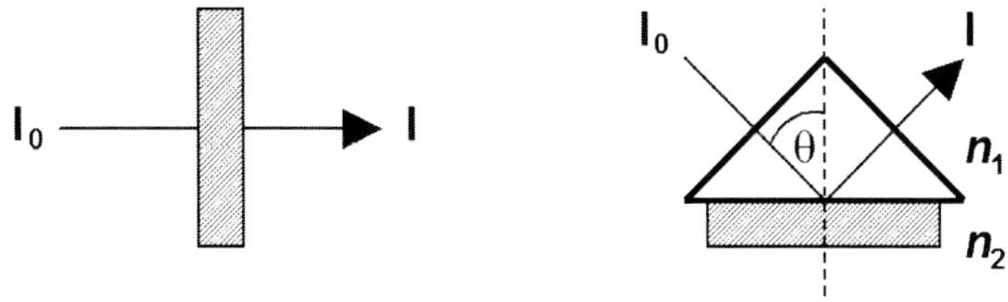

FIGURE 1 **Experimental geometry for transmission (left) and attenuated total reflection (ATR) infrared spectroscopy (right). I_0 and I are the intensity of the incident and transmitted (reflected) radiation, respectively. θ is the angle of incidence and n_1 and n_2 are the refractive indices of internal reflection element and sample, respectively.**

and less intensity is reflected, resulting in an *attenuated* total reflection (ATR). A schematic representation of the basic ATR-IR experimental setup is given in Figure 1 (right). A beam emitted from a light source is coupled into a so-called IRE (n_1). For the infrared, materials like ZnSe ($n_1 = 2.3$), Si ($n_1 = 3.4$) and Ge ($n_1 = 4$) are commonly used as IREs. On the way to the detector, the IR beam can undergo multiple internal reflections (several tens of reflections), which significantly increases the signal-to-noise ratio. Due to the evanescent nature of the field only a small volume is probed at the interface, which makes the ATR-IR technique surface sensitive and a powerful tool for the investigation of solid–liquid interfaces.

Description of the method

A light beam propagating in an optically dense medium with refractive index n_1 undergoes total reflection at the interface of an optically rare medium (n_2) when the angle of incidence exceeds the critical angle θ_c. Newton's experiments showed [5], and it follows from Maxwell's equations, that an electromagnetic wave propagates through the optical interface and generates an evanescent field in the rare medium. The electric field amplitude of the evanescent field falls off exponentially with the distance z from the surface as

$$E = E_0 e^{-z/d_p}. \tag{1}$$

Here, E_0 is the electric field amplitude at the interface, which depends on the angle of incidence, the refractive indices, and the polarization of the field. The depth of penetration d_p, defined as the distance required for the electric field amplitude to decrease to e^{-1} of its value at the surface, is given by

$$d_p = \frac{\lambda_1}{2\pi\sqrt{(\sin^2\theta - n_{21}{}^2)}} \tag{2}$$

where $\lambda_1 = \lambda n_1{}^{-1}$ is the wavelength in the denser medium and $n_{21} = n_2 n_1{}^{-1}$ is the ratio of the refractive indices of the rarer and denser medium.

Harrick introduced the concept of the effective thickness d_e, which is the equivalent path in a hypothetical transmission experiment that results in the same absorption signal as the ATR experiment under identical conditions [3]:

$$d_e = \frac{n_{21}E_0^2 d_p}{2\cos\theta} \quad (3)$$

For a Ge IRE (n_1=4) in contact with water (n_2 = 1.33) at an angle of incidence of $\theta = 45°$, the effective thickness at 1640 cm^{-1} is 0.44 μm for parallel polarized light (polarized in the plane of incidence of the light) and 0.22 μm for perpendicular polarized light. For a ZnSe IRE the corresponding values are 1.12 and 0.56 μm.

The incident plane wave has only field components perpendicular to the direction of propagation. In contrast, the evanescent wave has components along all directions x, y, and z of a Cartesian coordinate system attached to the IRE, as shown in Figure 2.

The direction of the incident field vector can be selected by use of a polarizer. The symbols $||$ and $\perp$ denote electric field vectors parallel and perpendicular to the plane of incidence (Fig. 2). A parallel polarized electric field $E_{||}$ results in field components E_x and E_z along the x and z axes. A perpendicular polarized incident electric field $E_\perp$ results in a y component E_y of the evanescent field. The relative electric field components in medium 2 at the phase boundary $z = 0$ (that is the ratio of the absolute field components in medium 2 at $z = 0$ and in medium 1) are given by Fresnel's equations [5]:

$$E_{0x2}^r = \frac{E_{0x2}}{E_{1||}} = \frac{2\cos(\theta)\left(\sin^2(\theta) - n_{21}^2\right)^{1/2}}{\left(1 - n_{21}^2\right)^{1/2}\left(\left(1 + n_{21}^2\right)\sin^2(\theta) - n_{21}^2\right)^{1/2}} \quad (4)$$

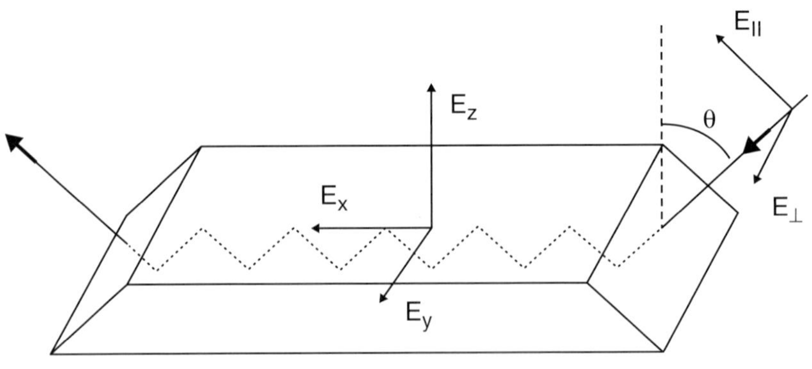

FIGURE 2 Electric fields in an ATR-IR experiment: θ is the angle of incidence. $E_{||}$ (parallel) and $E_\perp$ (perpendicular) denote the direction of the electric field components of the incident light with respect to the plane of incidence ((x, z) plane). E_x, E_y, and E_z denote the electric field components with respect to a coordinate system fixed on the internal reflection element.

$$E^{r}_{0z2} = \frac{E_{0z2}}{E_{1\parallel}} = \frac{2\sin(\theta)\cos(\theta)}{\left(1 - n_{21}^{2}\right)^{1/2}\left(\left(1 + n_{21}^{2}\right)\sin^{2}(\theta) - n_{21}^{2}\right)^{1/2}} \tag{5}$$

$$E^{r}_{0y2} = \frac{E_{0y2}}{E_{1\perp}} = \frac{2\cos(\theta)}{\left(1 - n_{21}^{2}\right)^{1/2}} \tag{6}$$

$$E^{r}_{02\parallel} = \left({E^{r}_{0x2}}^{2} + {E^{r}_{0z2}}^{2}\right)^{1/2} \tag{7}$$

$$E^{r}_{02\perp} = E^{r}_{0y2}. \tag{8}$$

The above equations are exact only for nonabsorbing rarer media, but are very good approximations for weakly absorbing samples. Note that E^{r}_{0x2}, E^{r}_{0y2}, and E^{r}_{0z2} are in general not equal!

Equations (4)–(6) hold for a bulk (non- or weakly absorbing) rarer medium 2. For a thin film with thickness $d \ll d_{\mathrm{p}}$ in contact with the IRE, the thin-layer approximation gives good results [3]. The latter one assumes that the electric field is determined by the optical constants of the IRE and the bulk medium above the thin film. The thin film is then considered as a dielectric in this field, and equations similar to equations (4)–(6) can be derived.

Equations (7) and (8) have to be used to calculate the effective thickness according to equation (3). The effective thicknesses for the two polarizations differ by the factor

$$\frac{d_{e_\parallel}}{d_{e_\perp}} = \frac{A_\parallel}{A_\perp} = \frac{E^{r2}_{0x} + E^{r2}_{0z}}{E^{r2}_{0y}} \tag{9}$$

which is called the dichroic ratio. This ratio is experimentally accessible by measuring the absorbance A of a vibrational band for the two polarizations of the incident light beam. The dichroic ratio is 2.0 for an isotropic bulk medium at an angle of incidence θ of 45°. Note that in a transmission experiment this quantity is unity for isotropic media.

Equation (9) forms the basis for orientation measurements in oriented samples. [6] In general, the intensity of a vibrational mode in an IR spectrum is proportional to the square of the scalar product of the electric field vector $\vec{E}$ and the transition dipole moment vector $\vec{\mu}$ associated with a normal mode of vibration:

$$I \propto (\vec{E} \cdot \vec{\mu})^{2} \tag{10}$$

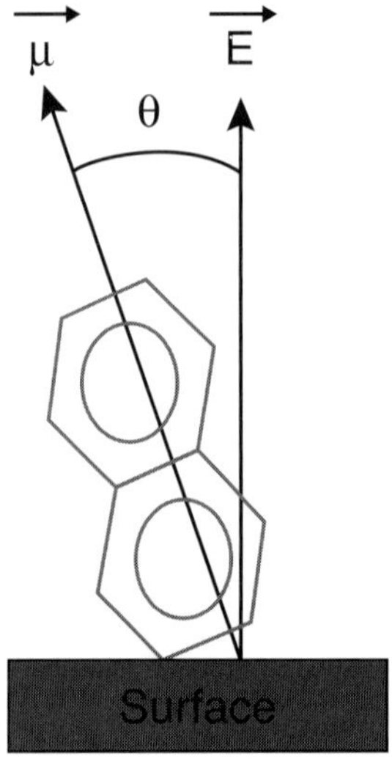

FIGURE 3 Schematic illustration of the electric field and transition dipole moment vectors for an adsorbed molecule. The orientation of the electric field vector with respect to the surface depends on the technique (geometry) used and the nature of the surface (e.g. metal vs. dielectric medium). The orientation of the transition dipole moment within the molecule depends on the vibrational mode.

which can be expressed as

$$I \propto \left|\vec{E}\right|^2 \cdot \left|\vec{\mu}\right|^2 \cos^2(\theta) \tag{11}$$

where θ is the angle between the two vectors. Obviously, equation (11) shows that the intensity of light absorption depends on the mutual orientation of these vectors (see Fig. 3).

ATR-IR spectroscopy can also be used for quantitative analysis analogous to the Lambert–Beer's law in a transmission experiment according to

$$c = \frac{\int\limits_{\text{band},k} A_{k,u} \mathrm{d}\tilde{\nu}}{N d_{e,k,u} \int\limits_{\text{band},k} \varepsilon_k(\tilde{\nu}) \mathrm{d}\tilde{\nu}}. \tag{12}$$

Here, c is the volume concentration, $\tilde{\nu}$ is the wavenumber, the numerator is the integrated absorbance of band k in the spectrum, and u stands for the polarization, that is, parallel or perpendicular. N denotes the number of active internal reflections, d_e is the effective thickness (see eqn. (3)) at band k and polarization u and finally $\int_{\mathrm{band},k} \varepsilon_k(\tilde{\nu})\mathrm{d}\tilde{\nu}$ is the integrated molar absorption coefficient. The situation is more complex for oriented samples [6].

For a thin layer ($d \ll d_\mathrm{p}$) it is adequate to quantify in terms of surface concentration Γ instead of volume concentration. The relationship between the two quantities is

$$\Gamma = cd \tag{13}$$

where d denotes the geometrical thickness of the sample. The surface concentration Γ is thus the number of molecules in the unit volume projected onto the unit area. The method outlined above is very well suited to quantify the adsorption of proteins as has been shown by Fringeli and coworkers [6–8].

Single beam sample reference (SBSR)

Most of the IR spectrometers used today are Fourier transform spectrometers. These have several advantages over dual-beam dispersive spectrometers, for example concerning light throughput, which generally results in lower noise. However, these spectrometers are single beam spectrometers, which also have disadvantages because most measurements such as transmittance and absorbance measurements need a reference spectrum. With a single-beam instrument, the reference has to be measured before (or after) the sample. For long measurements, this can be problematic, particularly if very small signals are to be measured and if the sample is subjected to some treatment, for example thermally. Instrument and sample drifts can cause unwanted artifact signals simply because the reference measured at the beginning of the experiment is not good anymore.

An elegant way to overcome these problems and to turn a FTIR instrument into a pseudo dual-beam instrument is the single beam sample reference (SBSR) method implemented by Fringeli [9]. The idea is to divide the IRE into a reference and a sample part. The element is then mounted on a lift that translates the two parts of the IRE alternating in and out of the IR beam, such that reference and sample are recorded quasi-simultaneously. A schematic of the experiment is given in Figure 4.

The SBSR technique has huge advantages for long-term experiments. For example think of an SBSR experiment to study the interaction of proteins or a lipopolysaccharide with a supported lipid bilayer. In such an experiment, the whole IRE is first covered by the lipid bilayer. Using a suitable flow through system the IRE can be divided into sample and reference compartments. Through the reference compartment a buffer solution is flowed, while the buffer contains the protein for the sample compartment. Assume now that slowly some of the lipid is leaking out of the double layer into the solution. In a normal experiment where the reference is measured before the sample this would show up in negative lipid signals due to the loss of lipid. However, it would be difficult

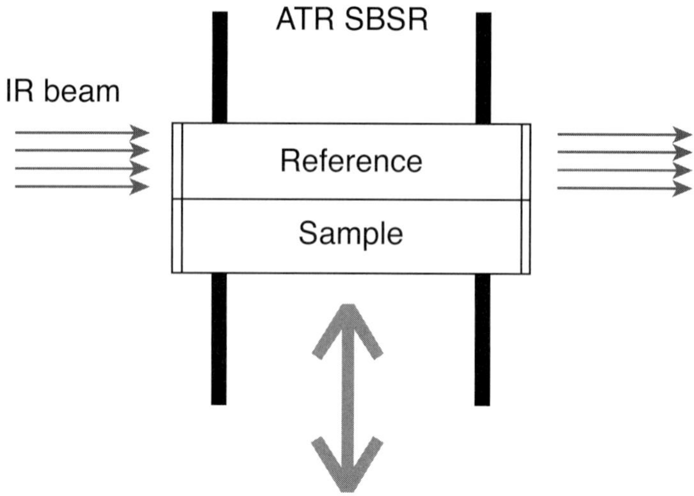

FIGURE 4 Principle of the single-beam sample reference (SBSR) technique. The ATR internal reflection element is mounted on a lift such that the reference and sample compartments can be moved into the infrared beam. Alternative change from sample to reference and vice versa is performed by a computer-controlled lifting and lowering of the ATR cell, which is synchronized with the data acquisition.

to decide if this loss is due to normal leaking or if the loss is associated with the protein adsorption. In an SBSR experiment normal leaking does not result in signals because sample and reference have the same history. Of course the single channel information (sample and reference) is also available such that the normal leaking would also be detected. Such an experiment will be shown below [8,10].

Examples

Interaction of a lipopolysaccharide (LPS) with a lipid bilayer Fringeli and coworkers studied the interaction of lipopolysaccharides (LPS) with lipid bilayers [8,10]. In the following, the interaction of negatively charged LPS from *Pseudomonas aeruginosa* with a positively charged lipid bilayer formed from a dipalmitoyl phosphatidic acid (DPPA) inner monolayer attached to a Ge IRE and an outer monolayer consisting of palmitoyl oleoyl phosphatidylcholine (POPC) and hexadecyl pyridinium (HDPyr) as positively charged component. Figure 5 shows the ATR spectra of the ν(CH) region that were obtained during the interaction of LPS at different concentrations with the bilayer. Three spectra are shown for each concentration. The lowest spectrum (sample S) is from the bilayer in contact with LPS. The middle spectrum (reference R) is the bilayer in

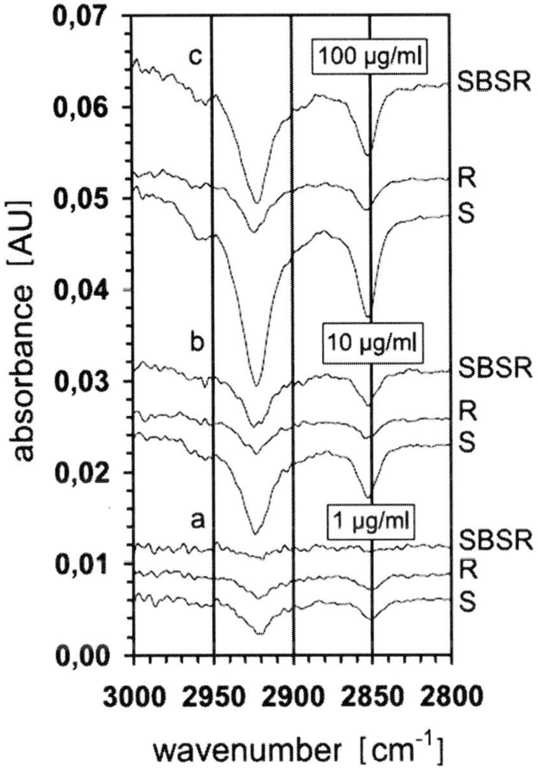

FIGURE 5 SBSR experiment revealing the interaction of a lipopolysaccharide (LPS) with a bilayer membrane consisting of dipalmitoyl phosphatidic acid (DPPA), palmitoyl oleoyl phosphatidylcholine (POPC), and the positively charged hexadecyl pyridinium (HDPyr). FTIR ATR absorbance spectra of a DPPA/(POPC: HDPyr) bilayer interacting with (a) 1 µg/ml, (b) 10 µg/ml, (c) 100 µg/ml LPS measured after 30 min of interaction with LPS in each case. Buffer: 20 mM phosphate pH 7.0, 100 mM NaCl. See the text for more details [10]. Reproduced with the permission of Chimia [10].

contact with pure buffer. The top trace corresponds to the difference spectrum S–R, which is the SBSR spectrum. For the R and S channels, a reference spectrum was measured after the preparation of the bilayer. Since both the R and S compartments could be accessed separately, the bilayer could be prepared simultaneously and in an identical way, whereas only the S compartment was then exposed to LPS.

The negative ν(CH) bands in the R spectra show that the bilayer was not completely stable during the experiment. However, interaction of LPS with the bilayer leads to further loss of lipid, which is obviously a consequence of LPS interaction with the positively charged bilayer. This example shows that more information can be obtained from a SBSR experiment than from a normal ATR experiment as outlined in the section above.

Orientation of N-acetyl-L-cysteine on gold In order to illustrate the power of ATR-IR spectroscopy to study particularly also the orientation of molecules and functional groups on surfaces, we discuss the case of *N*-acetyl-L-cysteine (NAC) on gold [11]. This is a special case due to the gold layer that was deposited onto the Ge IRE. Metals are strong absorbers and equations (4)–(6) above are not valid anymore. The electric field in this case is influenced by the metal layer and it has been shown that the metal surface selection rule, which is well-established for external reflection infrared spectroscopy, is also valid for this special case of internal reflection spectroscopy [12,13]. Locally, the electric field vector is oriented perpendicular to the metal surface, such as illustrated in Figure 3.

In order to determine the orientation of NAC on the Au surface, an ATR-IR spectrum of the randomly oriented molecule, that is, the dissolved molecule, is needed. Such a spectrum gives the intrinsic strengths of the vibrational bands, which are related to the lengths of the corresponding $\vec{\mu}$ vectors (see Fig. 3) for each of the vibrational modes. A comparison between the spectra of adsorbed and dissolved NAC is given in Figure 6.

Significant differences in band intensities with respect to the solution spectrum are observed, which strongly implies the presence of an oriented adsorbate layer on the gold surface. Specifically, the intensity ratios of the amide I (1665 cm^{-1})/amide II (1551 cm^{-1}) and amide II (1551 cm^{-1})/methyl (1375 cm^{-1}) bands change significantly upon adsorption of NAC on gold. Such a comparison between random and oriented molecule already allows one to draw some qualitative conclusions concerning the orientation of some functional groups with respect to the surface, since the orientation of the transition dipole moment within the molecule is known qualitatively for some vibrations. The analysis can however also be made in a more quantitative manner.

An experimentally accessible parameter directly related to the orientation is the intensity ratio of two vibrational bands for random (solution) and surface orientation, as expressed by the following equation [14]:

$$R_{\text{exp}} = \frac{(I_m^s / I_n^s)}{(I_m^r / I_n^r)} \tag{12}$$

where $I_{m,n}$ denotes the measured integrated intensity of vibrational bands m and n, respectively, and the superscripts s and r, respectively, refer to surface and randomly oriented (dissolved) species, respectively. The determination of the orientation, as described in the following, relies on the comparison between experimental and calculated R values. The calculated R values result from DFT calculations, which yield not only the frequencies of the vibrational modes but also the length and orientation of the corresponding $\vec{\mu}$ vectors. For different

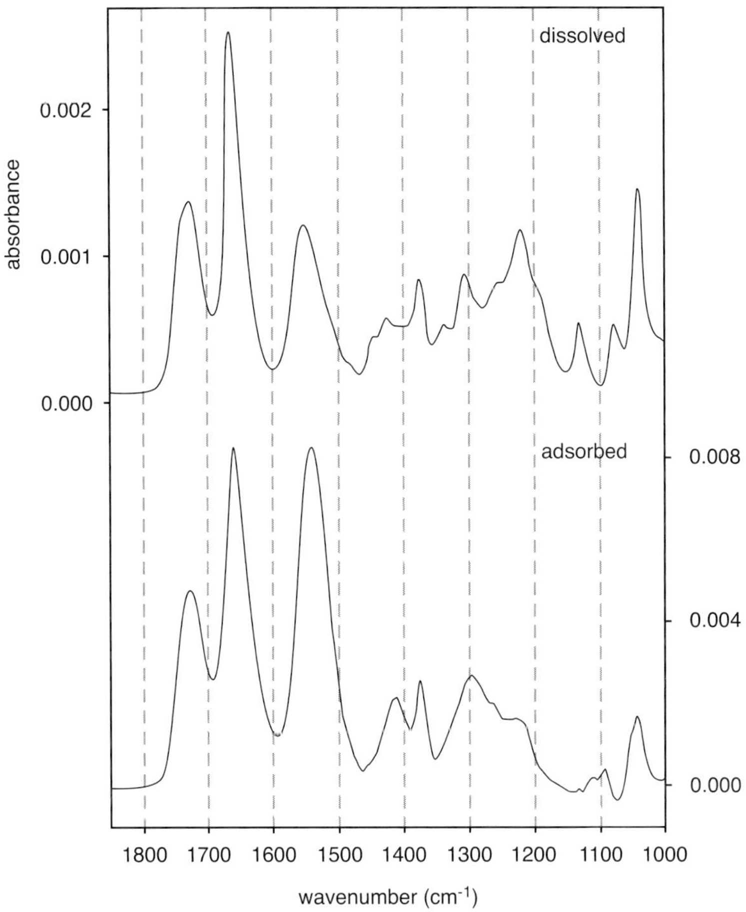

FIGURE 6 Top: ATR-IR spectrum of dissolved NAC corresponding to randomly oriented molecules. Bottom: ATR-IR spectrum of NAC adsorbed on a gold-coated Ge IRE (oriented) in the presence of 0.6 mM NAC in ethanol. Reproduced with the permission of the American Chemical Society [11].

orientations of the molecule with respect to the electric field vector (the surface normal in this case) different relative intensities of the vibrational bands are obtained. The goal is to find a corresponding value R_{calc}, analogously defined as R_{exp} in equation (12), for the calculated IR (randomly oriented molecules) and simulated ATR-IR (oriented molecules) spectra, using the orientation angles α and β, as defined in Figure 7, as adjustable parameters.

In principle, two values of R_{exp} (i.e., two pairs of vibrations) are sufficient to determine the orientation angles α and β. However, the amide I/amide II, amide I/ν(–COOH), and amide II/methyl pairs of modes were considered. The amide I,

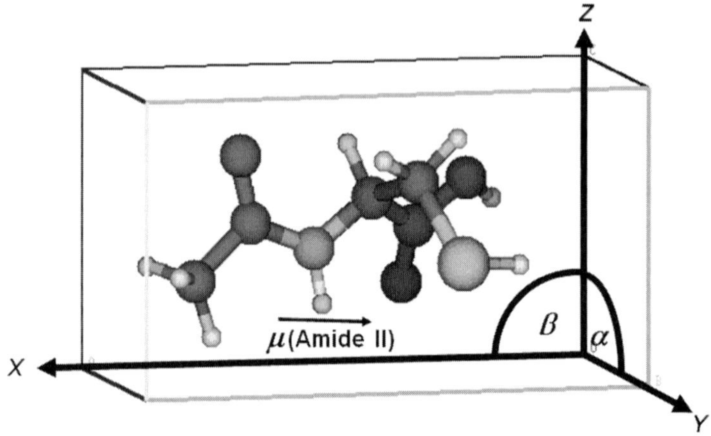

FIGURE 7 Optimized structure of NAC as found by density functional theory (DFT) calculations using a polarizable continuum model to take into account solvent effects. The molecular coordinate system is depicted as a box around the molecule with its axes parallel to the principal axes of inertia of the molecule. The fixed surface coordinate system (x,y,z) with the x–y plane defining the surface plane and z the surface normal has the same orientation as the molecular coordinate system for the reference orientation of the molecule. The molecular coordinate system is allowed to rotate around the angles α (x-axis) and β (y-axis), respectively, providing different orientations of the molecule with respect to the surface coordinate system. The orientation of the amide II transition dipole moment is also given. Reproduced with the permission of the American Chemical Society [11].

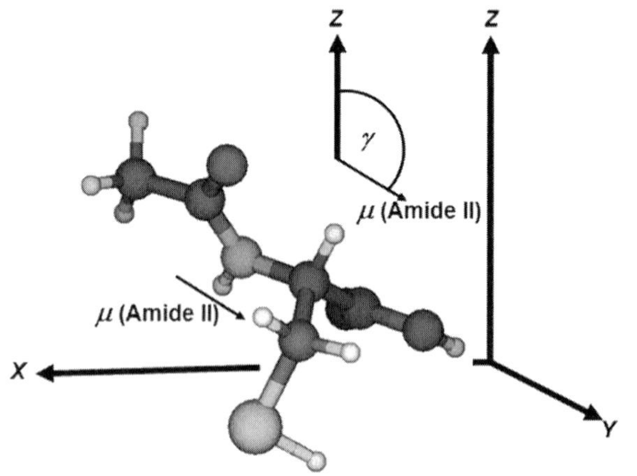

FIGURE 8 Orientation of NAC on the gold surface (x–y plane) emerging from the orientation analysis. The orientation of the molecule is indicated by the angle γ = 127.5° between the surface normal z and the amide II transition dipole moment vector (see the inset). Reproduced with the permission of the American Chemical Society [11].

amide II, and methyl vibrational bands of *N*-acetyl-L-cysteine are located within the *N*-acetyl part of the molecule, which is rather rigid and therefore serves as a pointer for the orientation analysis. It should be noted here that the angles that result from this analysis give an average value for the whole adsorbate layer, that is the method cannot distinguish between a system with a unique orientation of the adsorbed molecules and a system with molecules coexisting in different orientations. Furthermore, the described analysis assumes that the magnitude and orientation of the transition dipole moment vectors $\vec{\mu}$ within the molecule are the same in solution and in the monolayer. The result of such an analysis for NAC is given in Figure 8 [11].

In this orientation of NAC both the thiol and the carboxyl groups are close to the gold surface. This is in qualitative agreement with the analysis of VCD spectra of the same molecule adsorbed on small gold nanoparticles [15].

CONCLUSIONS

ATR-IR spectroscopy is a very powerful tool to study bio-interfaces. By choosing appropriate conditions (number of internal reflections *N*, material of IRE) bio-molecules at an interface can be probed efficiently in the presence of water, which is otherwise a big challenge for IR spectroscopy. The method has sub-monolayer sensitivity and is quantitative, thus allowing one to determine the surface concentration of molecules at the interface. For quantitative analysis, one needs to know the optical constants of IRE and sample as well as the absorption coefficient of at least one vibrational band of the analyte. ATR-IR is also very well suited to study kinetics on the time scale of seconds to minutes, for example of adsorption/desorption processes and structural changes of molecules at interfaces. By measuring spectra with parallel and perpendicular polarized light, the orientation of molecules can be determined, which requires knowledge of the direction of the transition dipole moments of vibrations within the molecule, which are accessible by DFT calculations.

One inherent disadvantage of the method is the fact that it relies on suitable optical elements (IREs). The latter have to be transparent in the spectral region of interest and be characterized by a high refractive index. This combination limits the suitable materials. If one wants to study a surface made from a specific material the IRE can be coated by a thin film of the respective material. Although this is possible it may require special techniques such a physical vapor deposition. In conclusion, ATR-IR is a powerful tool for the investigation of bio-molecules at interfaces and especially also for *in situ* studies in water. ATR-IR is a relatively straightforward technique and the necessary equipment is affordable, which makes the technique furthermore attractive.

Modulation Excitation Spectroscopy (MES)

Introduction and historical background

MES relies on the stimulation of the system under investigation by the periodic variation of an external parameter, such as concentration, pH, temperature, mechanical force, or any other parameter that influences the system. The periodic system response is detected and further processed by a PSD. The latter represents a narrow-band technique. This means that signals at frequencies different from the detection frequency are efficiently suppressed. Modulation techniques and PSD are widely applied in signal processing. In contrast, for spectroscopic investigations modulation excitation schemes are not yet extensively used although the technique can have striking advantages in specific situations as outlined in this chapter. The method was extensively used by Fringeli and Baurecht to study biological systems [16–18]. MES was however also applied in other fields like heterogeneous catalysis [19] photocatalysis [20], and enantiodiscrimination [21].

Description of the method

Infrared spectroscopy is not very specific. All species that are probed by the IR beam and that have IR active vibrations will give rise to signals in the spectrum, whether the species are involved or not in the process of interest. One may use difference spectroscopy to partly overcome this problem [22]. However, in order to further extract important information, such as, for example the kinetics of surface reactions, more sophisticated techniques are needed. Time-dependent stimulation is one of these methods that can be applied if the system under observation responds to an external parameter, such as, for example concentration, pH, or temperature.

The external stimulation can be applied in the form of a parameter jump [16] or a parameter modulation [17]. A prerequisite of the latter technique is a reversible system response to a periodic stimulation. Figure 9 shows the principle of MES.

The sample is stimulated periodically by an external parameter at the stimulation frequency ω (the stimulation is represented as a green sinusoidal line). The periodic system response (reversibility provided) due to the stimulation is shown as red line in the figure. The response can be delayed with respect to the stimulation (phase lag), but, importantly, the frequency of the response will be identical to the stimulation frequency (or harmonics thereof). Knowing the latter allows one to apply a PSD, which converts time-resolved to phase-resolved spectra.

Theory of phase-sensitive detection

If a system is disturbed periodically by an external parameter, all the species addressed by the stimulation will also change periodically at the same

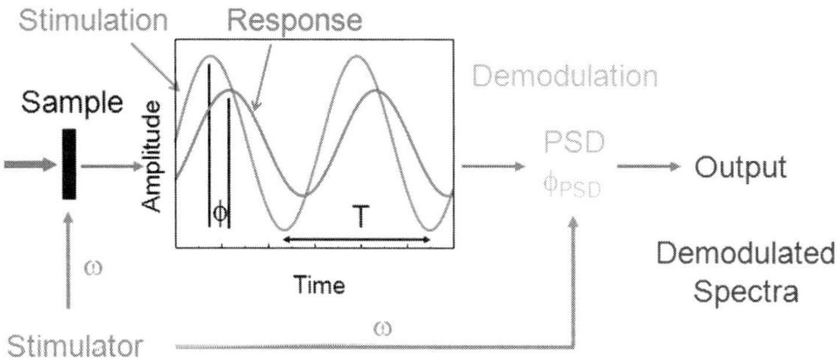

FIGURE 9 Principle of modulation spectroscopy.

frequency as the stimulation or harmonics thereof [23]. As is indicated in Figure 9, there may be a phase lag ϕ between the stimulation and system response. A phase lag is observed when the time constant of a process giving rise to a signal is on the same order as the time constant $T = 2\pi\omega^{-1}$ of the stimulation. After reaching a new quasistationary state at the beginning of the modulation, the system oscillates at a frequency ω around this state. The resulting absorbance variations $A(\tilde{\nu}, t)$ are followed by measuring spectra at different times during the modulation period T. The signal can be averaged over many periods. The set of time-resolved spectra $A(\tilde{\nu}, t)$ is then transformed by means of a digital phase-sensitive detection (PSD), according to equation (13), to a set of phase-resolved spectra:

$$A_k^{\phi_k^{PSD}}(\tilde{\nu}) = \frac{2}{T}\int_0^T A(\tilde{\nu}, t)\sin(k\omega t + \phi_k^{PSD})\mathrm{d}t \tag{13}$$

where $k = 1,2,3,\ldots$ determines the demodulation frequency, that is, fundamental, first harmonic, and so on, T is the modulation period, $\tilde{\nu}$ denotes the wavenumber, ω is the stimulation frequency, and ${\phi_k}^{PSD}$ is the demodulation phase angle. With a set of time-resolved spectra $A(\tilde{\nu}, t)$, equation (13) can be evaluated for different phase angles ${\phi_k}^{PSD}$ resulting in a series of phase-resolved spectra $A_k^{\phi_k^{PSD}}$. Equation (13) is closely related to a Fourier analysis [17]. According to the Fourier theorem, every periodic function $f(t)$ can be expressed as a Fourier series in the form

$$f(t) = a_0 + \sum_{k=1}^{\infty} a_k\cos(k\omega t) + \sum_{k=1}^{\infty} b_k\sin(k\omega t) \tag{14}$$

where a_k and b_k, are the orthogonal cosine and sine Fourier coefficients, respectively, and the integer k is the frequency multiplier, that is, fundamental, first harmonic, and so on. Now the Fourier series represented by equation

(14) can also be expressed as a sine series with an additional phase angle ϕ_k according to

$$f(t) = a_0 + \sum_{k=1}^{\infty} c_k \sin(k\omega t + \phi_k) \tag{15}$$

with $c_k = \sqrt{a_k{}^2 + b_k{}^2}$ and tan $\phi_k = a_k b_k{}^{-1}$. The Fourier coefficients of the periodic function described by equation (15) are given by

$$a_0 = \frac{1}{T}\int_0^T f(t)\mathrm{d}t \tag{16}$$

$$c_k = \frac{2}{T}\int_0^T f(t)\sin(k\omega t + \phi_k)\mathrm{d}t. \tag{17}$$

By replacing *f*(*t*) with $A(\tilde{\nu}, t)$ and ϕ_k with $\phi_k{}^{PSD}$ equation (17) yields the demodulation transformation given in equation (13). In other words, a demodulated spectrum $A_k^{\phi_k^{PSD}}$ represents the Fourier coefficient of the time varying signal $A(\tilde{\nu}, t)$ for a given frequency (usually fundamental, $k = 1$) and a given user-defined phase lag $\phi_{\mathrm{k}}{}^{\mathrm{PSD}}$.

What are the benefits of MES? The signals arising in demodulated spectra are exclusively related to the stimulation of the external parameter, that is, MES selectively highlights the species that are affected by the external parameter. Furthermore, demodulated spectra are of much higher quality with a better signal-to-noise ratio. The kinetic information of the investigated system is contained in the frequency-dependent amplitude of the response and the phase lag between stimulation and response (see Fig. 9). Consequently, species with different kinetics can be separated by setting the demodulation phase angle $\phi_k{}^{PSD}$ accordingly. Furthermore, MES helps the detection of minor species that are otherwise spectrally buried within the strong signals of the dominant spectral features. This is possible by choosing the demodulation phase angle such that the signals of the dominant species vanish. This works, provided that the dominant and the minor species have different kinetics in their response toward the stimulation. More information on MES can be found elsewhere [16,17,24].

Examples

Temperature modulation excitation of a hydrated poly-L-lysine film [25]

Temperature-induced structural changes are important in biological systems. The following example describes the application of MES to study temperature-induced structural changes in poly-L-lysine. For the study, a film of poly-L-lysine (PLL) was cast onto a Ge IRE for ATR spectroscopy. The film was hydrated with D_2O (80% relative humidity) and then the temperature was modulated ± 2°C around 28°C with a period T of 14.7 min. This causes a conformational change that can be followed by MES [25]. Figure 10 shows the stationary spectrum of the PLL film as well as a series of demodulated spectra that were obtained at different demodulation phase angles (see eqn. (13)). Each species i contributing to the spectrum can be associated with a phase angle ϕ_i. All the spectral bands associated with the same species will have the same phase angle. The amplitude of the bands becomes maximum when the difference between ϕ_i and the demodulation phase angle ϕ_k^{PSD} becomes zero, that is, for $(\phi_i - \phi^{PSD}) = 0°$, minimum (negative) for $(\phi_i - \phi^{PSD}) = 180°$, and zero for $(\phi_i - \phi^{PSD}) = 90°$ and $(\phi_i - \phi^{PSD}) = 270°$. Since the demodulation phase angle ϕ^{PSD} can be controlled one can easily determine the angle ϕ_i for a given species.

By comparison of the stationary spectrum and the demodulated spectra it becomes clear that the latter are better resolved although the resolution of the spectrometer was identical in both cases. In the demodulated spectra the spectral overlap is reduced because these spectra contain only absorption bands from species that are affected by the temperature stimulation. Furthermore, bands can be positive and negative, depending on the phase ϕ_i of the associated species. It is furthermore evident that the shape of the phase-resolved spectra (relative intensity of bands) changes with the demodulation phase angle ϕ^{PSD}. This is clear evidence of the existence of conformational populations of PLL with different phase angles ϕ_i. Thus, there are transient intermediate species for the structural change of PLL induced by the temperature modulation. Through further analysis, the spectra of the different populations can be extracted from the demodulated spectra as detailed elsewhere [25].

Enantiodiscrimination at chiral self-assembled monolayers

MES can also help to study specific interactions as described in the following example on enantioidiscrimination. Chirality plays a crucial role in biology and pharmaceutical sciences. Chiral structures can discriminate between enantiomers of a chiral molecule, which has important consequences since the behavior of the two enantiomers can be completely different. This is particularly also true for pharmaceuticals. At the origin of such dissimilar behavior is the different interaction of the two enantiomers with the chiral environment, for example in a receptor. In order to get molecular-level insight into the mechanism of enantioselection we have proposed a method that combines ATR-IR with MES [21] as schematically depicted in Figure 11.

In principle, the method can be applied to any enantiodiscriminating system. Here the interaction of a chiral self-assembled monolayer (SAM) with a chiral (bio-) molecule is described. An IRE for ATR-IR is first coated with a thin gold layer that hosts the SAM. The self-assembly of a chiral thiol on the gold surface

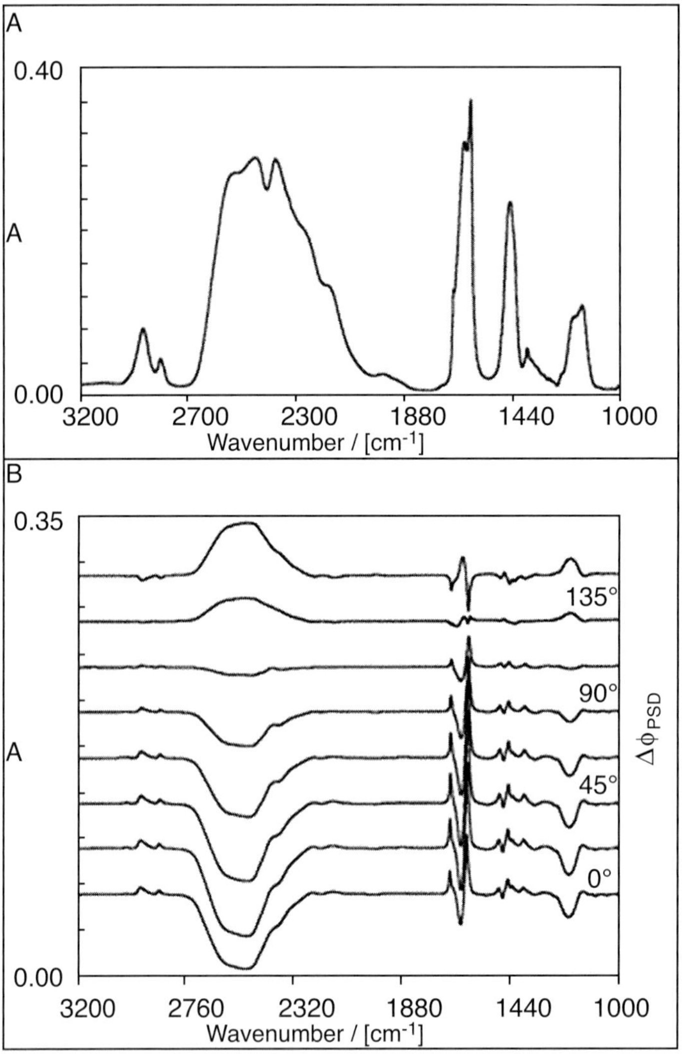

FIGURE 10 **Parallel polarized temperature-modulated ATR-IR spectra of a poly-L-lysine film hydrated with D_2O (80% relative humidity). The film was deposited onto a CdTe internal reflection element. A rectangular temperature modulation of +/– 2°C around 28°C with a period T of 14.7 min was applied. The top spectrum represents the stationary part of the PLL spectrum. In the bottom part, demodulated spectra at different demodulation phase angles as indicated are shown. Reproduced with the permission of the American Chemical Society [25].**

is then directly followed by ATR-IR. After the preparation of the SAM, solutions of the two enantiomers of the chiral probe molecule (selectand) are periodically flowed over the chiral self-assembled monolayer (SAM, selector) by using a specially designed flow-through cell, as schematically depicted in Figure 11. In

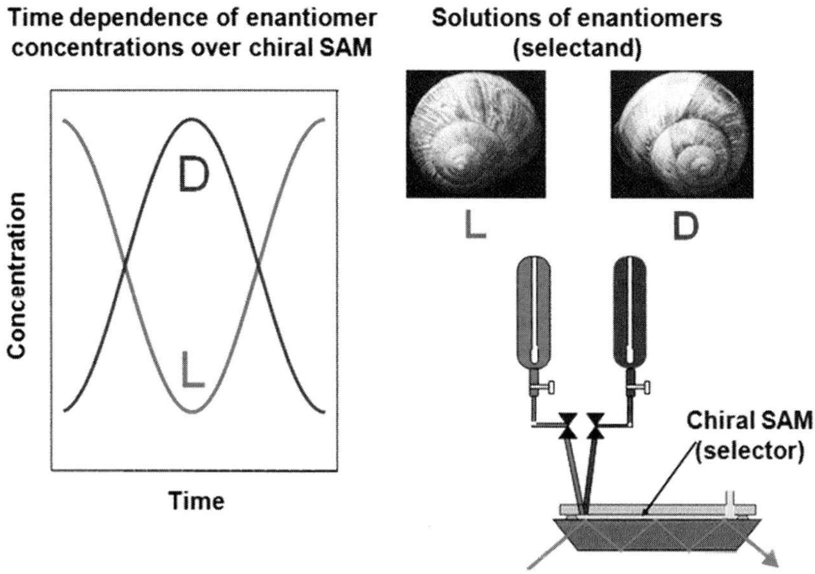

FIGURE 11 Principle of absolute configuration modulation combined with attenuated total reflection IR spectroscopy for investigating the interaction of chiral molecules with chiral surfaces. Left: schematic representation of time-dependent concentration of individual enantiomers in an absolute configuration modulation excitation experiment. The total concentration of both enantiomers remains constant. The stimulation is sinusoidal in the depicted example, which is however not a necessary requirement. Right: two reservoirs containing the solutions of the two enantiomers are connected to the flow-through cell. The solutions can be alternately flown over the sample (''modulated'') by switching two computer controlled pneumatically actuated valves.

this way the concentration of the two enantiomers is by turns high and low, respectively, while the total concentration stays constant throughout the modulation experiment (see concentration profile on the left in Fig. 11). During the modulation, ATR-IR spectra are recorded. These time-resolved spectra are subsequently demodulated, that is, transformed into phase-resolved or demodulated spectra according to equation (13). Only the periodically varying signals are thus detected and therefore no dissolved species are observed, because the IR spectra of the dissolved enantiomers are identical. Unspecific adsorption of the two enantiomers of the chiral molecule on the surface (on nonchiral sites) is also suppressed in the demodulated spectra because the resulting spectra remain identical for both enantiomers. Adsorption of the enantiomers on chiral sites that can distinguish between them leads to diastereomeric complexes (selector–selectand complexes) with different IR spectra. These differences might be small but are amplified by the PSD. The advantages of the method has been demonstrated by studying the interaction between chiral molecules and chiral stationary phases used in chromatography [21]. The technique was also applied to study the enantiodiscrimination of a chiral SAM of L-glutathione (GSH, γ-*glu-cys-gly*) on gold [26]. The latter SAM could quite efficiently discriminate between the enantiomers of proline. In the following, the interaction between SAMs of *N*-acetyl-L-cysteine (NAC) on gold and proline is described [27].

ATR-IR spectra of the interacting species, NAC adsorbed on gold (selector, bold solid line), and dissolved proline (selectand, thin solid line) are shown in Figure 12, bottom. The most prominent vibrational modes for NAC are the

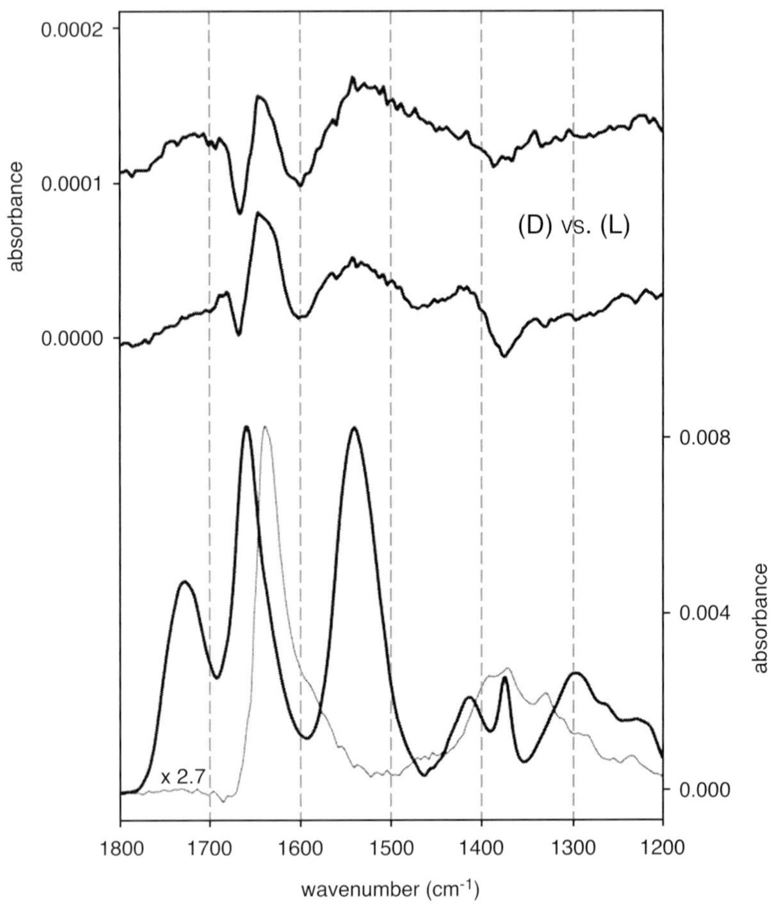

FIGURE 12 **Bottom: ATR-IR spectra of N-acetyl-L-cysteine (NAC, selector, bold solid line) adsorbed on a gold surface and proline (selectand, thin solid line, scaled by a factor of 2.7) dissolved in ethanol. Top: demodulated, that is, phase-resolved ATR-IR spectra. In this type of modulation experiment, D- and L-proline were periodically flowed over a chiral NAC SAM. The two demodulated spectra highlight the signals arising due to enantiospecific interactions between NAC and proline. The spectra were measured on two different gold surfaces illustrating the reproducibility of the absolute configuration modulation experiments. Reproduced with the permission of the American Chemical Society [27].**

ν(COOH) mode at 1727 cm^{-1}, amide I at 1661 cm^{-1}, and amide II at 1539 cm^{-1}. The proline ATR-IR spectrum is dominated by the $\nu_{as}(COO^-)$ vibrational mode at 1637 cm^{-1} with the δ_s (NH_2) scissoring mode appearing as a shoulder at 1570 cm^{-1}. On top in Figure 12, two phase-resolved, that is, demodulated spectra are presented (label "(D) vs. (L)") that refer to absolute configuration modulation experiments performed on two different gold surfaces (gold-coated Ge IREs). These phase-resolved ATR-IR spectra highlight the differences in intermolecular interactions between the two proline enantiomers and the NAC SAM. Note that the demodulated spectra (Fig. 12, top) are characterized by signals that are more than one order of magnitude weaker than the ones of the static spectra (Fig. 12, bottom). The demodulated spectra show a rather complex pattern with sharp positive and negative bands. Although the signals are considerably above the noise level (which is about 5×10^{-6} in the spectral region around 1700 cm^{-1}) the ATR-IR signals are small revealing a rather weak enantiodiscrimination between the NAC SAM and proline. The dispersive bands in demodulated spectra arise due to slight frequency shifts induced by the stimulation, that is, the change in absolute configuration of proline. The demodulated spectra can be viewed as difference spectra between the relevant diastereomeric complexes. The position of the dispersive bands suggests frequency shifts of the proline carboxylate $\nu_{as}(COO^-)$ (possibly also $\nu_s(COO^-)$) and NH_2 scissoring $\delta(NH_2)$ and NAC amide and carboxylate vibrations [27].

Density functional theory (DFT) calculations of an NAC–proline complex using a polarizable continuum salvation model (PCM) were also performed in order to further understand the nature of the interaction. Among different NAC–proline intermolecular forces, the ionic interaction between the positively charged proline NH_2^+ group and the negatively charged NAC carboxylate COO^- turned out to dominate the interaction. Structure optimization was performed for the NAC-(D)-proline and NAC-(L)-proline diastereomeric complexes and a pictorial representation is given in Figure 13. The difference in potential energy between the two chiral complexes was finally found to be very small (about 0.2 kJ mol^{-1}).

The calculated IR spectra for the NAC-D-proline complex, denoted by "(D)", and for the NAC-L-proline complex, denoted by "(L)", together with the corresponding difference spectrum (bold solid line, "L–D") is displayed in Figure 13, bottom. At first glance, the calculated spectra of the (D) and (L) complex seem to be identical, but the resulting difference spectrum, scaled by a factor 5, clearly reveals significant dispersive signals that originate from slight frequency shifts of vibrational bands in the order of one to several wavenumbers. This difference spectrum is directly comparable to the experimental spectra shown in Figure 12, top. The calculations are in qualitative agreement with the experiment. In particular, the observed dispersive line pattern between 1700 and 1550 cm^{-1} is very well reproduced by the calculations. This example demonstrates that even such weakly enantiodiscriminating

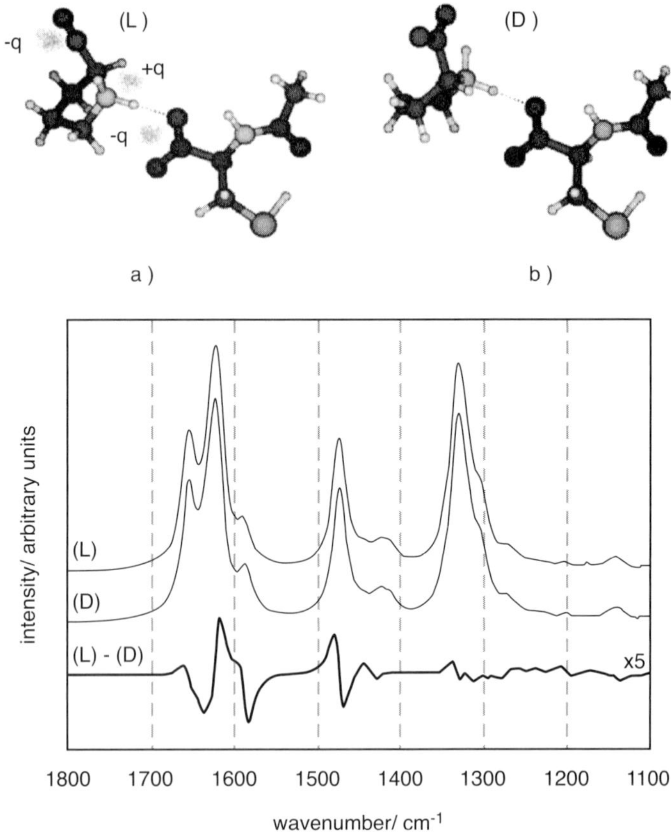

FIGURE 13 **Top: pictorial representation of optimized structures of the NAC-L-proline (a) and NAC-D-proline (b) complexes. The structure of the complexes is determined by ionic proline $NH_2^+\cdots COO^-$ NAC intermolecular interactions. In (a) the charges are schematically indicated by clouds. Bottom: calculated vibrational spectra of NAC-D-proline ("D") and NAC-L-proline ("L") complexes. The corresponding difference spectrum "(L)–(D)", scaled by a factor of 5, is presented at the bottom as a solid bold line. Note that the calculated difference spectrum is experimentally accessible by ATR-IR and corresponds to the phase-resolved spectra on top in Figure 12. Reproduced with the permission of the American Chemical Society [27].**

interactions lead to detectable signals in the difference spectrum of the two diastereomeric complexes as revealed by absolute configuration modulation and that the technique is a sensitive tool for the study of enantiodiscrimination at chiral solid–liquid interfaces. The PSD efficiently suppresses the signals arising from non-specific interactions, which makes it possible to detect even subtle spectral features of the specifically bonded species.

CONCLUSIONS

MES combined with phase-sensitive detection is a powerful technique for the investigation of reversible systems. It has several advantages over conventional spectroscopy with regard to sensitivity, specificity, and resolution of overlapping bands. The PSD scheme allows the separation of small varying signals from a large static background. Only those species give rise to signals that are affected by the external parameter that is periodically varied. Analysis of the phase-resolved spectra largely helps to disentangle crowded spectra. The spectrum of a species exhibiting distinct kinetics can be unequivocally subtracted by choosing the appropriate phase-setting. This then also allows the observation of minor species in the system, which are otherwise buried by the dominant (most abundant) species. Moreover, rigorous kinetic data can be gained from the analysis of phase angles and amplitudes measured as a function of modulation frequency, as has been shown elsewhere [16].

Experimentally, a modulation experiment is usually relatively easy to realize and not much special equipment is needed in addition to the conventional spectroscopy. Also, the phase-sensitive detection, that is, the application of equation (13), is straightforward.

A prerequisite for MES is the reversibility of the system under investigation, which may hinder the application of the technique. A critical parameter is the modulation frequency. Ideally, the modulation period is on the same order of magnitude as the timescale of the process of interest. In such a situation the most information can be gained from MES, by performing experiments at different modulation frequencies. In case the physical–chemical process is much faster than the modulation, the system response will just follow the stimulation and the demodulated spectra correspond to difference spectra between two states of the system. One might think, for example of a photo-sensitive sample that switches extremely fast and that is probed by a slow MES experiment. The demodulated spectra then correspond to the difference spectra between the two states of the system, that is, the light on and the light off state. In case the stimulation is much faster than the system response, the system cannot follow the stimulation and the amplitude of the response will be zero or at least very small.

An important issue is the maximal modulation frequency that can be realized in an experiment. This depends of course on the concrete system. One limiting factor is the spectrometer itself. Modern FTIR spectrometers can record up to 100 spectra per second (depending on the spectral resolution), which is limited by the mechanical movement of the mirror on the interferometer. In a MES experiment several time-resolved spectra have to be recorded during one modulation period. As a lower limit, 16 spectra per period can be considered. This means that the spectrometer limits the accessible modulation frequency to a few Hertz in the most optimistic case, 1 Hz or even less being more realistic. In the example discussed above, a concentration modulation was in principle applied in the presence of a liquid phase. In such a case, mass transport of the molecules (convection and diffusion) may be a limiting factor for the applicable modulation frequency [28].

VIBRATIONAL CIRCULAR DICHROISM (VCD)

Introduction and historical background

VCD is the differential absorption of left and right circularly polarized infrared light [29]. The signals are relatively small compared to the corresponding IR signals. Anisotropy factors $\Delta A/A$ of 10^{-4} to 10^{-5} are typical, where A is the absorbance and ΔA is the difference in absorbance for left- and right-circularly polarized light. With technological developments such as lock-in amplifiers, liquid-nitrogen-cooled semiconductor detectors and IR polarization modulators for the generation of left- and right-circularly polarized radiation the measurement of such small signals is nowadays possible [30–32]. VCD is usually measured in transmission mode and the sample is prepared as liquid film of the neat compound or as a solution. However, also films of solid compounds were measured [33].

VCD has been successfully measured in the mid seventies by Holzwart and coworkers [34] and Stephens and coworkers [35]. At that time dispersive infrared spectrometers were used. Shortly after that, Nafie used for the first time a FTIR spectrometer for VCD experiments [36]. Nowadays, several VCD spectrometers are available in the market.

Description of the method

The heart of the experiment is a photoelastic modulator (PEM) that modulates the polarization state of the light usually at several tenths of kilo hertz. This PEM generates left and right circularly polarized light, which leads to a modulated signal at the detector provided the sample absorbs left and right circular polarized light to a different extent. This modulated signal, after lock-in amplification and Fourier transformation of the interferogram, contains essentially the VCD spectrum.

VCD of biomolecules in aqueous environment is challenging. Water is a strong IR absorber and therefore only very short path lengths can be used, which means that the concentration of the biomolecule has to be quite high in order to obtain sufficiently large signals. An interesting development for the study of biomolecules is the extension to the near infrared region NIR [37]. We should also mention the differential scattering analog to VCD, Raman optical activity (ROA), which is very well suited for studies of biomolecules in aqueous solution [38, 39]. A further challenge in VCD is artifacts originating from linear birefringence due to optical elements in the beampath between the PEM and detector. A design with two PEMs and two lock-in amplifiers was demonstrated to cancel out birefringence effects [40].

VCD spectra contain substantial structural information. First of all, the technique can be used to determine the absolute configuration of a chiral molecule without the need to grow single crystals [41–43]. Enantiomers have opposite signs for corresponding vibrational bands. Second, and perhaps even more important, the technique yields information on the conformation of dissolved molecules [44–46]. This information has to be extracted from the experimental spectrum through comparison with theory. For this, the possible conformations of the molecule have to be determined, their structure refined, and their VCD spectrum calculated. Generally, the VCD spectra of conformers differ significantly. This strategy for structure determination of dissolved molecules obviously relies on the quality of the calculation. Today, methods are available, notably relying on density functional theory (DFT), which have predictive character for VCD spectra of small molecules.

VCD is also sensitive toward the secondary structure of proteins and peptides [47,48]. Particularly, the amide vibrations can be used to get information on the secondary structure as is illustrated in Figure 14, which shows the VCD spectra of the amide regions of several polymers differing in the content of secondary structure elements.

Example

VCD of molecules adsorbed on metal nanoparticles VCD of flat interfaces and adsorbed molecules on such interfaces is very difficult and has not been reported yet, to the best of our knowledge. The challenge is at least twofold. First, VCD signals are small even for bulk samples. In the latter case ΔA values of 10^{-5}–10^{-4} are typically measured for vibrational bands giving rise to absorbances of A of 0.1–1. For monolayer films of molecules absorbance signals A of $10^{-2} - 10^{-4}$ are typical, depending on the method used. The VCD signals are then expected to be smaller than 10^{-6} for a monolayer of molecules. Second, the interface itself may change the polarization state of the light and lead to polarization artifacts. Most IR techniques to probe interfaces rely on reflection geometries that strongly alter the polarization of the incident light. For the investigation of chiral molecules at flat interfaces other techniques might be more suitable, for example second harmonic generation looks promising [50]. Furthermore, on surfaces molecules are very often oriented, which can complicate the analysis of (linear or nonlinear) circular dichroism spectra.

As will be shown below, VCD can still be used to study adsorbed molecules. The trick is to use a high surface area material that can be dissolved, that is a nanoparticle. The application of VCD spectroscopy to nanoparticles (NPs) has been demonstrated recently by studying the conformation of the adsorbed chiral molecules [51]. For example the cysteine derivatives *N*-acetyl-cysteine and *N*-isobutyryl-cysteine adsorbed on gold NPs of about 2 nm core diameter were studied. VCD spectra of gold NPs covered by the two enantiomers of *N*-isobutyryl-cysteine (NIC) were measured [51]. Whereas the IR spectra were

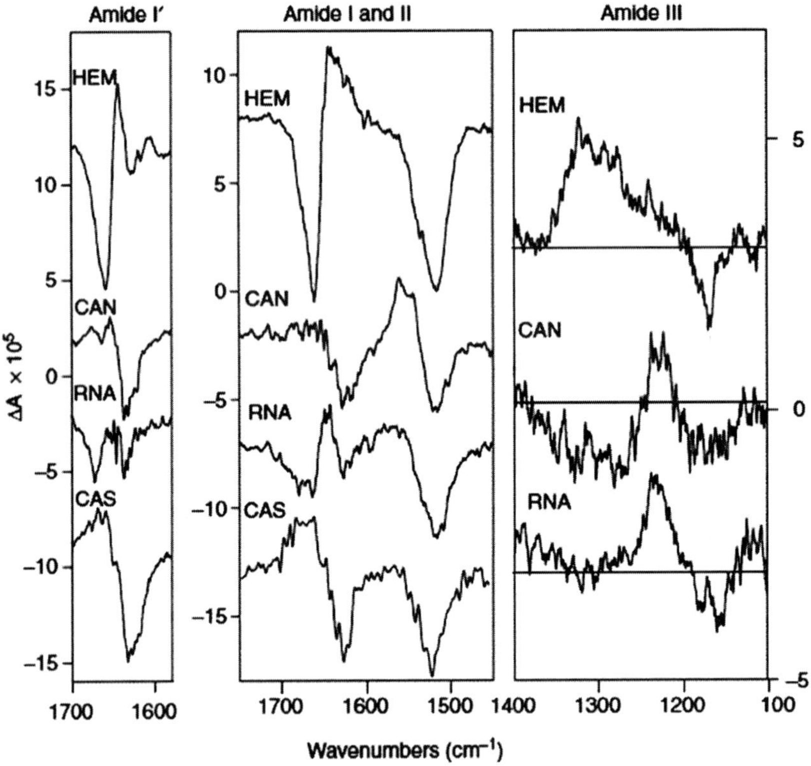

FIGURE 14 **Comparison of typical amide I′ (N-deuterated), amide I & II (middle), and amide III (right) VCD spectra for proteins in a solution. The proteins have different dominant secondary structures. Hemoglobin (HEM) is highly helical, concanavalin A (CAN) contains no helix, but β-sheet, ribonuclease A (RNA) contains a mixture of sheet and helix and casein (CAS) is a ''random coil'' protein with no extensive secondary structure. Reproduced with the permission of Elsevier [49].**

identical for the two enantiomers, the VCD spectra showed mirror image relationship as can be seen in Figure 15.

In order to extract structural information from the VCD spectra density functional theory (DFT) calculations were performed. Different conformers and different adsorption modes of the molecule adsorbed on a small gold cluster were considered for these calculations. The calculations showed that the conformer of the adsorbed molecule had a large influence on the calculated VCD spectra.

The simulated VCD spectrum of the conformer of *N*-isobutyryl-cysteine adsorbed on a small gold cluster that matches best the experimental data is shown in Figure 16[51]. This conformation is characterized by an interaction of the carboxylate with the gold particle. Thus, the carboxylate group seems to be a second anchoring point beside the strong Au–S bond.

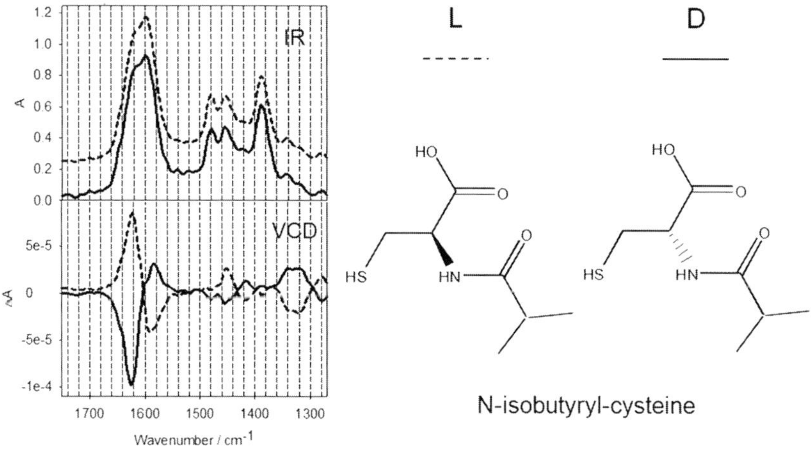

FIGURE 15 IR and VCD spectra of gold nanoparticles of about 2 nm in diameter covered by the two enantiomers of N-isobutyryl-cysteine. The particles were dissolved in basic D_2O. Reproduced with the permission of the American Chemical Society [51].

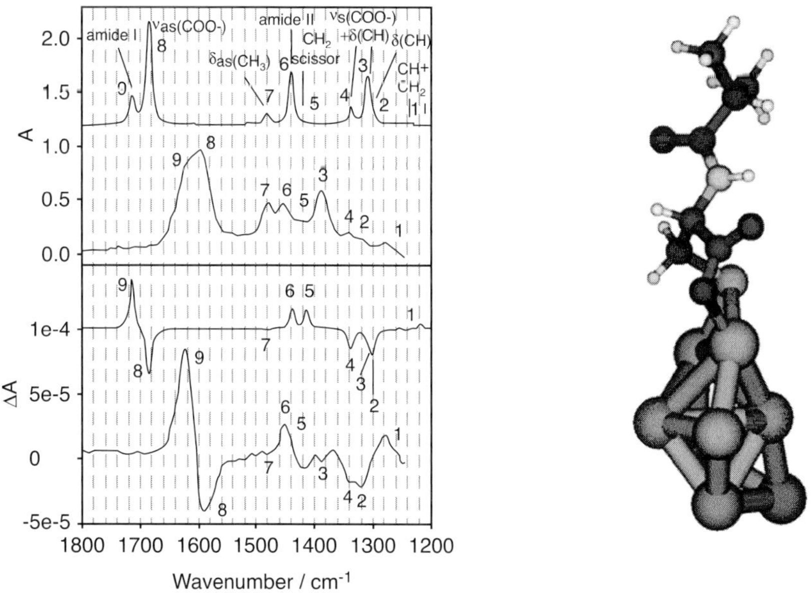

FIGURE 16 Comparison between calculated and experimental IR and VCD spectra of N-isobutyryl-L-cysteine on gold nanoparticles (left). The calculated spectra are shown for the conformer on the right. Corresponding bands are numbered and the assignment is given in the upper part of the Figure. The experimental spectra were recorded for gold nanoparticles covered by N-isobutyryl-L-cysteine dissolved in D_2O [51].

Most importantly, it has been shown that VCD is a promising tool to study the structure of adsorbed chiral molecules on small metal nanoparticles. Such information is relevant because the structure (conformation) of the adsorbed molecules can have a pronounced effect on surface properties and function such as molecular recognition. It should be noted that the system described above is challenging for mainly two reasons. First of all, *N*-isobutyryl-cysteine is a quite flexible molecule with considerable conformational freedom. Therefore, several conformers might contribute to the experimental spectrum. Second, the particles were studied in water. It is very likely that water specifically interacts with *N*-isobutyryl-cysteine via hydrogen bonding and this will influence the VCD spectra at least to a certain extent. This should be included in the calculation of the VCD spectra. Taking into account this potential for improvement, the approach for studying the structure of adsorbed chiral (bio-) molecules on metal particles outlined above can be considered very promising although up to now only a few examples were reported on the use of VCD for the investigation of molecules adsorbed on metal particles.

CONCLUSIONS

VCD is sensitive toward the absolute configuration and the conformation of molecules and the secondary structure of proteins. Molecules adsorbed on flat surfaces and interfaces are difficult to be studied by VCD spectroscopy, due to the weakness of the signals. However, VCD is possible for molecules adsorbed on soluble nanoparticles, which are high specific surface area materials. One interesting avenue to consider in future work is surface enhancement effects similar to what is explored for Raman optical activity, where the study of the interaction of proteins with metal colloids seems within reach [53].

ACKNOWLEDGMENT

I would like to thank Professor Dr. Urs Peter Fringeli, who passed away much too early, for sharing his large knowledge and experience in ATR-IR and modulation excitation spectroscopy with me, but also for his support and for his exemplary function as a scientist.

REFERENCES

[1] Newton, I. *Optics*; 2nd Edition, 1717; Vol. Book III.

[2] Harrick NJ. *Phys. Rev. Lett.* 1960;4:224.

[3] Harrick NJ. *Internal Reflection Spectroscopy*. New York: Interscience; 1967.

[4] Fahrenfort J. *Spectrochim. Acta* 1961;17:698.

[5] Born M, Wolf E. *Principles of Optics*. 3rd Edition Oxford, UK: Pergamon; 1979.

[6] Fringeli, U. P.; Baurecht, D.; Siam, M.; Reiter, G.; Schwarzott, M.; Bürgi, T.; Brüesch, P. In: *Handbook of Thin Film Materials* (H. S. Nalwa, Ed.). New York: Academic Press, 2001; Vol. 2, 191.

[7] Reiter G, Hassler N, Weber V, Falkenhagen D, Fringeli UP. *Biochim. Biophys. Acta* 2004;1699:253.
[8] Reiter G, Siam M, Falkenhagen D, Gollneritsch W, Baurecht D, Fringeli UP. *Langmuir* 2002;18:5761.
[9] Fringeli, U. P.; Goette, J.; Reiter, G.; Siam, M.; Baurecht, D. *AIP Proceeding of 11th International* Conference on *Fourier Transform Spectroscopy* (J. A. deHaseth, Ed.).: Woodbury, NY: American Institute of Physics, 1998, Vol. 430.
[10] Baurecht D, Reiter G, Hassler N, Schwarzott M, Fringeli UP. *Chimia* 2005;59:226.
[11] Bieri M, Burgi T. *J. Phys. Chem. B* 2005;109:22476.
[12] Osawa M, Ataka K-I, Yoshii K, Yotsuyanagi T. *J. El. Spectr. Rel. Phenom.* 1993;64/65:371.
[13] Hutter E, Assiongbon KA, Fendler JH, Roy D. *J. Phys. Chem. B* 2003;107:7812.
[14] Street SC, Gellman AJ. *J. Phys. Chem. B* 1997;101:1389.
[15] Gautier C, Burgi T. *Chem. Commun. 2005.*, 5393.
[16] Fringeli, U. P.; Baurecht, D.; Günthard, H. H. In *Infrared and Raman Spectroscopy of Biological Materials* (H. U. Gremlich, B. Yan, Eds.). New York, Dekker, 2000, 143.
[17] Baurecht D, Fringeli UP. *Rev. Sci. Instrum.* 2001;72:3782.
[18] Baurecht D, Porth I, Fringeli UP. *Vib. Spectr.* 2002;30:85.
[19] Burgi T, Baiker A. *J. Phys. Chem. B* 2002;106:10649.
[20] Dolamic I, Bürgi T. *J. Catal.* 2007;248:268.
[21] Wirz R, Burgi T, Baiker A. *Langmuir* 2003;19:785.
[22] Strehlow H, Knoche W. *Fundamentals of Chemical Relaxation.* 1977;Vol.10 :Weinheim: Verlag Chemie; 1977.
[23] Fourier J. *The Analytical Theory of Heat.* Cambridge: Cambridge University Press; 1878.
[24] Bürgi T, Baiker A. *J. Phys. Chem. B* 2002;106:10649.
[25] Muller M, Buchet R, Fringeli UP. *J. Phys. Chem.* 1996;100:10810.
[26] Bieri M, Burgi T. *J. Phys. Chem. B* 2005;109:10243.
[27] Bieri M, Bürgi T. *Chem. Phys. Chem.* 2006;7:514.
[28] Urakawa A, Wirz R, Burgi T, Baiker A. *J. Phys. Chem. B* 2003;107:13061
[29] Nafie LA. *Annu. Rev. Phys. Chem.* 1997;48:357.
[30] Freedman TB, Cao X, Dukor RK, Nafie LA. *Chirality* 2003;15:743.
[31] Herse C, Bas D, Krebs FC, Bürgi T, Weber J, Wesolowski T, Laursen BW, Lacour J. *Angew. Chem, Int. Ed.* 2003;42:3162.
[32] Naubron JV, Giordano L, Fotiadu F, Bürgi T, Vanthuyne N, Russel C. GB. *J. Org. Chem.* 2006;71:5586.
[33] Lombardi RA, Cao XL, Kim SS, Dukor RK, Nafie LA. *Biochemistry* 2003;42:195.
[34] Holzwart G, Hsu EC, Mosher HS, Faulkner TR, Moscowit A. *J. Am. Chem. Soc.* 1974;96:251.
[35] Nafie LA, Keiderling TA, Stephens PJ. *J. Am. Chem. Soc.* 1976;98:2715.
[36] Nafie LA, Diem M. *Appl. Spectrosc.* 1979;33:130.
[37] Ma SL, Freedman TB, Dukor RK, Nafie LA. *Appl. Spectrosc.* 2010;64:615.
[38] Barron LD, Buckingham AD. *Annu. Rev. Phys. Chem.* 1975;26:381.
[39] Hug W, Kint S, Bailey GF, Scherer JR. *J. Am. Chem. Soc.* 1975;97:5589.
[40] Nafie LA. *Appl. Spectrosc.* 2000;54:1634.
[41] Su CN, Keiderling TA. *J. Am. Chem. Soc.* 1980;102:511.
[42] Schweitzer-Stenner R, Eker F, Griebenow K, Cao XL, Nafie LA. *J. Am. Chem. Soc.* 2004;126:2768.
[43] Freedman TB, Cao XL, Dukor RK, Nafie LA. *Chirality* 2003;15:743.

[44] Bour P, Navratilova H, Setnicka V, Urbanova M, Volka K. *J. Org. Chem.* 2002;67:161.
[45] Bürgi T, Vargas A, Baiker A. *J. Chem. Soc. Perkin Trans.* 2002;2:1596.
[46] Devlin FJ, Stephens PJ, Cheeseman JR, Frisch MJ. *J. Phys. Chem. A* 1997;101:9912.
[47] Baumruk V, Keiderling TA. *J. Am. Chem. Soc.* 1993;115:6939.
[48] Schweitzer-Stenner R. *J. Phys. Chem. B* 2004;108:16965.
[49] Keiderling TA. *Curr. Opin. Chem. Biol.* 2002;6:682.
[50] Han SH, Ji N, Belkin MA, Shen YR. *Phys. Rev. B.*, 66.
[51] Gautier C, Bürgi T. *J. Am. Chem. Soc.* 2006;128:11079.
[52] Bieri M, Bürgi T. *J. Phys. Chem. B* 2005;109:22476.
[53] Brazhe NA, Brazhe AR, Sosnovtseva OV, Abdali S. *Chirality* 2010;21:E307.

Chapter 6

Synchrotron infrared interface science

P. Dumas[1] and V. Humblot[2]
[1]*SOLEIL Synchrotron, BP48, Saint Aubin, F-91192 Gif sur Yvette Cédex, France*
[2]*Université Pierre et Marie Curie, Paris 6, Laboratoire de Réactivité de Surface - UMR CNRS 7197, 4 place Jussieu Paris VI, 75252 Paris Cedex 05, France.*

INTRODUCTION

Infrared (IR) spectroscopy forms part of "biophotonics," a field of research that merges biology and optics and covers an ensemble of techniques that exploit light or more generally all forms of photonic energy in the field of life sciences. IR spectroscopy of biomolecules has been a very active research area over the last 15 years or so, and it has made a tremendous impact in this branch of molecular spectroscopy. Applications of this vibrational technique to proteins, lipids, enzymes, nucleic acids, carbohydrates, and biomembranes are numerous [1]. With the addition of a microscope as an accessory to conventional Fourier Transform InfraRed (FTIR), spectrometers have brought the potential to examine tissues at cellular resolution. It has led to IR imaging, where biochemical and spatial information are combined.

Biointerface Characterization by Advanced IR Spectroscopy. DOI: 10.1016/B978-0-444-53558-0.00006-0

This book is dedicated to biointerfaces, a field which is exhibiting an increasing interest over the past few years. IR spectroscopy has become an important approach for the characterization of biomolecule–surface interactions at a molecular level, and several chapters of this book are demonstrating such capabilities. Often, the sensitivity of the technique for such a complex interface has built upon the previous knowledge and expertise of IR spectroscopy at model systems adsorbate substrate. There have been, undoubtedly, very important achievements using conventional spectrometers, complemented by news ways of applying IR spectroscopy.

Apart from the sophisticated techniques associated with IR spectrometers, exemplified in this book, the heated filament that is the IR thermal source in spectrometers has been challenged when it was realized that synchrotron radiation could be a much brighter source in the IR region [2].

Synchrotron radiation is the light produced by a beam of relativistic charged particles, most often electrons, in synchrotron storage rings. A synchrotron facility or center is a large machine (about the size of a soccer field, or even bigger) that accelerates electrons to almost the speed of light. These storage rings involve combinations of magnets that bend (and focus) the electron beam onto an approximately circular orbit. They can accelerate and store hundreds of milliamperes of current for hours at a time. As electrons are deflected through magnetic fields they create extremely bright light. Replenishing the beam on a regular basis compensates the slow loss of electron beam, due to scattering. Nowadays, a refilling mode of the synchrotron storage ring, named top-up, is maintaining an almost constant intensity of the electron beam over days, with less and less beam perturbation during reinjections. The light produced is channeled down to beamlines and to experimental workstations where it is used for research [3].

Although most synchrotron light sources are optimized to produce X-ray and vacuum ultraviolet radiation, they also produce broadband radiation in the IR region. The primary advantage of synchrotron IR radiation is its brightness (number of photons per unit area and per unit solid angle). Because the light originates from a small bunch of electrons, the source can be treated as a point source. Thus, IR light from a synchrotron can be easily collimated and/or focused to diffraction limited spot sizes ($\sim$1–10 μm), allowing high spatial resolution (IR spectromicroscopy) and high spectral resolution both the far-IR to near-IR energy domain.

IR spectroscopy and microscopy beamline end stations have flourished during the last 10 years, and almost all synchrotron facilities worldwide have, or are planned to have, IR beamlines. If there was a strong motivation and scientific activity in surface science, in the 1990s, the motivation has evolved, nowadays toward biology.

The far-IR domain has provided, undoubtedly, the strong initial boost for exploiting the synchrotron radiation. In solid-state physics, phonons and other collective modes have resonant excitations in the so-called far-IR and THz spectral range. These can involve transport phenomena and the various interactions between electrons and other excitations in solid, making IR spectroscopy an important tool for studying complex electronic systems (such as excitons in semiconductor nanoparticles, and ferroelectricity, magnetoresistance, and superconductivity in complex metal oxides). In many cases, the information complements that obtained from higher energy photons (e.g., photoemission, X-ray spectroscopy, diffraction, and scattering), yielding a more complete picture of a material's structure and behavior.

The subject of interface bonding and intermolecular dynamics has attracted considerable attention with much of the information about the nature of the bonding being inferred from measurements in the intramolecular vibrational modes in the 600–4000 cm^{-1}. With a synchrotron radiation, one can measure at lower frequency, and study not only these intramolecular motions, but also the bonding modes in the frequency domain 20–600 cm^{-1}. This is crucial in any attempt to understand the many macroscopic phenomena involving interfacial processes such as adhesion, adsorption, desorption, friction, lubrification, catalysis, epitaxial growth, and electronic properties.

The high brightness of IR synchrotron radiation is crucial for a variety of low-throughput techniques. These include microprobes and imaging microspectroscopy, for micron-sized samples such as individual cells and tissues at subcellular resolution. This is exemplified in this chapter. Dynamical processes can be investigated using any of these techniques by exploiting the pulsed nature of the synchrotron source. Though lasers can probe shorter time scales, no other source allows for such a wide spectral coverage.

SYNCHROTRON INFRARED EMISSION

Synchrotron radiation is the light produced by a beam of relativistic charged particles, most often electrons, in synchrotron storage rings. One of the key parameters defining the synchrotron emission is the Lorentz factor, a convenient way to define a relativistic quantity for the accelerated electrons (also called relativistic mass enhancement factor):

$$E = \gamma \cdot E_0 \tag{1}$$

where $E_0 = m_0c^2 = 0.511$ MeV is the rest energy of the electron and $\gamma \equiv \frac{1}{\sqrt{\left(1-\left(^{v}/_{c}\right)^2\right)}}$, where v is the velocity, in unit in c, the speed of light.

The synchrotron light is pulsed (hundreds of picoseconds) and is emitted in a narrow cone with an half-angle of $1/\gamma$. For example, modern synchrotron facilities are operating at energy of 3 GeV or so. This gives a value of $\gamma = 5871$, and an emission angle of 0.170 mrad (0.0095°).

Relativistic effects cause the emitted radiation to span a large spectral range, often reaching into the hard X-ray spectral range. In contrast to free electron lasers (FELs), synchrotron radiation is a spectrally continuous (i.e., "white") source of light. For a more detailed description, we report readers to [4].

Exploiting the IR energy domain from synchrotron radiation is quite recent. IR emission from synchrotron radiation was early proposed as a useful source for long wavelength studies of throughput limited experiments [2,5–8].

The first studies utilizing such a source were carried out at Tantalus, Daresbury, and Bessy I, starting in the 1970s [9]. The real boost in IR experimental activities in synchrotron facilities, started right after, at UVSOR [10], NSLS VUV [11–13], later LURE SuperACO [14]. All beamlines were designed with reasonably large apertures, in order to extract more efficiently the IR emitted photons (see below). However, at the early time, the electron beam motions in synchrotron facilities were generating very noisy photon beams, and this was quite detrimental for the spectral quality. Moreover, since the emitted IR flux is proportional to the electron beam current, and this later could be barely made higher than few tens of millamperes, the collected flux was not as high as originally expected due to the limited electron current in the storage rings. Despite this, the initial science program with such IR beamlines was mostly concentrated on the far-IR wavelength domain, and in application domain such as gas phase, solid-state physics and surface science [15–17]. These scientific fields have remained very active over the years and represent, even today, an important research activity in the far-IR (also named THz) using the synchrotron radiation.

As mentioned in the introduction, the real boost of interest in using synchrotron radiation as an IR source arose when it was demonstrated that the high brilliance advantage is a key issue in microanalysis. This has led to a bigger impact in applications, and for wide recognition of the usefulness of the IR synchrotron source [18,19].

Infrared emission in storage rings

IR radiation is generated by electrons traveling at relativistic velocities, either in a curved path through a constant magnetic field (i.e., bending magnet radiation [2]) or when their trajectories encounter variable magnetic fields, for example, at the edges of bending magnets (i.e., edge radiation (ER)) [20,21] (Fig. 1). In the latter, the ER is generated by a high-energy charged particle when it passes the entrance or exit through the region of a rapid change of magnetic field amplitude of a bending magnet. The photons emitted at two adjacent bending magnets bounding a straight section appear in the same cone and are subsequently synchronized by the electron itself. This leads to an interference effect, which manifests itself in oscillations of radiation intensity [20,21].

In synchrotron world, spectral flux is more traditionally expressed as the number of photons per time unit and per relative spectral bandwidth,

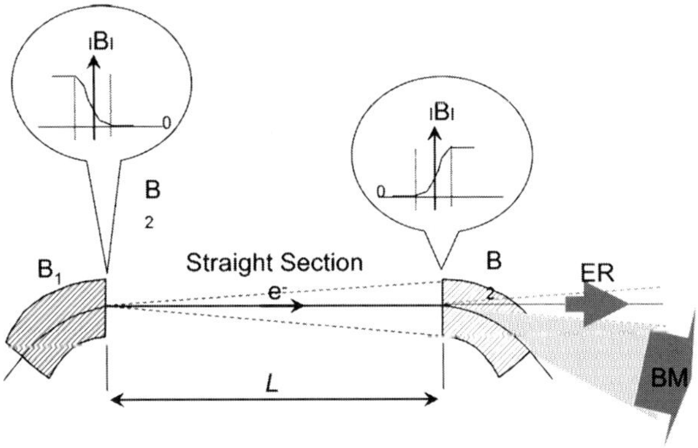

FIGURE 1 **Schematic drawing of the infrared emission from edge radiation (ER) and bending magnet radiation (BM). L is the length of the straight section, and this influences the interference effects with the upstream ER dipole emission.**

while in the spectroscopy world, this is expressed in terms of the emitted power per unit wave number. Both quantities are related by the following formulas [22]:

$$F_{src} \equiv \frac{\mathrm{d}W}{\mathrm{d}(1/\lambda)} = hc\frac{\mathrm{d}N}{\mathrm{d}t(\mathrm{d}\lambda/\lambda)}, \tag{2}$$

where h is the Planck constant and c is the speed of light.

In practical units

$$\frac{\mathrm{d}W}{\mathrm{d}(1/\lambda)}\left(\frac{W}{\mathrm{cm}^{-1}}\right) \approx 2\times 10^{-20}\frac{\mathrm{d}N}{\mathrm{d}t\,(\mathrm{d}\lambda/\lambda)}\left[\frac{\text{Photons}}{s\,(0.1\%\,bw)}\right]. \tag{3}$$

The flux emitted by the two main sources of IR emission can be calculated using the following equations [22].

From constant field of bending magnet

Assuming that vertical angular aperture of an IR beamline $\Delta\theta_y$ is larger than the natural opening angle of synchrotron radiation, that is, $\theta_y \gg (\lambda/\rho)^{1/3}$, where λ is the radiation wavelength and ρ the magnet radius, the spectral flux of synchrotron radiation can be estimated as

$$\left(\frac{\mathrm{d}W}{\mathrm{d}(1/\lambda)}\right)_{SR} \approx \frac{\sqrt{3}}{4\pi}\frac{e\gamma I}{\varepsilon_0}\theta_x G(\lambda_c/\lambda), \tag{4}$$

with

$$G(x) \equiv x \int_{x}^{+\infty} K_{5/3}(x')\,dx',$$

where ε_0 is the permittivity of free space, e is the charge of electron, $\gamma = E/m_0c^2\gamma \gg 1$ is the electron relativistic mass enhancement factor, I is the electron beam current, θ_x is the horizontal angular aperture, $\lambda_c = 4\pi\rho/(3\gamma^3)$ is the critical synchrotron radiation wavelength for the bending magnet, and $K_{5/3}$ is the modified Bessel function. In practical units, the above equation becomes

$$\left(\frac{dW}{d(1/\lambda)}\right)_{SR}\left(\frac{W}{cm^{-1}}\right) \approx 4.88 \times 10^{-7} E(\text{GeV}) I[A] \theta_x(\text{mrad}) G(\lambda_c/\lambda) \qquad (5)$$

where E is electron energy (in GeV).

Taken as example, the French synchrotron facility SOLEIL, which has a storage ring with parameters $E = 2.75$ GeV, $I = 0.5$ A, $\lambda_c = 1.43$ Å, a beamline with horizontal angular aperture $\theta_x = 40$ mrad, at the wavelength $\lambda = 10$ µm (1000 cm^{-1}), the emitted flux is

$$\left(\frac{dW}{d(1/\lambda)}\right)_{SR} \approx 1.40 \times 10^{-6} \frac{W}{cm^{-1}}.$$

Edge radiation from extremities of bending magnets

In the near field observation region, the spectral flux per unit surface of the radiation emitted at two edges of bending magnet limiting one straight section of a storage ring is approximately

$$\left(\frac{dW}{dSd(1/\lambda)}\right)_{ER} \approx \frac{2eI}{\pi^2\varepsilon_0 r_\perp^2}\sin^2\left[\frac{\pi L r_\perp^2}{2\lambda z(z+L)}\right] \qquad (6)$$

where $r_\perp$ is the distance from observation point to the straight section axis ($r_\perp^2 = x^2 + y^2$), z is distance from downstream bending magnet edge to observation plane, assumed to be $r_\perp \ll z \ll \lambda\gamma^2$, and L is the distance between bending magnet edges (i.e., the straight section length). This intensity distribution is the result of interference of the emission from two bending magnet edges, and the near-field effect.

Equation (6) allows us to estimate the angular size of the first interference ring of the ER intensity distribution, as observed at a distance $z = r_{\perp 1}/z \approx 2[2\lambda(z+L)/(zL)]^{1/2}$. At $L = 10$ m, $z = 3$ m, and $\lambda = 10$ µm, this gives $r_{\perp 1}/z \approx 6$ mrad. This is much less than the natural opening angle of the SR from constant field of the bending magnet. One should note, however, that the first interference ring does not contain all the flux of the ER.

After integration of Equation (6) over $r_\perp$ within a circular aperture $z\theta_r$ centered on the axis of the straight section (θ_r is the angular radius of the aperture, assuming the origin at the edge of the downstream magnet), we get

$$\left(\frac{dW}{d(1/\lambda)}\right)_{ER} \approx \frac{eI}{\pi\varepsilon_0} H\left[\frac{\pi z L \theta_r^2}{\lambda(z+L)}\right], \tag{7}$$

where $H(x) \equiv \ln(x) - ci(x) + C$, and $\mathrm{ci}(x) \equiv -\int_x^{+\infty} \cos(t)t^{-1}dt$ is the cosine integral function, C $\approx$ 0.577216 is the Euler constant. In practical units, equation (7) can be rewritten as

$$\left(\frac{dW}{d(1/\lambda)}\right)_{ER}\left(\frac{W}{cm^{-1}}\right) \approx 5.76 \times 10^{-7} I(A) H\left[\frac{\pi \cdot \theta_r^2(\mathrm{mrad})}{\lambda(\mu m)} \frac{zL}{z+L}(m)\right]. \tag{8}$$

Taking the following parameters from SOLEIL synchrotron: $I = 0.5$ A, $L = 10$ m, $z = 5$ m, $\theta_r = 10$ mrad, and $\lambda = 10$ μm, we obtain from equation (8)

$$\left(\frac{dW}{d(1/\lambda)}\right)_{ER} \approx 1.5 \times 10^{-6} \frac{W}{cm^{-1}}.$$

Therefore, these two IR sources are almost equivalent in terms of flux as well as in brightness. However, for completion, it is interesting to know the intensity distribution profiles of both sources and their polarization properties.

In Figure 2, the distribution profiles at two wavelengths (10 and 100 μm) have been calculated, using SRW [23] for an energy storage ring of 3.0 GeV and for a prototypical opening angle of the dipole chamber of 20 × 40 mrad for the collection of bending magnet IR emission, and 20 × 20 mrad for the collection of the edge radiation IR emission. These profiles illustrate that, for bending magnet emission, the angle of emission is larger than that of the ER. However, it should be noted that bending magnet emission can be increased if the horizontal opening angle is increased (which is the case actually in several new IR ports at synchrotron facility, for example, 78 mrad at SOLEIL (France), 50 mrad at DIAMOND (UK), 60 mrad at Australian Synchrotron, and even 320 mrad at SRC, (Wisconsin (USA)).

A marked difference between the two sources is their polarization properties (Fig. 3). Compared to bending magnet emission, which is strictly horizontally polarized in the plane of the electron trajectory, ER has a radial polarization. Circular right and left polarization, as well as vertical polarization can be selected by collecting the appropriate portion of the IR beam emission, as seen in Figure 3.

The apparent source size is usually not determined by the physical electron beam dimensions (few tens of microns) but rather by the limits imposed by diffraction. When diffraction is limited, the rms source dimensions are

$$\sigma_{\text{bend magnet}} \approx 1.6\left(\rho\lambda^2\right)^{1/3} (\text{dipole bend source})$$

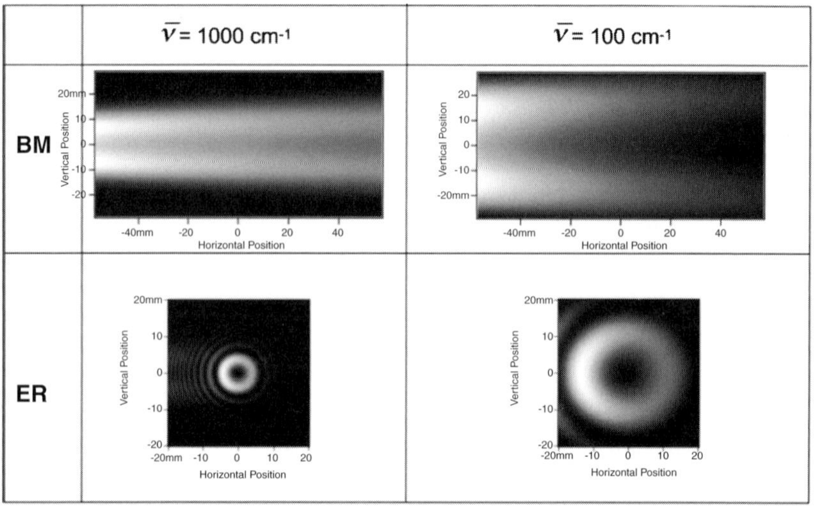

FIGURE 2 Intensity distribution for two wavelengths, l = 100 mm (left column) and l = 10 mm (right column). The distribution has been calculated for a bending magnet (BM) emission, and an ER emission. This calculation has been carried out for the case of a storage ring of 3 GeV, a vertical aperture of 20 mrad, and a horizontal aperture of 20 mrad for ER and 40 mrad for BM.

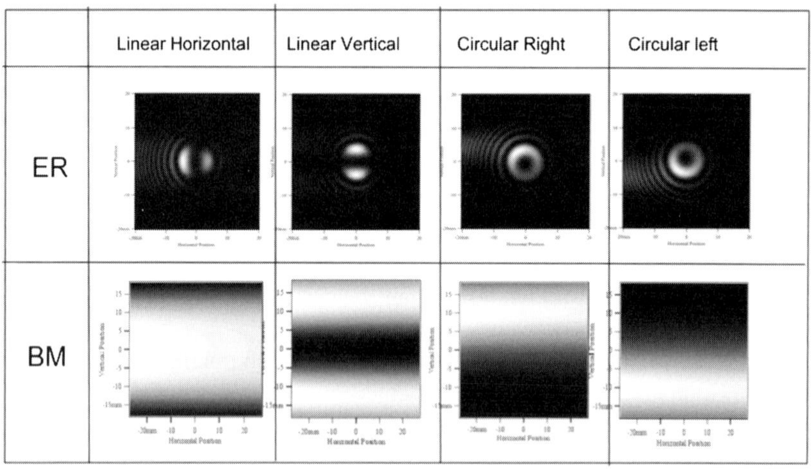

FIGURE 3 Intensity distribution for various polarization states for ER and bending magnet radiation (BM).

where λ is the wavelength of interest and ρ is the bend radius in the dipole magnet. Taken, as an example, a storage ring having an electron beam energy of 3 GeV ($\gamma = 5871$) and a bending radius on the order of 7 m, the dipole bend rms source size at $\lambda = 10$ μm is about 1.42 mm.

These source dimensions pertain to optical extractions limited to the natural opening angle of the emitted light, measured in milliradians (mrad). The rms half-angles for emission are

$$\theta_{\text{bend}} = 0.6(\lambda/\rho)^{1/3} \quad \text{(dipole bend source)}$$

$$\theta_{\text{edge}} = {}^{1}/_{\gamma}. \quad \text{(edge source)}$$

Taken, as the above example of a synchrotron facility at 3 GeV, the dipole bend emission angle at $\lambda = 10$ μm is 6.7 mrad while for ER the emission is below 0.2 mrad. Either of these sources can be demagnified to produce a small spotsize for surface science studies in grazing incidence or in the microscope, where the rms spot size can be made on the order of λ. It should be noted that the $1/\gamma$ emission angle for ER depends on having the full source dimensions extracted. In practice, the source is often truncated by the storage ring's metal vacuum chamber, resulting in larger opening angles that also depend on the wavelength. A horizontal collection angle greater than these values can result in additional horizontal source segments, turning the point-like, diffraction-limited source into an extended source. One could imagine collecting and extracting the entire horizontal swath of light from a bending magnet source, but the mechanical design of the dipole magnets and the electron beam vacuum chamber set a practical limit of ~30° or less. Note that this only pertains to the dipole bend source as the edge source is, by definition, limited to the fringe field region of the magnet (typically less than 1° into the dipole). Though infrared ER by itself nearly always behaves as a point-like source; it evolves into bending magnet radiation that can be a part of the extracted flux.

Therefore, the important characteristic of the synchrotron IR source, such as other wavelengths, is its very high brightness, the result of the intrinsically small source size and narrow emission angle for synchrotron radiation.

Williams [24] has shown that in the IR spectral region, where the synchrotron radiation source is typically diffraction limited, the brightness is given by

$$B_{\text{sr}}(\bar{\nu}) \cong 10^{-8} I\, \bar{\nu}^2 \ \text{W/mm}^2/\text{sr/cm}^{-1} \tag{9}$$

where I is the stored beam current in amperes and $\bar{\nu}$ the frequency in wavenumbers (cm^{-1}).

A blackbody radiator has a brightness, which can be expressed as is

$$B_{bb}(\omega) = \frac{\hbar\omega^3}{(2\pi)^2 c^2} \frac{d\omega}{e^{\hbar\omega/kT} - 1} \tag{10}$$

Assuming emission into 2π steradians (plus ignoring any obliquity factors), substituting $T = 2000$ K (the temperature of a typical thermal source), and converting to wavenumber yields

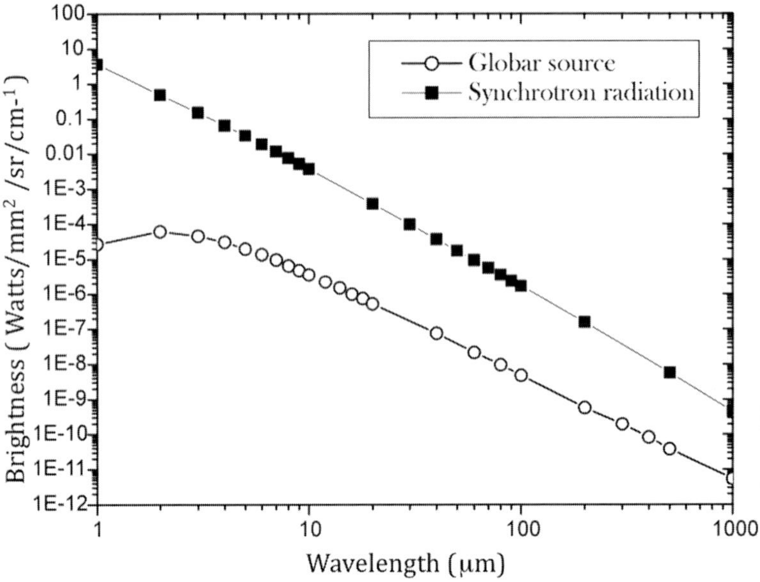

FIGURE 4 Brightness comparison between a synchrotron source and a thermal source (at 2000 K).

$$B_{bb}(\bar{\nu}) = 6 \times 10^{-15}\bar{\nu}^3 \frac{d\bar{\nu}}{e^{\bar{\nu}/1400} - 1} \text{ W/mm}^2/\text{sr/cm}^{-1} \tag{11}$$

Figure 4 compares the synchrotron versus blackbody brightness according to equations (9) and (11). Clearly, the synchrotron offers approximately three orders of magnitude advantage over the thermal source.

It is important to note that synchrotron IR source is an absolute source since its power depends strictly on the electron beam current stored in the ring, which can be measured accurately. This has opened a completely new field in surface dynamics, as will be illustrated in the next section.

Today, most synchrotron radiation facilities offer a port dedicated to IR spectroscopy and spectromicroscopy. While early synchrotron sources had beam stability issues, the high electron beam stability of modern third-generation machines and/or the use of optical beam stabilizing techniques [25] provide the required beam stability for high-quality FTIR spectroscopy.

SYNCHROTRON INFRARED STUDIES IN SURFACE SCIENCE

Most of the studies have been performed on simple molecules (CO, NO) with few of them related to organic molecules.

Identification of low frequency modes of adsorbed species

The main initial objective at the beginning of the exploitation of the synchrotron far-IR photons was to study the bonding between adsorbates and substrate (mainly metallic). Not surprisingly, the first system studied was the interaction of the carbon monoxide (CO) with monocrystalline surfaces (Cu, Pt) [26–30]. The C–Cu vibrational mode was found at ~340 cm^{-1} while the C–Pt vibrational mode appears at ~470 cm^{-1}. It is appealing to note that the same system was studied using a thermal source at low temperature [31] and the C–Pt signal was recorded after 3 days of data accumulation, (10^5 s) compared to the synchrotron-based spectra obtained in 100 s in [30]. Interestingly, the C–Pt band position not only depends on the coverage, but also exhibits an upward frequency shift at 300 K, and indicates a temperature onset for the lifting of the surface reconstruction [32].

Perhaps the most illustrative example of the potential for synchrotron far-IR approach for adsorbate–substrate interaction is the study of the absorption of NO on Cu [28]. After completion of a monolayer of NO, no IR signal was observed in the stretching mode region of NO, suggesting that the molecules have reacted and desorbed. However, the far-IR spectra revealed one band, which, upon isotope labeling, was assigned to N_2O, bonded to the copper via the oxygen atom, and with a N–N axis parallel to the surface (Fig. 5).

Several studies have identified the bonding nature of organic molecules on the metallic surface [27]. Adsorption of the diketones, methylacetoacetate, and acetylacetone, on Cu(1 1 0) yields a pure νCu–O vibration at <300 cm^{-1}, while the monotartrate form of (R, R) tartaric acid adsorbed via its carboxylate group

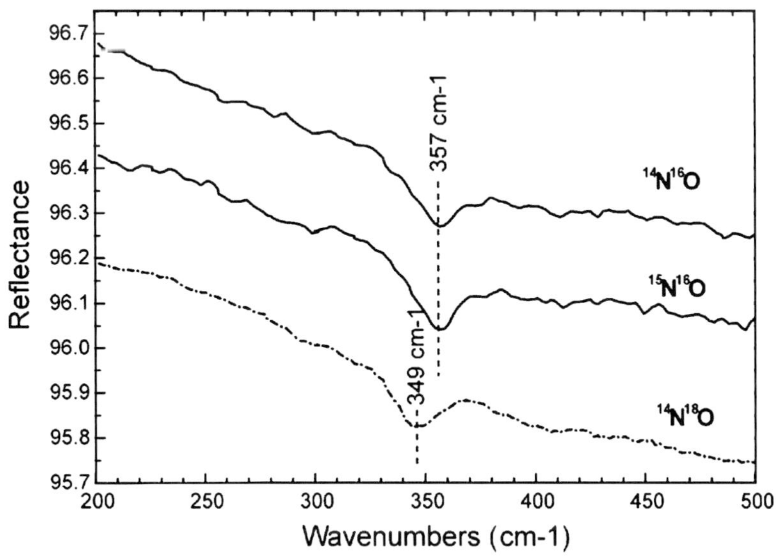

FIGURE 5 **Frequency dependence of the low frequency mode, for 5 L exposure of NO, on Cu (111), at T = 90 K. The low frequency mode is downward shifted upon oxygen labeling (3%). This has been interpreted as a Cu–ON$_2$ peak, as the Cu–O peak would have shifted by 6%. From [28].**

on Cu(1 1 0) shows a νCu–O stretching vibration at 378 cm^{-1}. This is almost identical to the νCu–O stretching frequency of formate adsorbed via its carboxylate unit at Cu(1 0 0) and Cu(1 1 0) surfaces. For formate, the high resolution of the IR technique has enabled the asymmetry of the absorption band to be detected, suggesting the presence of heterogeneity in the adlayer.

Vibrational dynamics

Low frequency vibrational modes of adsorbates are precursors to motion, about or away from the surface, as their energies are directly related to the potential energy surfaces of the molecule or atom bonded. Due to their low energies, they are easily thermally populated and contribute predominantly to the adsorbate entropy. Therefore, energy transfer through these modes is important in various processes, such as surface diffusion and reactions. Since the beginning of the 1990s, the interest of the study of the low energy modes was renewed when Persson suggested that frustrated translational modes of molecules or atoms are the key to the frictional force induced by the adsorbate on the metallic surface [33–34]. A direct relation between the lifetime of the frustrated translation and the friction coefficient, responsible for the diffuse scattering of the conduction electron of the substrate by the adsorbate (resistivity change) has been suggested [35].

Vibrational dynamics at surfaces is manifested in two major observations in IR spectra. First, a broadband reflectance change is observed upon adsorption, and the overall reflectivity change provides a value of the electron hole lifetime (which is related to the variation of the density of states at the Fermi level induced by the adsorbate [36,37]). Simultaneous measurements of dc resistivity and IR reflectivity changes were performed on the CO/Cu system, confirming the model (Fig. 6). The strict dependence of the synchrotron far-IR photons on

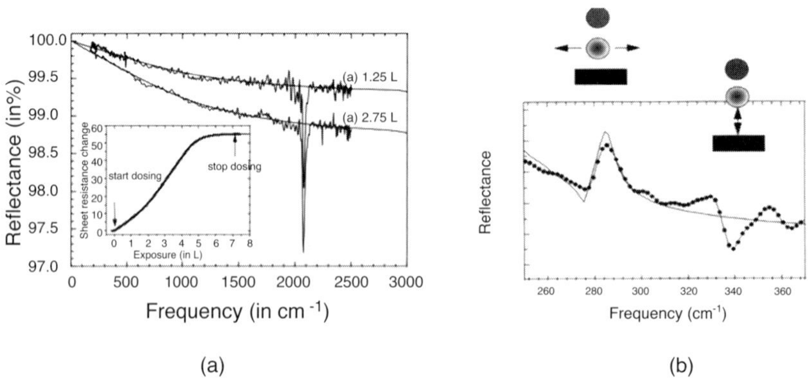

FIGURE 6 (a) Broadband IR reflectance change for CO adsorbed on a Cu (111) film (67 nm thick), at T = 90 K. The solid lines through the spectra are fits according to Persson and Volokitin's model [34]. The insert shows the change and shape of the resistance during CO exposure. (b) C–Cu vibration band and anti-absorption mode arising for the frustrated rotation of CO on the copper surface.

the electron beam current is mandatory to record such a low reflectivity change upon molecular adsorption, and was used to calculate the electron hole pair lifetime for other interfaces such as C_{60} on Ag(1 1 0) and Cu(1 0 0), providing values of 480 and 602 ps, respectively, and indicating a high metallic character of C60 on these metallic surfaces [38].

Such studies have not been undertaken on biomolecular interfaces yet.

SYNCHROTRON INFRARED APPLICATIONS IN BIOLOGY

It is well recognized that IR microspectroscopy is providing a direct indication of a sample's biochemistry [39]. The brightness advantage of the synchrotron source is being increasingly used for subcellular studies, since, with diffraction limited spot size, the spectral are obtained with high (S/N) quality. Armed with information on sample histology and pathology, variations in nucleic acid, protein, and lipid content or protein secondary structure have been scrutinized as potentially providing important details about the chemistry of diseased states [40].

It is therefore not surprising that, up to now, the large majority of synchrotron mid-IR microspectroscopy studies have concentrated on isolated cell and tissues, for biomedical applications. However, until recently, no spatially resolved studies at a biomolecular interface have been carried out. The recent achievement using a synchrotron-powered IR microscope, and exemplified in this section, is opening new avenues in the research field.

Single cells and tissues studies

It is expected that the nucleus of isolated cell might possess a biochemical different spectra form the cytoplasm. From the changes occurring in the nucleus, phenomena such as apoptosis, necrosis can be identified and are often characterized, in an IR spectrum, by a change in a secondary structure of the protein (Amide I band) [41] and consequently the effect of external agent, drug can be addressed by following the biochemical signature of these nucleus [40,41]. As demonstrated in the previous reference, synchrotron IR microscopy is mandatory to follow the subtle changes specifically in the nucleus. In addition, the spectral changes have to be recorded on an important number of different cells, in order to have a statistical relevancy. Therefore, statistical methods, such as principal component analysis (PCA), hierarchical cluster analysis (HCA), and K-means clustering, among others, are used routinely in biological studies using IR microspectroscopy. Accuracy of the statistical treatment of spectra tremendously relies upon the high spectral quality obtained. Figure 7(a) illustrates the spectrum obtained from a nucleus of a cancer cell (HELA), using a spot size of $6 \times 6\ \mu m^2$, as well as the PCA loadings of the spectra (Fig. 7(c)) recorded in the nucleus and in the cytoplasm of isolated cells displayed in Figure 7(b). It is quite obvious that spectra belong to two separate groups, and the PCA score displayed

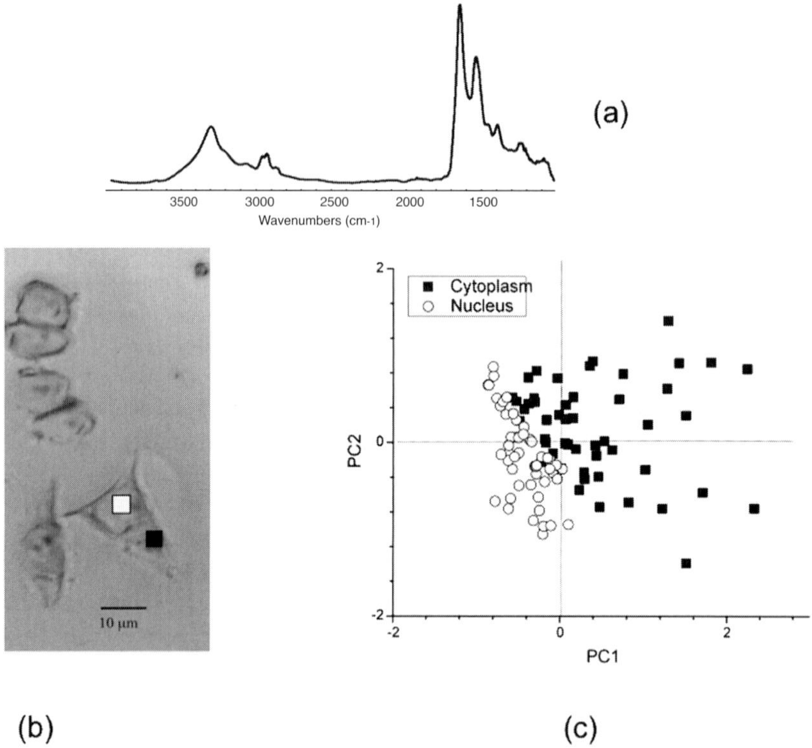

FIGURE 7 **Infrared spectrum of a single cell taken with an aperture of 6×6 mm^2 (16 scans, 4 cm^{-1} resolution). The light micrograph showing the area probed inside one single cell using synchrotron infrared microscopy. The dimension of the projected aperture on the sample was set to 6×6 µm^2 (represented as a dashed square), allowing analysis of either the nucleus or the cytoplasm of each cell. (B): PCA analysis of the whole set of spectra taken inside the nucleus and the cytoplasm of about 100 cells. There is a clear separation of the spectra into two clusters, where the discrimination is 92% along the PC1 axis.**

in Figure 7(c) shows that a different secondary structure of the proteins is identified in the nucleus. Subsequent efficient drug treatment of cancer cells will induce a further change in the proteome of the nucleus [42]. It should not be ignored that the thorough analysis of single cell spectra is facing challenges, since physical phenomena is adding on the spectral shape. This is currently a subject of intense work, and important understanding and progresses in data treatment is being made, to disentangle spectral variation due to biochemical changes from those coming for physical artifacts [42].

Figure 8 illustrates the power of the synchrotron beam in imaging, with subcellular resolution, the distribution of the main components of individual cells. The chemical images have been generated by raster scanning a single

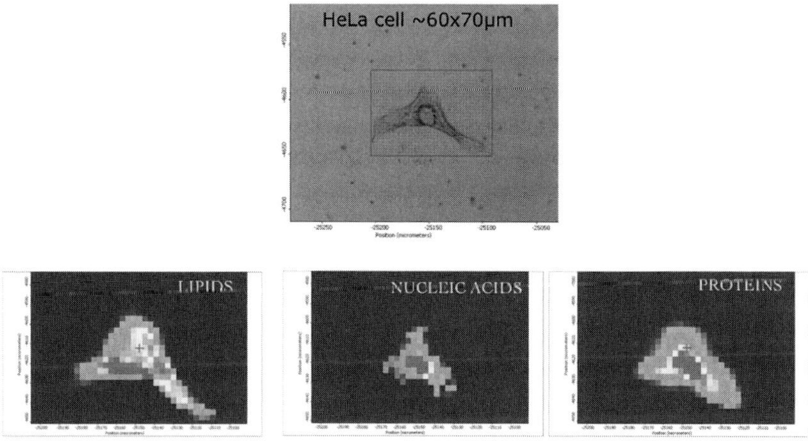

FIGURE 8 Chemical image of the distribution of lipids, nucleic acids, and proteins across one individual cancer cell (HELA). The step size for analysis is 1 μm. The acquisition parameters are as follows: 32 scans, aperture = 4 × 4 μm^2, resolution = 8 cm^{-1}, electron beam current= 400 mA. Courtesy: M.J. Tobin

HELA cancel cell, with a step size of 1 μm, and with 32 scans per spectrum. One can easily see that RNA and proteins are located in majority in the cell nucleus, while the lipids are distributed mainly outside the nucleus. This is often observed in cancer cells, and there is not consistent explanation yet for this phenomenon.

Many human tissues have been studied with high spatial resolution, and univariate (single band mapping) and multivariate (cluster mapping) imaging has been generated: Alzheimer, bones. (For a review, see [39].)

Biomolecular interface studies should build on the actual studies on biological cells and tissues. In a near future, with the improvement of the technical approaches, of the sensitivity of the detection, and of the brightness of the synchrotron source, high quality spectra can be acquired and subtle changes in conformation could be much better determined using statistical approaches, at biomolecular interfaces systems.

Spatially resolved biomolecular interface study

Providing that we can now benefit for a brighter IR source, in the extended frequency domain between the far-IR (THz) and the mid-IR, one can envision initiating spatially resolved studies at biointerfaces, in the mid- and far-IR region. One of the limiting factors has remained, until recent years, the beam stability, since the signal strength expected at biointerfaces is lower than that for the model system studies, and reported in this chapter.

The first step toward the application of spatially resolved studies at biointerfaces was carried out by the group of C. M. Pradier and V. Humblot at the

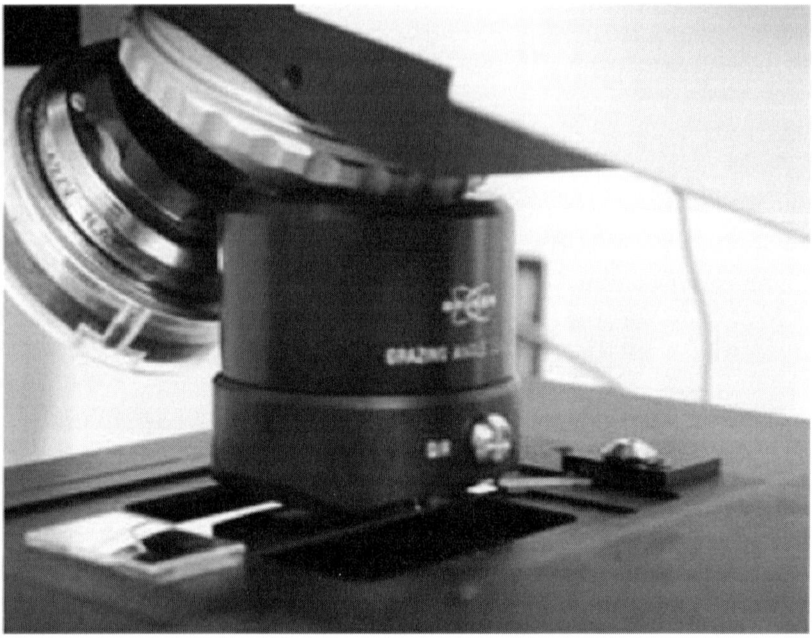

FIGURE 9 Grazing incidence objective (Bruker) mounted on a microscope (NICPLAN from Thermo Fischer), for performing biomolecular interface analysis of one monolayer of GSH on Au(111).

SMIS beamline at SOLEIL, the French synchrotron facility. The essential motivation was the high stability of the electron beam of this facility (one of the latest synchrotron facility made operational), who could now pave the way for such feasibility study.

Equipped with a grazing incidence objective (Fig. 9), a better spatial resolution is obtained when the synchrotron source is used instead of the thermal source. In Figure 10, we have reported the reflectivity signal obtained, on a 1951 USAF resolution target, at two different wavenumbers (1000 and 3000 cm^{-1}). Clearly, the use of the synchrotron source allows achieving a much better resolution, in the grazing incidence mode, compared to the thermal source. It can also be noted that the resolution degrades with the lower wavenumber (higher wavelength) frequency, due to diffraction effect.

Besides improved spatial resolution, minimizing noise is an important factor prior to any attempt to record an IR spectrum for a biomolecule monolayer. This was successfully achieved for the study of an adsorbed monolayer of polypeptide (glutathione or GSH(γ-glu-cys-gly)) on Au(1 1 1) [44]. The study of GSH was motivated by the strong affinity of the thiol group of the cystein for metal atoms and, from the fact that GSH is the major low-molecular-mass thiol compound in plants and animals in which it plays an important role in metabolic mechanisms, as natural reducing and the nucleophilic molecule.

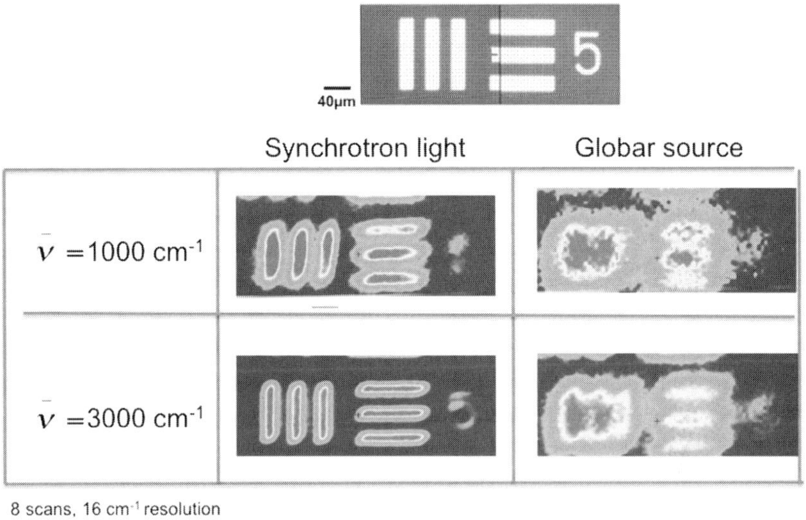

FIGURE 10 Resolution test performed on a 1951 USAF target, at two wavenumbers: 1000 and 3000 cm^{-1} (10 and 3.3 μm wavelength, respectively), and using either the synchrotron source (left) or the thermal source (right). The spectra have been acquired at 16 cm^{-1} resolution.

Little was known about the long-range scale homogeneity of the protein layer on the gold substrate. For that purpose, IR microscopy, using a grazing incidence objective, was performed to image its distribution. The experiment was carried out at the SMIS beamline at SOLEIL synchrotron (France), with a Magma 560 spectrometer to which a NICPLAN (Thermo Nicolet) microscope was coupled. A grazing incidence objective (Bruker) was used, as well as a liquid nitrogen cooled mercury cadmium telluride detector (50 μm). The spectra were collected in the 4000 – 800 cm^{-1} mid-IR range at a resolution of 4 cm^{-1} with 1024 coadded scans; the signal-to-noise ratio was measured around 3×10^{-4} peak-to-peak for band height in the range of 0.05 to 0.1%. The scanned areas obtained with the microscope objective were ~150 × 600 μm^2 with the internal thermal source and 20 × 80 μm^2 with the synchrotron light. Data analysis of IR spectra and chemical images was performed using the OMNIC software (Thermo Scientific).

Figure 11 reports the spectrum of a monolayer of L-glutathione or GSH on a Au(1 1 1) surface, while Figure 12 displays the spatial distribution of the protein over a surface of ≈ 100 × 300 μm^2 [44]. This figure shows how one can image the heterogeneous distribution of a protein overlayer on a metallic substrate.

Armed with such equipment, spatially resolved IR spectra and imaging has been shown to be feasible and one can expect a further development in this field in a near future.

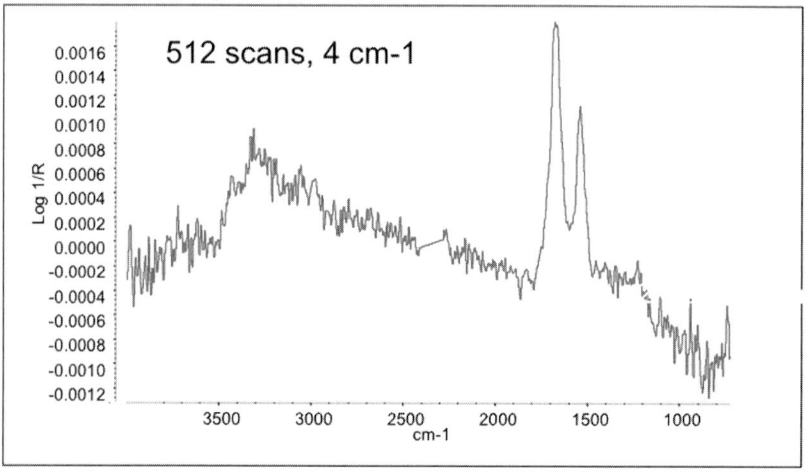

FIGURE 11 **Infrared spectrum of the one monolayer of GSH deposited on the Au(111) surface. The spectrum has been recorded by the synchrotron source, as the infrared beam, using a grazing incidence objective (512 scans, 4 cm^{-1} resolution).**

PERSPECTIVES IN SYNCHROTRON INFRARED FOR BIOINTERFACES

Many synchrotrons worldwide have active programs involving biomedical applications of synchrotron IR microscopy. The motivation for far-IR studies of such biointerfaces on extended surfaces has to be revived, since the number of

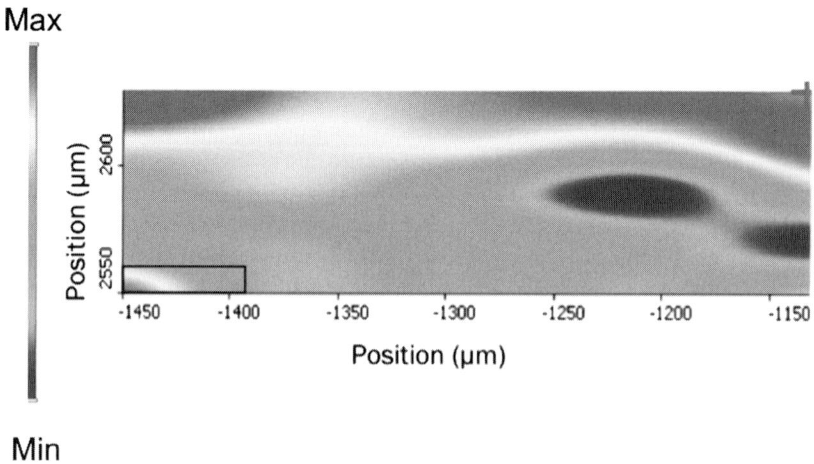

FIGURE 12 **Chemical image of the spatial distribution of adsorbed GSH on Au(111). Due to the grazing incidence geometry, the projected spot size onto the sample is $\approx 20 \times 80\ \mu m^2$. The image has been acquired with a step size of 80 μm in one direction and 20 other in the perpendicular direction. Each spectrum has been recorded with 1024 scans at 8 cm^{-1} resolution. The synchrotron beam has remained stable during all the course of the data recording.**

surface science-dedicated station in the far-IR is reduced at the moment, after the recent closing of the surface science station at NSLS and Daresbury. The only operational surface science dedicated end station is located at Spring-8 [45]. A new surface science equipment is being developed at SOLEIL, for such studies, and will be accessible to the users soon. For scientists that are not familiar with the operation of these centers access to these facilities may seem restrictive. However, almost all synchrotron IR beamlines worldwide offer beam time free of charge based on a peer-reviewed proposal system. In many cases, researchers have active programs using a conventional IR microscope, equipped either with a single detector or an array detector.

In synchrotron IR microspectroscopy, one of the pitfalls remains the diffraction limited spot size and slow data collection rates. Nowadays, focal plane array detectors are being implemented in microscopes that use a global source, and the performance of these detectors has not yet been exploited with a synchrotron source. Clearly, the size of the detector array has to be adapted to match the projected size of the synchrotron beam, in order to keep the brightness advantage of this source. Smaller detector arrays might well be available soon, and will allow faster acquisition of data, and a slightly improved lateral resolution using point-spread function deconvolution. Readers interested in the details of future improvements with such a combination of bright IR sources with up-to-date bidimensional detectors can find them in [22].This is a direction that can be complemented by the collection of many more IR photons from a bending magnet with much large horizontal opening angles [46].

Another upcoming area in biomolecular interface research is the use of very intense THz beams that can be generated with so-called coherent emission. Coherent synchrotron radiation is produced when the relativistic electron bunches have longitudinal density variations on a scale comparable to or smaller than the wavelength. Synchrotron and transition radiation emitted from a bunched electron beam becomes coherent and highly intense at wavelengths about or longer than the electron bunch length. The radiation has a continuous spectrum in the submillimeter to millimeter wavelength range. One of the main factors determining the intensity of the coherent radiation is the electron bunch shape. The intensity of the coherent radiation emitted from an electron bunch is becoming proportional to the square of the electron intensity. But very recently, significant progress has been made toward stable operation of storage rings with a small momentum compaction factor, which results in a short (submillimeter) longitudinal size of the electron beam, and as a consequence, a high flux of coherent synchrotron emission in the far-IR spectral range [47]. It is also possible that ultrashort pulses can also be generated with the assistance of laser slicing techniques [48]. One can expect that if this progress continues, generation of coherent SR may be possible also in the mid-IR range. If this happens, the synchrotron sources will be much more advantageous over black-body sources not only in terms of brightness, but also in terms of flux.

Other exciting new directions in synchrotron IR spectroscopy are emerging. The pulsed nature of the synchrotron source has been exploited in few cases [49] but synchronization with external lasers will certainly see more applications in the future. FELs are also being developed in many countries. Production of intense IR pulses, with very short duration, is currently possible. There are new projects combining FELs and continuous emission (synchrotron) from a storage ring.

Therefore, synchrotron IR spectroscopy and microscopy are very attractive for many scientific disciplines, in a wavelength energy domain extending from the very far- to the near-IR. We hope that readers of this chapter have found motivation to utilize synchrotron infrared beamlines for undertaking new approaches in biomolecular interface studies.

REFERENCES

[1] Mantsch HH, Chapman D. *Infrared Spectroscopy of Biomolecules.* New York: Wiley; 1996.

[2] Duncan W, Williams GP. Infrared synchrotron radiation from electron storage rings. *Appl. Opt.* 1983;22:2914–2922.

[3] Several synchrotron facilities websites offers a practical guide to synchrotron production and exploitation, such as, among others: http://www.esrf.eu/AboutUs/AboutSynchrotron.

[4] Duke P. *Oxford Series on Synchrotron Radiation.* Oxford: Oxford University Press; 2008.

[5] Einfeld, D; Nagel, J. BESSY Report No. TB34, 1981.

[6] Lagarde P. Infrared spectroscopy with synchrotron radiation *Infrared Phys* 1979;18:395–460.

[7] Stevenson JR, Ellis H, Bartlett R. Synchrotron radiation as an infrared source. *Appl. Opt.* 1973;12:2884–2889.

[8] Stevenson JR, Lathcart JJ. *Nucl. Instr. Math. Phys. Res. B* 1980;172:367.

[9] Schweizer E, Nagel J, Braun W, Lippert E, Bradshaw AM. The electron storage ring as a source of infrared radiation. *Nucl. Instr. Math. Phys. Res. A* 1985;239:630–634.

[10] Nanba T. *Int. J. Infrared Millimeter Waves* 1986;7:759–770 .

[11] Williams GP. The initial scientific program at the NSLS infrared beamline. *Nucl. Instrum. Methods Phys. Res. A* 1990;291:8–12.

[12] Williams GP, et al. Infrared synchrotron radiation measurements at Brookhaven. *Rev. Sci. Instrum.* 1989;60:2176.

[13] Williams GP, Takacs PZ, Klaffky RW, Schleifer M. The infra-red beam line IR4 at the NSLS. *Nucl. Instr. Math. Phys. Res. A* 1986;246:165.

[14] Roy P, Mathis YL, Lupi S, Nucara A, Tremblay B, Gerschel A. SIRLOIN, the infrared beamline at LURE. *Synchrotron Radiat. News* 1995;8:10.

[15] Forro L, Carr GL, Williams GP, Mandrus D, Mihaly L. Far infrared transmission study of single crystal $Bi_2Sr_2Ca1Cu_2O_x$ superconductors. *Phys. Rev. Lett.* 1990;65:1941.

[16] Hirschmugl CJ, Williams GP, Hoffmann FM, Chabal YJ. Adsorbate-substrate resonant interactions observed for CO on Cu(100) in the Far-IR. *Phys. Rev. Lett.* 1990;65:408.

[17] References for gas phase with synchrotron.

[18] Carr. GL, Hanfland M, Williams GP. Mid-infrared beamline at the national synchrotron light source port U2B. *Rev. Sci. Instr.* 1995;66:1643–1645.

[19] Carr GL, Reffner J, Williams GP. Performance of an infrared microspectrometer at the NSLS. *Rev. Sci. Instrum.* 1995;66:1490.

[20] Bosch RA. *Nucl. Instrum. Math. Phys. Res. A* 1997;386:525.
[21] Matthis YL, et al. Magnetic field discontinuity as a new brighter source of infrared synchrotron radiation. *Phys. Rev. Lett.* 1998;80:1220.
[22] Carr, G. L.; Chubar, O.; Dumas, P. Multichannel *Detection with a Synchrotron Light Source: Design and Potential, in Spectrochemical Analysis Using Infrared Detectors*, (I.W.L.R., Bahrgava, Ed.)., Blackwell Publishing, Oxford, 2006.
[23] Chubar O, Elleaume P. Accurate And Efficient Computation Of Synchrotron Radiation In The Near Field Region. 1998; EPAC98 Conference p. 1177–9.
[24] Williams GP. Synchrotron and free electron laser sources of infrared radiation. J. M. Chalmers., P. Griffiths (Eds.), *Handbook of Vibrational Spectroscopy*. 2001;Vol.1:New York: Wiley; 2001 pp. 341 .
[25] Scarvie T, Anderson N, Baptiste K, Byrd JM, Chin MJ, Martin MC, McKinney WR, Steier C. *Infrared Phys. Technol.* 2004;45:403–408.
[26] Hirschmugl C, Williams GP, Hoffmann FM, Chabal YJ. Adsorbate-substrate resonant interactions observed for CO on Cu(100) in the far infrared. *Phys. Rev. Lett.* 1990;65:480–483.
[27] Humblot V, Bingham C. JA, Le Roux D, Mateo ME, McNutt A, Nunney TS, Ortega LM, Roberts AJ, Williams J, Surman M, Raval R. Synchrotron far-infrared RAIRS studies of complex molecules on Cu(110). *Surf. Sci.* 2003;537:253–264.
[28] Dumas P, Suhren M, Chabal YJ, Hirschmugl CJ, Williams GP. Adsorption and reactivity of NO on Cu(111) a synchrotron infrared reflection absorption spectroscopy study. *Surf. Sci.* 1997;371:200–212.
[29] Williams GP. Infrared spectroscopy at surfaces with synchrotron radiation. *Surf. Sci.* 1996;368:1–8.
[30] Surman M, Hagans PL, Wilson NE, Baily CJ, Russell AE. Adsorption of CO on Pt(111): a synchrotron far-infrared RAIRS study. *Surf. Sci.* 2002;511:L303–L306.
[31] Engström E, Ryberg. *J. Chem. Phys* 2001;115:519.
[32] Baily CJ, Surman M. Russell Investigation of the CO induced lifting of the (1x2) reconstruction on Pt(110) using synchrotron far-infrared RAIRS. *Surf. Sci.* 2003;523;111–117.
[33] Volokitin AI, Personn B. NJ. Quantum theory of IR spectroscopy for dipole-forbidden vibrational modes of adsorbed molecules on the surface of a metal in the frequency range of the anomalous skin effect. *J. Exp. Theor. Phys.* 1995;81:545–552.
[34] Personn B. NJ, Volokitin AI. Infrared reflection-absorption spectroscopy of dipole-forbidden adsorbate vibrations. *Surf. Sci.* 1994;301:314–336.
[35] Persson B. NJ, Schumacher D, Otto A. Surface resistivity and vibrational damping in adsorbed layers. *Chem. Phys. Lett.* 1991;178:204–212.
[36] Hirschmugl CJ, Dumas P, Chabal YJ, Hoffmann FM, Suhren M, Williams GP. Dipole forbidden vibrational modes for NO and CO on Cu observed in the far IR. *J. Electron. Spectrosc. Relat. Phenom.* 1993;64-65:67–73.
[37] Hein M, Dumas P, Otto A, Williams GP. Friction of conduction electrons with adsorbates: simultaneous changes of DC resistance and broadband IR reflectance of thin Cu(111) films exposed to CO. *Surf. Science.* 1999;419:308–320.
[38] Rudolf P, Raval R, Dumas P, Williams GP. *Appl. Phys. A, Mater. Sci. Process.* 2000;75:147–153.
[39] Miller LM, Dumas P. *Biochim. Biophys. Acta* 2006;1758:846–857.
[40] Dumas P, Sockalingum GD, Sulé-Suso J. Adding synchrotron radiation to infrared microspectroscopy: what's new in biomedical applications? *Trends Biotechnol.* 2007;25:40–44.

[41] Dumas P, Miller LM, Tobin MJ. *Acta Phys. Pol.* 2009;115:446–454.

[42] Chio-Srichan S, Refregiers M, Jamme F, Kascakova S, Rouam V, Dumas P. Photosensitizer effects on cancerous cells: a combined study using synchrotron infrared and fluorescence microscopies. *Biochim. Biophys. Acta, –Gen. Subj.* 2008;1780:854–860.

[43] Bassan P, Byrne HL, Bonnier F, Lee J, Dumas P, Gardner P. Resonant Mie scattering in infrared spectroscopy of biological molecules: understanding the dispersion artifact *Analyst* 2009;134:1586–2159.

[44] Vallée A, Humblot V, Méthivier C, Dumas P, Pradier CM. Modifying protein adsorption by tripeptide preadsorbed layers on Au(111). Submitted to J. Physics: Condensed Matter.

[45] Sakurai M, Moriwaki T, Kimura H, Nishida S, Nanba *T. Nuclear instruments and methods, section A*: accelerators, spectrometers, detectors and associated equipment. 2001, *467-468*, 1481–1484.

[46] Nasse MJ, Reininger R, Kubala T, Janowski S, Hirschmugl C. *Nucl. Instrum. Methods A* 2007;582:107.

[47] Williams GP. *Rep. Prog. Phys.* 2001;69:301.

[48] Schoenlein RW, Chattopadhyay S, Chong H. HW, Glover TE, Heimann PA, Shank CV, Zholents AA, Zolotorev MS. *Science* 2000;287:2237.

[49] Carr GL, Lobo R, LaVeigne J, Reitze D. *Phys. Rev. Lett.* 2000;85:3001–3004.

Chapter 7

IR spectroscopy for biorecognition and molecular sensing

C.M. Pradier[1], M. Salmain[2] and S. Boujday[1]
[1]*UPMC, Université Paris VI, Laboratoire de Réactivité de Surface, F-75005 Paris, France*
[2]*ENSCP, Chimie ParisTech, Laboratoire Charles Friedel, F-75005 Paris, France*

Biointerface Characterization by Advanced IR Spectroscopy. DOI: 10.1016/B978-0-444-53558-0.00007-2

INTRODUCTION

Many if not all biological processes essential to life are mediated by a sequence of individual events relying on the molecular recognition between two partners. In a most simplistic way, these molecular recognition processes stem from a complementarity of shape and function between the host and the guest that associate with high affinity and specificity. For example, this kind of process is encountered between enzymes and substrates that are then converted into products in a chemo/stereospecific fashion. The same kind of association is found between receptors and ligands and between antibodies and antigens.

Physical methods to detect biorecognition events have been mostly implemented for bioanalytical purposes, that is, to detect and quantify one of the partners of the association (the host or the guest) but also for more fundamental purposes such as the study of molecular mechanisms underlying the interaction between therapeutic agents and their biological targets.

Surface IR spectroscopy appears nowadays as a powerful technique to characterize the adsorption and reactivity of molecules on various material surfaces. This chapter, aiming at demonstrating the interest of using IR spectroscopy to monitor the functionalization of surfaces, as well as their use for biorecognition monitoring, will be divided into two parts.

In the first part, we will describe how the immobilization of bioreceptors on materials can be studied by various IR surface techniques for the elaboration of biosensors and recognition of target molecules. Emphasis will be put on three of these techniques, which benefit from a surface-enhanced sensitivity, namely, attenuated total reflection infrared (ATR-IR) spectroscopy, surface enhanced infrared spectroscopy (SEIRAS), and eventually infrared reflection absorption spectroscopy (IRRAS). The last technique allows us to reach low detection limits, thanks to the use of modulation of the polarization. Several examples of application of these IR techniques will be presented. Recent examples of very sensitive and selective mid-IR biosensors operating in direct or competitive format will be given, using IR as a transduction technique to detect several targets going from small molecules to pathogenic bacteria.

The second part will be devoted to the use of transition metal carbonyl and some other chemical probes in conjunction with IR spectroscopy for molecular sensing and biorecognition. Having in mind that, for biosensing applications, analytical sensitivity is of critical importance, the use of metal carbonyl probes, as well as some other good IR reporters, will be explained and illustrated by some examples of biorecognition events.

PART I: SURFACE IR SPECTROSCOPY FOR THE LABEL-FREE DETECTION OF BIORECOGNITION EVENTS

The ATR technique

FTIR spectroscopy, in the attenuated total reflection configuration, relies on the total reflection of the IR beam within a high refraction index crystalline material, occurring at a certain angle of incidence; the sample which is in intimate contact with the ATR crystal is then probed by the evanescent wave, with a sensitivity decaying with the distance (Z) from the crystal interface [1].

$$E = E_0 e \left[\frac{-2}{\lambda_1} \pi (\sin^2\theta - n_{21}^2)^{1/2} z \right] = E_0 e^{-\gamma Z}$$

where $\lambda_1 = {}^{\lambda}/_{n_1}$ is the wavelength in the sample, n_{21} is the refractive index ratio, n_2/n_1, θ, is the angle of incidence, and Z is the distance from the surface (all notations are those written in Fig. 1).

The penetration depth, and thus the length of the probed sample, depend on the sample density and on the wavenumber, which makes the sensitivity varying with the spectral region; as an example, the probed distance is of the order of a few micrometers for a layer of oxide nanoparticles, and equal to 1.39 μm in water (at 1100 cm^{-1}). The principle of the technique, as well as experimental description of the setups and practical considerations, are extensively described in several reviews or books [3, 4]. We only recall that, in an ATR setup, the sample to be analyzed must be in close contact with the so-called ATR element, and IR in the ATR mode is particularly adapted to probe liquid media.

Flow-through cells have been built to monitor changes in adsorption or adhesion phenomena in real time. One of the first examples of such a setup was presented by Fleming *et al.* in 1995 [5]. In this very pioneer study, the authors monitored the growth and physiological changes within a *Pseudomonas putida* biofilm on a germanium ATR crystal; the four spectral regions correlate to various functional groups and molecules; as an example the appearance of an IR signal at 1085 cm^{-1} indicates the formation of exopolysaccharides (EPS) upon toluene addition in the medium. Moreover, monitoring the adhesion and growth of the *P. putida* film makes clear the progressive increase of the amide bands and the irregular evolution of the EPS characteristic bands, showing their complex role in biofilm growth (Fig. 2).

Importantly, since crystals with high refraction indexes are a prerequisite for internal reflection, ATR is often used to investigate surface reactions on Si, Ge,

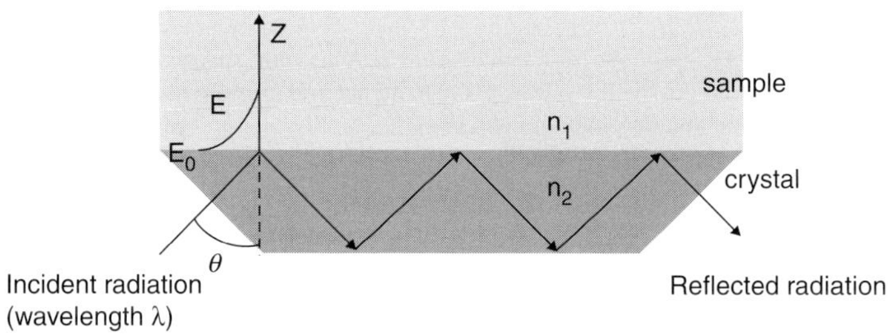

FIGURE 1 Schematic representation of the optical characteristics for attenuated total reflection IR (taken with permission from Ref. [2]).

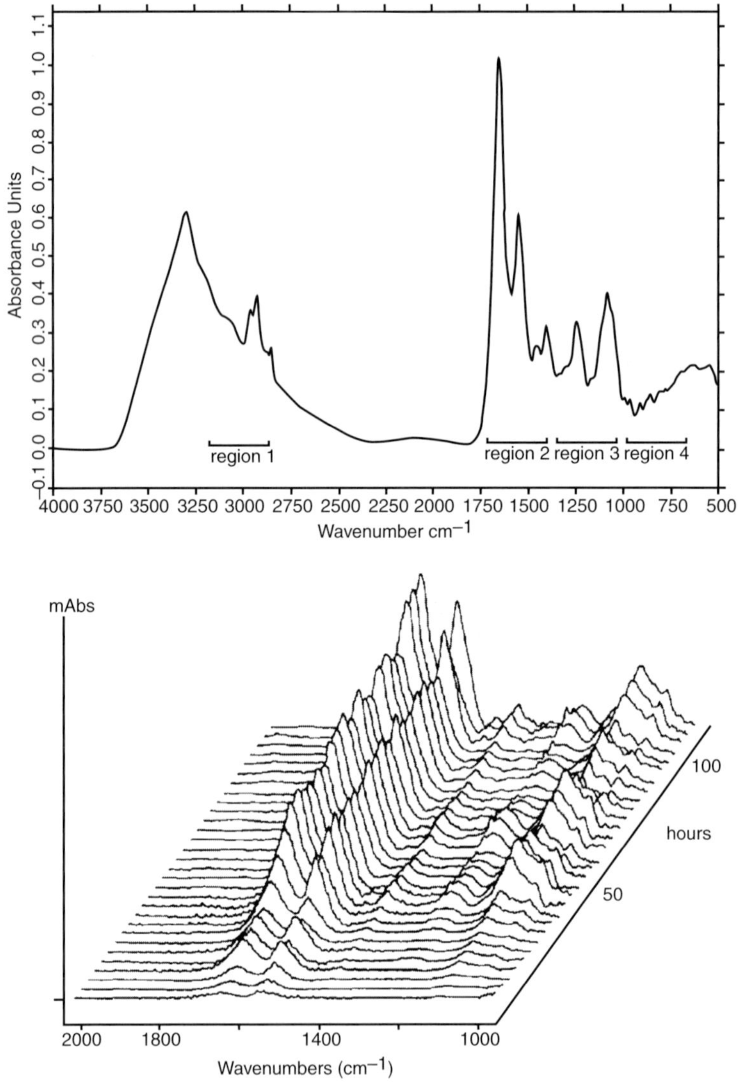

FIGURE 2 Up: FTIR spectrum of *Pseudomonas putida*; four spectral regions have been identified corresponding to the fatty acid, protein, mixed and polysaccharide regions, respectively. Down: adhesion and development of a *P. putida* biofilm monitored in an ATR flow cell. Taken with permission from Ref. [5].

ZnS, or ZnSe surfaces. Of course, the use of FTIR in the ATR mode for monitoring biorecognition requires the ATR optical element to be functionalized with the adequate receptors. This can be easily done on Si surfaces, the drawback being its IR opacity below 1550 cm^{-1} [6]. The use of Ge instead of Si not only enables a broader IR analytical range, but also changes the functionalization procedures; this was exemplified by Marchand-Brynaert *et al.* who setup an original functionalization method of Ge, building a thin organic layer made of PEG-silanes and photografting of an NHS-terminated azide, ready for the covalent anchoring of any biomolecule; they monitored this chemical activation of the Ge crystal surface, and then the binding and stability of a protein by in situ FTIR-ATR analyses. The stability of the so-functionalized Ge sensor surface was checked by performing immersion in PBS and monitoring the amide band intensity as a function of time; no intensity changes were observed after 5 h under a PBS flow, that is, under conditions mimicking physiological ones commonly used for monitoring biorecognition phenomena [7].

Detection of ligand–receptor interactions was tested and compared on two Ge ATR crystals, having different geometries, one being trapezoidal and yielding 25 internal reflections and the other using triangular-shaped crystal and relying on one single reflection (see Fig. 3) [8].

After activation and silanization of the Ge ATR elements, a thin phospholipids membrane was deposited and submitted to a specific or nonspecific binding of coagulation factor VIII in the presence of other proteins. The authors demonstrated that building a homogeneous lipidic layer on the second type of the ATR element (much smaller surface corresponding to one single reflection) is much easier than on the classical large area Ge crystals. This enables a better control of the nonspecific interactions; in this example, the decrease in the number of reflections being compensated by the better quality of the surface functionalization.

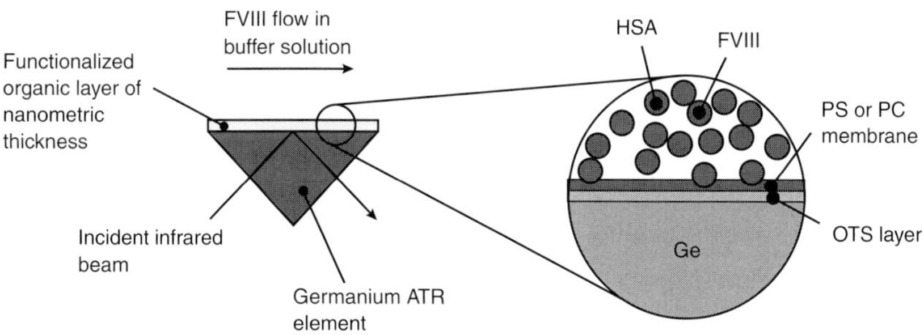

FIGURE 3 Schematic view of the Ge triangular bar used for ATR–FTIR. The principle of FTIR sensors for factor VIII (FVIII) detection and details of the functionalized interface (taken with permission from Ref. [8]).

Moreover, by combining FTIR spectroscopy and second-derivative analysis of the IR spectra, the specific character of the protein–protein interaction can be checked [9]. As an example, an immunosensor was built up on a GaAs ATR element schematized on Figure 4.

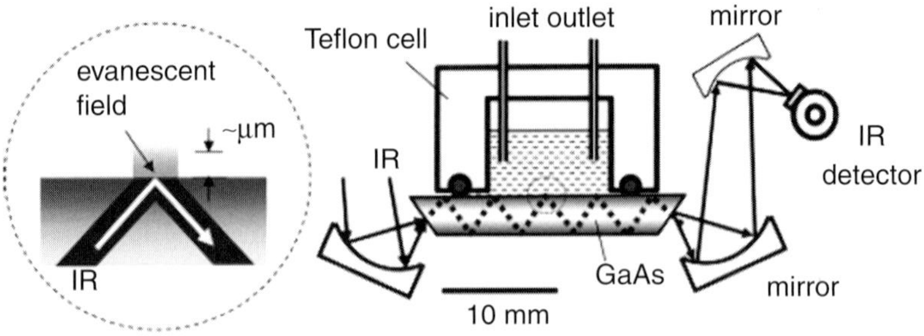

FIGURE 4 Optical configuration and flow cell used for the mid-IR ATR analysis (taken with permission from Ref. [9]).

To do so, a synthetic peptide antigen was immobilized on a SiO_2 film deposited on the GaAs prism; the latter configuration was preferred, compared to the use of a pure Si prism due to the better transparency of GaAs compared to Si [10]; the quantitative analysis of antibody recognition was then achieved by integrating the amide I and amide II bands while specific and nonspecific interactions were discriminated, thanks to the second-derivative analysis of the amide I band [11]. As a matter of fact, the major fraction of β-sheets in antibodies [12] makes them relevant targets to probe the method. Figure 5 shows typical ATR-IR spectra recorded after interaction of the antigen-coated (a) or uncoated (b) ATR crystal with the antiserum (containing the specific antibody) diluted in a buffer. Bands at 1548 and 1639 cm^{-1} indicate the binding of antibody, in a much higher amount on the antigen-modified surface.

Moreover, the second derivative spectrum, corresponding to the upper spectrum, in the amide I region, exhibited one main peak at 1638 cm^{-1}, showing the

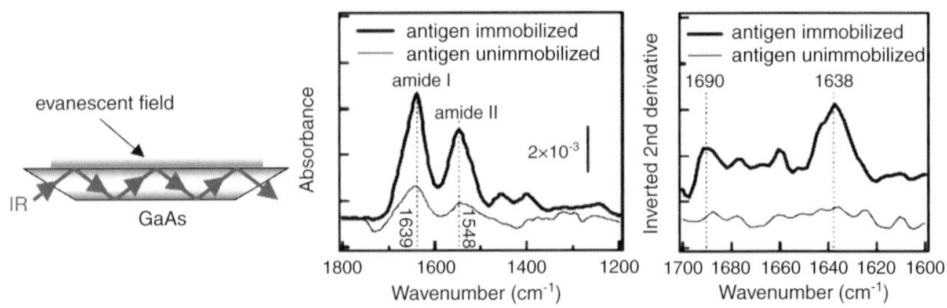

FIGURE 5 Left: schematic layout of the optical configuration used in this experiment. Middle: typical IR absorption spectra of antigen-immobilized (thick line) and uncoated ATR crystals (thin line), measured after 1 h contact with antiserum diluted 1:1000 in the reaction buffer. Right: inverted second derivatives of the previous figure spectra (taken with permission from Ref. [9]).

predominance of β-sheet-folded proteins, that is, antibodies. Conversely, the second derivative of the lower spectrum (blue line) indicates no, or a very weak amount of specific antibodies, thus demonstrating that specific interactions can be discriminated via this spectral analysis. Eventually, a quantitative analysis of the protein concentration was achieved, based on the amide II intensity, because the amide I region is perturbed by the strong absorption of water. A detection limit (1:36,000 dilution of antiserum), lower than that reached by classical ELISA tests, was obtained showing the potential of this label-free IR-based immunosensor.

A similar FTIR ATR-based sensor was setup and utilized by Reiter *et al.* to monitor the specific recognition of a monoclonal antibody by human tumor necrosis factor-α (TNFα) [13]. The latter receptor was simply adsorbed on a Ge prism; after washing with BSA to saturate the remaining binding sites of the Ge surface and thus prevent nonspecific interactions, a flow of anti-TNFα antibody was pumped over the Ge element. The resulting protein surface concentration was determined, taking into account the refractive indices, the molar absorption coefficient of the amide II band and the number of internal reflections, following the procedure described in Ref. [14]. This analysis gave evidence that the TNFα receptors were adsorbed as a full monolayer and one antibody molecule was adsorbed per ca. two TNFα receptors, corresponding to a binding efficiency of 38%. In this work, parallel and perpendicular polarized IR lights were used; the calculation of the dichroic ratio, from the integrated absorbance of the amide II bands corresponding respectively to parallel and perpendicular polarized incident lights, and comparison to that of an isotropic arrangement of molecules, gave information about a preferential orientation of some groups; this was the case for TNFα molecules which tend to align their amide groups normal to the Ge surface, likely favored by the interaction of the other side of the molecule with the surface. This phenomenon was not observed with larger proteins such as antibodies.

More recently, Gosselin *et al.* reported results concerning the detection of a small molecule, namely 2,4-dinitrophenol (DNP), using ATR FTIR spectroscopy [15]. They attached a DNP-HSA (HSA: carrier protein) conjugate covalently to the silanized surface of a Ge ATR element and carried out competitive immunoreactions using several anti-DNP antibodies, in the presence of either free or coupled DNP. To do so, solutions containing monoclonal antibodies and DNP or DNP-HSA inhibitors were injected over the DNP-HSA-functionalized ATR surface. The measured surface absorbance, in the amide II region, was then converted into percentage of inhibition (see following equation):

$$\mathrm{I} = 100(1 - (A_i - A_0/A_{\max} - A_0)),$$

where A_i is the measured absorbance, A_0 is the absorbance before contact with the antibody solution, and $A_{\max}$ is the absorbance measured in the absence of inhibitor in solution.

Figure 6 displays the schematic representation of the sensors and plots of the inhibition curves for DNP-HSA (open symbols) and DNP (filled symbols). The

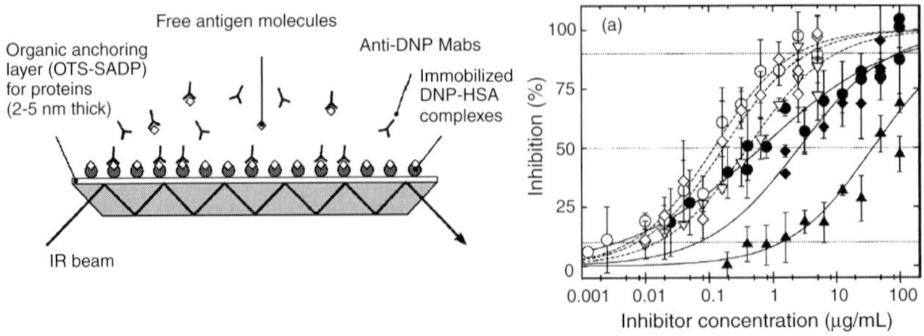

FIGURE 6 Left: schematic representation of the ATR sensor; right: percentage of measured inhibition using DNP-HSA (open symbols), or free DNP (filled symbols) (taken with permission from Ref. [15]).

same procedure was applied to perform antibody detection with ELISA technique, in order to compare the sensitivity of the two methods. The limits of detection were equivalent for both techniques, ca. 10–40 μg/mL. Interestingly, the sensitivity was significantly increased when using HSA-DNP as inhibitor.

Another type of biosensor, based on DNA recognition, may also take full advantages of IR spectroscopy. The process of DNA immobilization and subsequent characterization on Si surfaces was investigated *in situ* by using IR absorption spectroscopy with a multiple internal reflection setup [16]. The technique enabled to assess the covalent binding of DNA to the terminal amine functions of a silane, and of course the hybridization of a complementary oligonucleotide yielding a specific signal at ca. 1685 cm^{-1}. This signal has been attributed to the C=O stretching vibration mode which appears due to conformational changes of the immobilized DNA molecule. As a matter of fact, interaction of a noncomplementary oligonucleotide did not give rise to this absorption signal. Note that, for the sake of spectral clarity in the 1600–1700 cm^{-1} region, all experiments were conducted *in situ* in a D_2O solution.

The authors presented a careful analysis of the conformation of DNA and its relation with the IR spectra. Figure 7 shows the ATR-IR spectra of two complementary DNA strands, comprising 30 bases, and yielding expected absorption bands at 1580, 1620, and 1680 cm^{-1} (C=O and C=N stretching vibrations). These experiments clearly showed that DNA hybridization induces IR spectral changes, in particular in the C=O stretch region; this has been attributed to the hydrogen bonding between A and T or G and C bases as represented in Figure 8.

Label-free detection of proteins and other small molecules could also be achieved by FTIR-ATR through DNA conjugation [17]. In the experiment reported by Liao *et al.*, the authors used 20 nucleotide-long aptamers, specific or not to thrombin; those were covalently immobilized on hydrogen-terminated Si surfaces; incubation with solutions of thrombin (10^{-8} to 10^{-6} M) was then

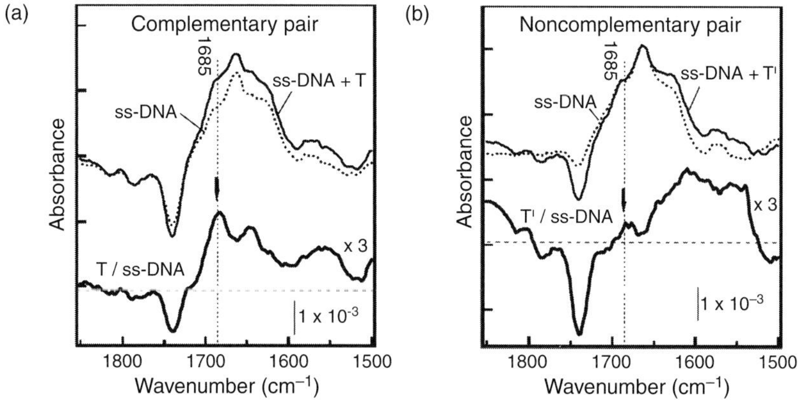

FIGURE 7 (a) ATR-IR spectra of DNA-modified Si surfaces before (dotted lines), and after interaction with complementary ss-DNA (a), or noncomplementary ssDNA (b). The thicker lines are the difference spectra (solid thin line minus dotted line) (taken with permission from Ref. [16]).

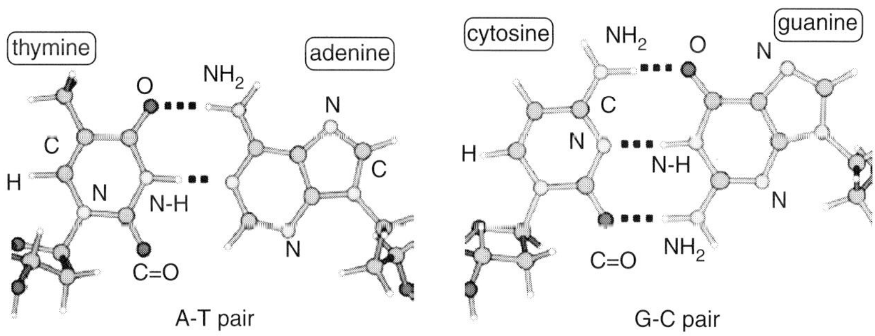

FIGURE 8 Base paring in DNA, thymine-adenine and cytosine-guanine, showing the H-bondings (taken with permission from Ref. [16]).

achieved; the binding of protein was monitored by the increase of the IR bands in the 3550–2800 cm^{-1} region (N–H + CH_2 bands); the affinity constant between the aptamer and the protein could be determined by fitting the concentration-dependant IR intensity curves using a Langmuir isotherm equation.

Note eventually that measurements on gold or other metal surfaces may also be achieved using the ATR configuration; the metal film must be thin enough to let a high intensity of the evanescent wave, while the reflectivity of the metal/ATR crystal is high enough to let a significant IR intensity reach the detector. The optimal film thickness depends on the detector and number of internal reflections; it is usually of the order of a few nm. Such an ATR setup was applied to studying the formation of self-assembled monolayers of chelator alkanethiols, and subsequent binding of proteins, at the gold/water interface [18].

This configuration has also been widely exploited to combine optical and electrochemical measurements at an electrode surface. Practical, as well as theoretical considerations about the use of IR-ATR technique for electrochemical investigations are given in Ref. [19].

The SEIRAS technique

After the discovery of surface enhanced Raman scattering (SERS) in the early 1970s, a similar enhancement effect has been observed on nanostructured surfaces by IR spectroscopy. The enhancement factor of SEIRAS (surface enhanced infrared absorption spectroscopy) is weaker than the one of SERS, $10–10^3$ compared to $10^6–10^{12}$; however, thanks to the high IR cross-section of IR, SEIRA, such as SERS, provides significant and well detectable molecular information in many cases, in particular when biomolecules are targeted. Let us recall that, like IRRAS, surface selection rules imply that only vibrational modes involving a change in the dipole moment normal to the surface are active [20].

SEIRA signal intensity depends on the surface roughness, on the size and shape of nanoparticles in the case of colloids; and the SEIRAS technique may be applied with various geometries of the optical setup.

The preparation of the surface is critical for the success of SEIRAS analyses: SEIRA-active metal surfaces may be fabricated by vacuum evaporation at a controlled rate. The best SEIRA effects have been reported for Au or Ag rough films [21]. SEIRA signal enhancement has been observed on Cu or Pt. A reproducible and thus quantifiable SEIRA effect may be obtained by using nanolithography technique [22] or electroless chemical deposition [23] to prepare the adequate rough surfaces. Colloidal gold nanoparticles may also serve as SEIRA templates [24]. The most commonly adopted geometries for SEIRA measurements are the internal reflection (Kretschmann attenuated total-reflection geometry) or total reflection on a metal substrate.

The sensitivity of SEIRA spectroscopy is comparable to that of Surface Plasmon Resonance (SPR), but it has the great advantage of providing a chemical identification of the detected molecules.

The signal enhancement by SEIRA effect can be seen on Fig. 9 (upper one) which shows the IR spectrum of a layer of thymine, analyzed in two different modes, ATR-IR or SEIRA [25].

For the former, a 50 nm thymine film was deposited on a KBr plate while, for SEIRA experiments, a 7 nm film of thymine was evaporated onto rough silver. Peak intensities are on an average slightly more intense on the SEIRA spectrum, despite a much lower amount of thymine; the difference in intensities, for respectively the 7 nm (SEIRA) and 50 nm (RAIRS) films indicates an enhancement factor of at least 10 with the former technique. Moreover, some frequency shifts may be observed; for instance the C=O stretch mode, at 1750 cm^{-1} on the SEIRA spectrum is obviously shifted to lower wave-number on the ATR spectrum; this witnesses the loss of some intermolecular interactions (hydrogen-bonding network) which exist in polycrystalline thymine deposited on the ATR crystal.

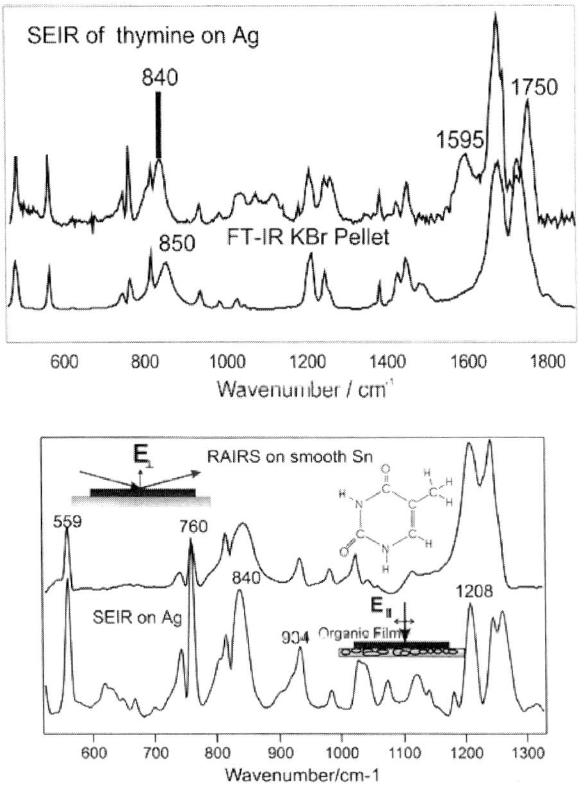

FIGURE 9 Up SEIRA, spectrum of thymine on rough silver, compared to the ATR spectrum of thymine powder deposited on a KBr plate; Down, transmission SEIR spectrum of thymine on rough surface, and reflection absorption spectrum of a 50 nm thymine film on smooth tin (taken with permission from Ref. [25]).

In Fig. 9 (lower one), the SEIRA spectrum of a thymine film on rough Ag is compared to its IRRAS one on a smooth tin surface. Having in mind the orientation-dependency of the IRRAS peak intensities, one observes a very weak out-of-plane mode of the thymine cycle, at 760 cm^{-1}, suggesting that the thymine molecules are, on an average, preferentially oriented in an edge-on way, as schematized on the figure.

In a second example, SEIRA spectroscopy was combined with visible microscopy to localize bacteria on geologic material surfaces and probe biomolecule interactions [26]. A SEIRA-active surface was obtained by evaporating a thin gold film on a rough rock sample, so as to acquire SEIRA spectra in the metal-overlayer configuration. Localized different chemical compositions were observed from the SEIRA spectra, thanks to different spectral features in the 1300–1000 cm^{-1} region. These may be used as biomarkers to define the spatial distribution of bacteria microcolonies on a basalt sample. The authors noticed a special sensitivity enhancement due to SEIRA effect around 1500 cm^{-1}, but not below 1200 cm^{-1}, where specifically differentiating bands were expected.

The SEIRA effect may easily be applied in the ATR configuration, by depositing a metal film on an ATR element; the presence of the metal film induces a SEIRA effect. Examples of characterization of redox processes in electroactive bacteria such as *Geobacter sulfurreducens* or *Pseudomonas fluorescens* by combining cyclic voltammetry and ATR-SEIRA have been recently described [27,28]. For such a device, the working electrode is made of a thin (ca. 35 nm) film of gold deposited by argon sputtering on a silicon prism, inserted in a glass electrochemical cell; the IR spectra were recorded with *p*-polarized light.

As an example, Figure 10 shows the ATR-SEIRA spectra acquired during cyclic voltammetry. Observing these two series of spectra upon potential cycling, witnesses the two redox pairs (low and high potential redox) on the *Geobacter sulfurreducens* bacteria adsorbed on gold. Observed spectral changes resemble the electrochemical behavior of c-type cytochromes, located in the outer sphere of the cell, suggesting their role in the electron transfer processes. Interestingly, shifts in the amide I band reveal some conformation transition of the cytochromes during the electron transfer.

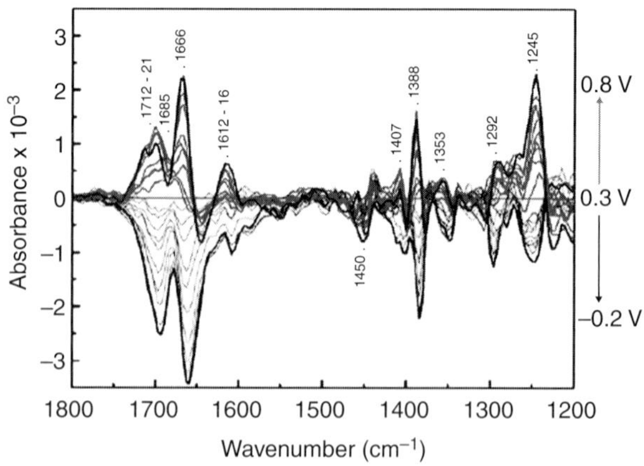

FIGURE 10 ATR-SEIRAS spectra of *G. sulfurreducens* acquired during repeated cyclic voltammetric analysis. Taken with permission from Ref. [28] (see details in the same reference).

The IRRAS and PM-IRRAS techniques

The use of Fourier-transform infrared reflection-absorption spectroscopy (FT-IRRAS or IRRAS), at grazing angle, provides a precise chemical and semiquantitative characterization of layers adsorbed on reflective surfaces; moreover, they can be used in the air, sometimes in liquid environment which make them of great interest for biointerface characterizations. The principle of IRRAS and PM-IRRAS techniques, dedicated to adsorption studies on metal surfaces in UHV or under low-pressure conditions, has been described in detail in Chapter 1. The PM-IRRAS technique, applied to the characterization of molecular phenomena at dielectric surfaces was the subject of Chapter 3.

In this chapter, we will first describe the experimental setups that can be used to characterize molecular films by IRRAS or PM-IRRAS at reflective (metallic) surfaces, either in the air or in a liquid environment. The interest of these techniques will then be exemplified in different applications: monitoring the elaboration of biosensors, or its use for the detection of various targets, toxins, DNA proteins, or cells.

IRRAS to monitor surface functionalization and biointerfaces

Gold surfaces and their modification by self-assembled monolayers (SAMs) for the immobilization of biomolecules constitute typical systems for which IRRAS is an attractive characterization technique. IRRAS has been widely used to characterize the growth and organization of thiol SAMs on gold [29]. Indeed, gold reacts with alkylthiols that spontaneously form more or less ordered self-assembled monolayers widely studied to controlling the amount and orientation of the biological elements [30–32]. The immobilization of various types of biomolecules can then also be controlled by IRRAS thanks to some very intense signals in the C=O stretch spectral region.

A very classical example is the characterization of SAMs of acid-terminated thiols by FT-IRRAS on Au films. The technique, performed after rinsing and drying the substrates, was also used to monitor activation of terminal acid groups into ester groups and the following binding of proteins [33]. Though measurements could be easily done in the air and exploited, one has to be careful about the background samples and possible variations in the humidity, or CO_2 concentration, in the air that may induce false signals or poor baselines. We will see later that PM-IRRAS brings much more rapid and satisfying results.

FT-IRRAS was also applied to characterize the functionalization of Si surfaces [34]. As an example, Figure 11 shows the IRRA spectra of a Si–H surface after immersion in solutions of 1-decene (lower spectrum) or *N*-hydroxysuccinimidyl undecyl-1-enate (NHS)-ester (upper spectrum); these results indicate that the reaction of hydrosilylation indeed takes place via the C=C double bond since no signal at ca 1700 cm^{-1} is present and, second, the presence of the NHS ester functions is confirmed by the three characteristic C=O stretch vibrations at 1817, 1788, and 1745 cm^{-1}, respectively; the position of the CH stretches of the methylene groups, 2854 and 2926 cm^{-1} suggest a poor crystallinity of both the alkene and NHS ester-terminated monolayers [29,35].

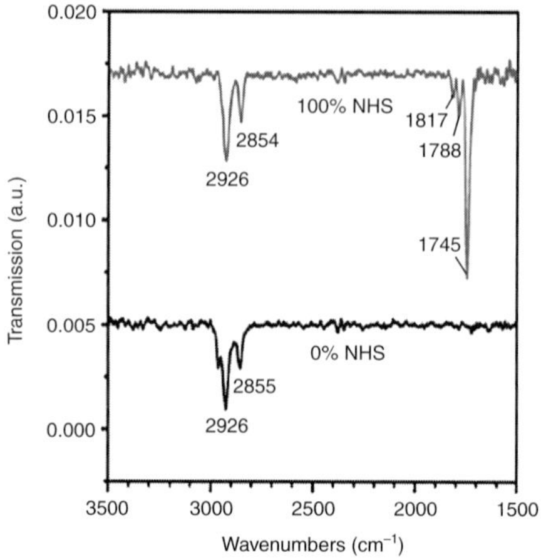

FIGURE 11 FT-IRRAS spectra of a pure 1-decene (lower spectrum), or N-hydroxysuccinimidyl undecyl-1-enate (NHS)-ester (upper spectrum) monolayer on a Si surface (taken with permission from Ref. [34]).

An IRRAS experimental system was even setup to probe solid–liquid interfaces for the observation of solid–solution interfaces on biosensor materials (see Fig. 12). It consisted in a cell containing a CaF_2 prism which ensures a very thin (1–2 μm) layer of solution between the sample and the prism surfaces, by pushing the sample with a spring; solution could be injected with a syringe pump. The prism surface was precovered with PEG (2-methoxy-(polyethylene)

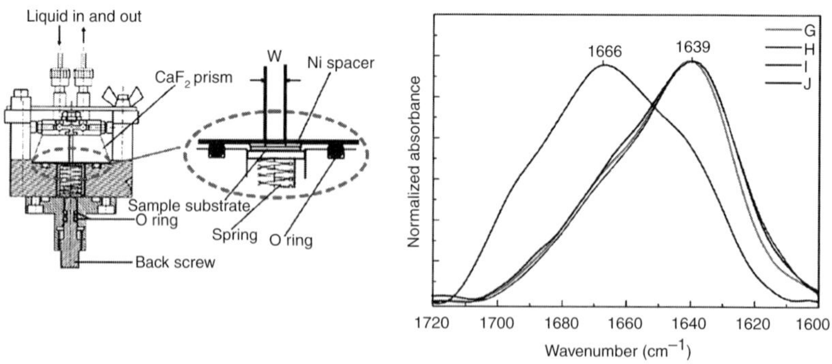

FIGURE 12 (a) Schematic drawing of the cell used for the IRRAS characterization of solid–liquid interfaces; (b) Amide I band of IgG dissolved (G) in a D_2O-PBS solution, (H) adsorbed on the nonactivated SAM layer and measured in the solution, (I) adsorbed on the activated SAM layer, and measured in solution, and (J) adsorbed on the activated SAM layer, and measured under vacuum; taken with permission from Ref. [36].

oxypropyltrimethoxysilane) to avoid adsorption of proteins, which would false the measurements. Gold-coated glass substrates were modified with a SAM of 16-mercaptohexadecanoic acid, which was activated with EDC/NHS mixture prior to covalent attachment of IgG. All measurements were performed in D_2O, in order to avoid perturbation of the spectrum in the region of interest, and background spectra were recorded in the presence of the protein-free solution. An original result, obtained thanks to this *in situ* IRRAS setup was the evidence that no conformational change of the protein occurs when covalently bound to an activated thiol layer; conversely, protein denaturation clearly took place when the cell was evacuated, that is, when the proteins were dried on the surface (spectrum J, blue shift of ca. 27 cm^{-1}) [36].

The performances of SEIRAS (transmission) and IRRAS were compared for the detection of specific antibody–antigen reaction [37]. SEIRA substrates were prepared by depositing Au nanoparticles ($d = 13$ nm) on a SiO_2/Si wafer surface. The Au Nps were then covered with one monolayer of antibody and submitted to an antigen solution. The procedure was repeated on a smooth Au film for IRRAS spectral analysis. Considering the accessible Au area on the nanoparticles, compared to that of the smooth film, the SEIRA enhancement factor was estimated to be close to 25; importantly, chemically prepared and deposited gold nanoparticle films were shown to be efficient SEIRA substrates.

PM-IRRAS to monitor the elaboration of biosensors and biorecognition events

For the molecular characterization of surfaces of silicon, gold, or any reflective material, using PM-IRRAS brings the great advantage of easy and rapid analyses in the air, without requiring a perfect repositioning of the sample or control of the surrounding atmosphere. As an example, Figure 13, left, shows the

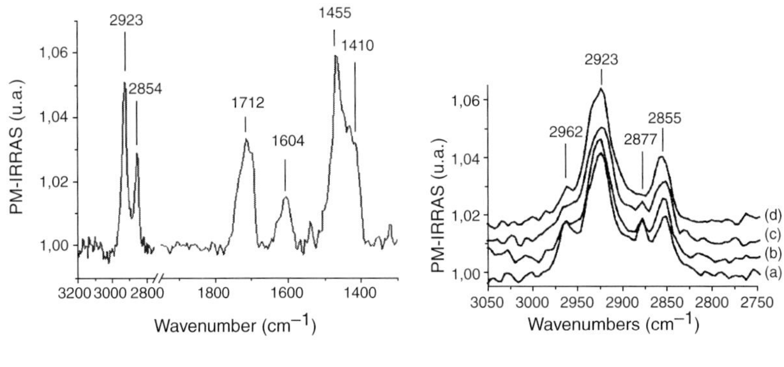

FIGURE 13 PM-IRRAS spectra of a gold surface. Left: after MUA chemisorption; right: after chemisorption of mixtures of MUA and mercaptohexanol, molar fraction of MUA equal to 0.25 (a), 0.5 (b), 0.75 (c), and 1 (d). Figure taken from Ref. [38] with permission.

PM-IRRAS spectrum recorded upon chemisorption of a carboxylic acid terminated thiol, mercaptoundecanoic acid (MUA), on a gold thin film, providing, in a few minutes, the following information

- The ν_{CH} vibrations at 2854 and 2923 cm^{-1}, proving the presence of the alkyl chains in a rather ordered state.
- The COOH signature at 1712 cm^{-1} and those of COO at 1604 and 1410 cm^{-1} indicating that protonated and nonprotonated forms of the carboxylic groups coexist on the surface.

Moreover, if dilution of the active terminal functions is desired, mixed SAMs can be built and PM-IRRAS used for a semiquantitative characterization of the SAM composition, as well as of the layer crystallinity. Figure 13, right, displays the ν_{CH} region of mixed layers of MUA and mercaptohexanol. One observes an increase of the ν_{CH} bands intensity when the MUA fraction increases, as expected from the number of methylene groups, and a progressive shift of the amide I band toward lower wavenumbers, suggesting that the layer crystallinity gets better and better when the fraction of long chain thiols increases. The progressive ordering of the layer was correlated to its further reactivity when antibodies were grafted; antibody interactions were then probed on these layers [39].

By combining PM-IRRAS and DFT calculation, characterization of mixed thiol SAMs on Au(1 1 1), constituted of mercaptohexanol and MUA, of varying concentrations, segregation was made clear; namely, the most stable surface configuration results in the coexistence of separate domains, some rich in alcohol functions and some rich in acids; the main information came from the IR feature in the C=O stretch region, the $\nu_{C=O}$ frequency being very sensitive to the existence of H-bonding with neighboring groups [40].

Activation of the terminal groups of a thiol, for example, activation of acid into ester groups, is a classical way to covalently attach proteins or DNA probes. As an example, Figure 14 shows the evolution of the IR spectra upon chemical activation followed by protein A immobilization. Spectrum a shows bands at ca. 1410 and 1720 cm^{-1} confirming the presence of acid-terminated thiols; on spectrum b, one can see the three C=O ester bands at 1745, 1788 and 1817 cm^{-1}, and spectrum c proves the efficient binding of a protein, by reacting with the ester groups.

Measurement of the amide band intensities provides semiquantitative information on the amount of adsorbed proteins, and thus on immunosensor efficiency when, for instance, antibodies and other proteins or cells are used as capture probes and target molecules, respectively. Plotting the IR band intensity has been used to calculate the affinity constant between an antibody and a target (see Fig. 15). Let us take the example of an allergen in milk, namely, β-lactoglobulin (β-LG), which was immobilized on gold transducers in order to build up a biosensor to detect allergen antibodies.

Two different SAMs were explored: an amine-terminated SAM and an acid-terminated one. β-Lactoglobulin was grafted via its acid or amine groups on the amine or acid-terminated layers, respectively. Even though both acid and

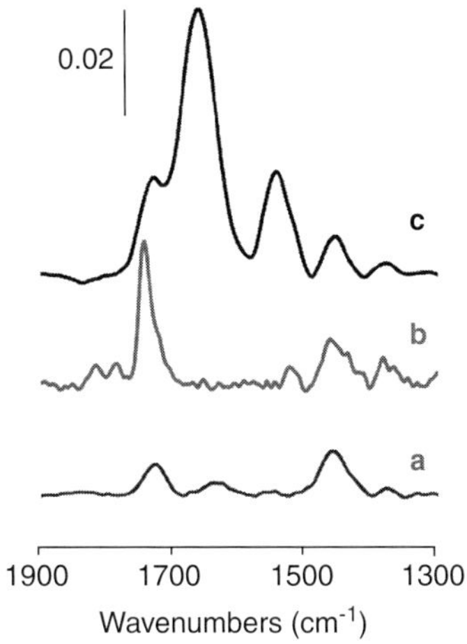

FIGURE 14 PM-IRRAS spectra of a gold surface (a) after mercaptoundecanoic acid grafting, (b) after its activation, and (c) after β-lactoglobulin immobilization.

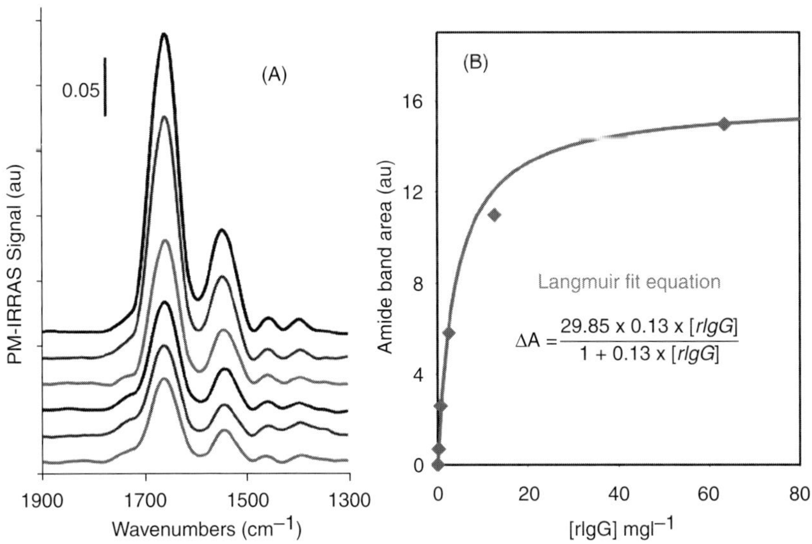

FIGURE 15 (A) PM-IRRAS spectra obtained on amine terminated layers on which β-LG was covalently bound upon interaction with rabbit antiserum, with increasing concentrations of anti-βLG IgG, from bottom to top: 0, 0.2, 0.6, 2.5, 12.7, and 63.3 μg/mL. (B) Amide band area as function of rabbit IgG concentration, also shown are experimental data fit to the Langmuir isotherm equation. Figure taken from Ref. [41] with permission.

amine grafting led to a similar amount of β-lactoglobulin, an interesting influence of the immobilization technique on the molecular recognition of specific rabbit anti-β-LG IgG was observed. Immobilization of β-LG via its acid functions led to a better affinity, 2.5 times higher, than immobilization via its amine groups with respect to the recognition of its specific antibody. In both cases, the affinity constants were deduced from PM-IRRAS and SPR experimental data. A very good agreement between the two techniques was observed [41].

Interestingly, a similar system was then used to detect a small chemical molecule, benzo[a] pyrene (BAP), a carcinogen pollutant representative of the polyaromatic hydrocarbons (PAH) contaminants, whose detection at trace amount level is a current challenge. IR is a perfect technique to detect these PAH compounds thanks to their specific ν_{CH} aromatic signals at ca 3050 cm^{-1} [42,43]. The PM-IRRAS-based sensor was able to detect the target with a limit of detection slightly below 5 μM (Fig. 16).

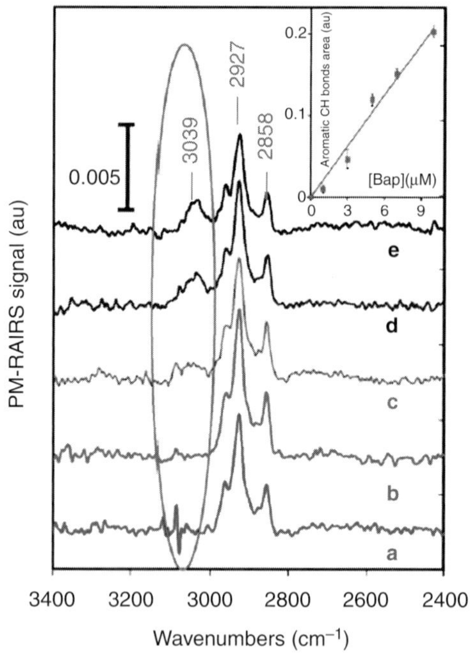

FIGURE 16 PM-RAIRS spectra recorded after immunodetection of benzo[a]pyrene at 1, 3, 5, 7, and 10 × 10^{-6} M for a, b, c, d, and e, respectively; taken with permission from Ref. [43].

PM-IRRAS to monitor DNA hybridization

DNA hybridization on gold surfaces can be also characterized by PM-IRRAS, implying the preliminary binding of a single-stranded DNA to the surface; detection of DNA hybridization by PM-IRRAS was described for the first time by Brewer *et al.*, the target was the exact complementary ssDNA of an immobilized SH-modified ssDNA [44]. The difficulty was, in this type of experiment, to find the right compromise between the optimal temperature for hybridization (60–80°C) and the possible desorption of thiolated ssDNA from the surface. Note that the spectrum of the gold surface after ssDNA immobilization exhibited signals characteristic of the DNA phosphodiester backbone (broadband at 1188 cm^{-1}), and the asymmetric PO_2^- stretching mode at 1238 cm^{-1} (Fig. 17). Here again, the hybridization was made clear from the appearance of a new band at ca 1670 cm^{-1}, attributed to changes in hydrogen bonding between complementary bases, that is, conformation changes in the adsorbed ssDNA. DFT calculation permitted to assign the position of this signal to a blue shift of the NH_2 bending mode, which becomes more constrained upon base pair formation.

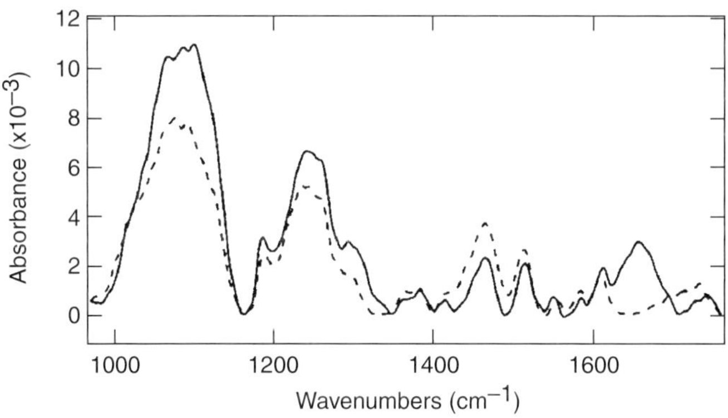

FIGURE 17 **PM-IRRAS spectra of a mixed monolayer of mercaptohexanol and ssDNA before (dashed line) and after (solid line) hybridization with the complementary DNA; taken with permission from Ref. [44].**

The hybridization IR signature was fully interpreted using DFT, by calculating the vibration frequencies for each individual nucleotide and modeling the difference spectrum obtained by subtracting the ssDNA from the dsDNA spectra (see Fig. 18) [45]. These DFT calculations confirm that the signals in the 1550–1780 cm^{-1} region are mainly due to the shifts of the NH_2 bending and C=O stretch, involved in Watson–Crick hydrogen bonds formed upon base pairing.

Base-pairing between DNA strands was also detected by PM-IRRAS, by using oligonucleotide-derivatized nanoparticles self-assembled onto gold films, thus reaching an enhancement of the sensitivity [46]. Au and Ag nanoparticles were decorated with oligonucleotides by reaction with disulfide-protected oligonucleotides. In parallel, thiolated oligonucleotides were self-assembled on a gold film and then exposed to the Au or Ag nanoparticles functionalized by the complementary oligonucleotide; note that a mixed monolayer of DNA

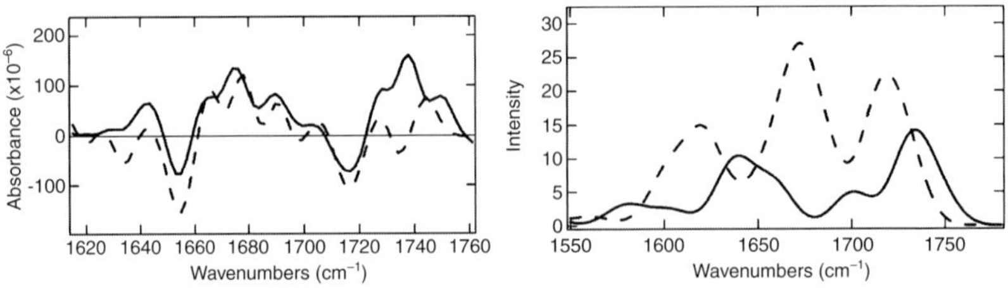

FIGURE 18 **Left: difference obtained by subtracting the PM-IRRAS spectra of ssDNA from that of dsDNA; The solid and dotted lines correspond to measurements performed at 25°C and 45°C respectively. Right: DFT-calculated spectra of SSDNA (solid) and dsDNA (dotted line) (taken with permission from Ref. [45]).**

(containing only adenine bases) and mercaptohexanol was formed in order to block direct interaction of the complementary DNA with uncovered gold areas. Figure 19 shows the PM-IRRAS spectrum of the single strand oligonucleotide (thymine only) coadsorbed with mercaptohexanol on gold (spectrum 1), after hybridization of the complementary DNA (adenine only) (spectrum 2), or after hybridization of the same complementary DNA, but grafted on either Au or Ag nanoparticles (spectra 3 and 4).

Important to note are the spectral changes observed upon hybridisation, (i) the hybridisation signature at 1550–1650 cm^{-1} and (ii) the signal enhancement when complementary DNA is coupled to nanoparticles. An increase of the bands characteristic of the DNA backbone, between 800 and 1300 cm^{-1} is also observed.

PM-IRRAS is thus capable of detecting the formation of thin films of ssDNA, but also the hybridization with complementary DNA.

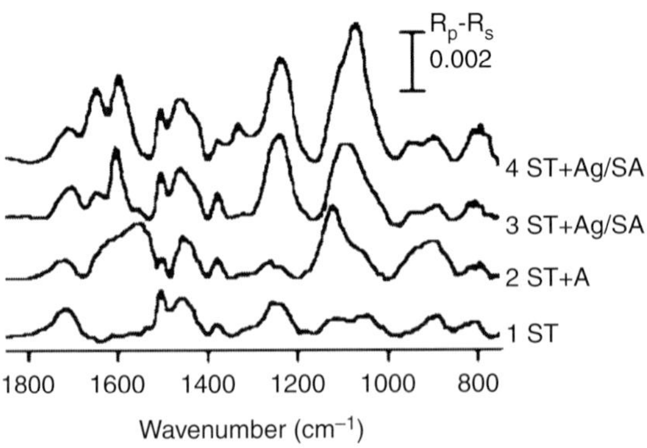

FIGURE 19 **PM-IRRAS spectra of a 12 mer ssDNA (spectrum 1), after hybridization with complementary DNA (spectrum 2); after hybridization with complementary DNA grafted to Au (spectrum 3) or Ag (spectrum 4) nanoparticles (taken with permission from Ref. [46]).**

Another efficient way of detecting DNA hybridization consists in immobilizing peptide nucleic acids, PNAs, on gold surfaces; and this was advantageously done by using by PM-IRRAS and taking advantage of the PNA/DNA IR specific signatures [47]. Cysteine-derivatized PNA fragments were grafted onto gold surfaces, leading to intense bands at 1549–1270 cm^{-1}; those have been attributed to the in-plane ring vibrations, thus supporting the idea that the PNA chains stand upright, almost normal to the surface, favorable to the hybridization with complementary DNA [48]. Hybridization of complementary DNA implies the presence of new phosphate chemical groups, not present on the PNA backbone, yielding indeed the characteristic IR signature of the PNA-DNA heteroduplex formation (P–O stretch vibration, in PO_2^-, at 1237 and 1081 cm^{-1}) [49]; the PNA-based biosensors, coupled to the PM-IRRAS technique enables the detection of label-free nucleic acid targets in biological solutions (Fig. 20).

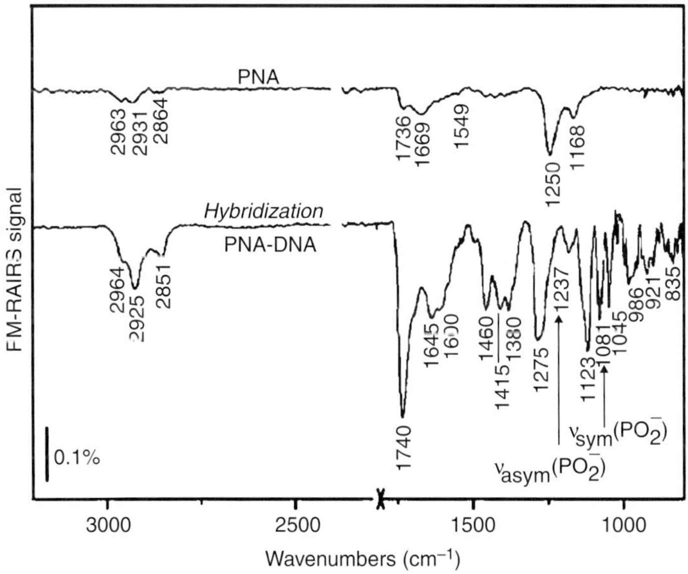

FIGURE 20 PM-IRRAS spectra of gold surface modified by PNA (upper spectrum) and after hybridization with complementary DNA sequence (lower spectrum) (taken with permission from Ref. [49]).

Conclusion

FTIR, spectroscopy can be setup in various configurations, making the technique sensitive to very low amounts of molecules adsorbed on solid surfaces; it can be advantageously used to characterize biomolecular recognition phenomena. Taking advantage of the enhanced electric field in the vicinity of a reflective surface (IRRAS effect), or on a rough surface (mostly of Au or Ag; SEIRA effect), IR signals can be enhanced by one or several orders of magnitude; experimental setups have been optimized including cells for measurements under flowing conditions; some examples have been reported showing the successful detection of protein binding or DNA–DNA pairing, for example.

In the case of protein binding, FTIR also provides an accurate way to discriminate specific and nonspecific interactions by performing a second derivative spectral analysis and thus analyzing the secondary structure of the targeted proteins. Thus, in addition to their sensitivity, FTIR-based, label-free, sensors have indeed the great advantage in allowing the differentiation of bound molecules.

PART II: TRANSITION METAL CARBONYL (TMC) PROBES

IR spectroscopy of metal carbonyls

Transition metal carbonyls (TMC) complexes are among the longest known classes of organometallic compounds. Carbon monoxide bonds to transition metals at low oxidation states (< +2) via three modes of coordination, that is, terminal, doubly bridging (μ^2) and triply bridging (μ^3) (Fig. 21).

The metal-CO bond is classically described as a resonance hybrid leading to a bond order between 1 and 2 for the M–C band and 2 and 3 for the C–O bond (Fig. 22).

The most important contribution to M–CO bonding is provided by the σ-donor interaction where the less electronegative carbon atom donates lone pair electron density to a σ-oriented metal orbital. It is completed with one π-donor interaction where electrons from a filled π-orbital of the CO are transferred to an empty d-orbital of the metal. Additionally, another π-bonding interaction occurs in which filled metal orbitals back-donate electron density to the available π*-orbital of the carbonyl group (Fig. 23).

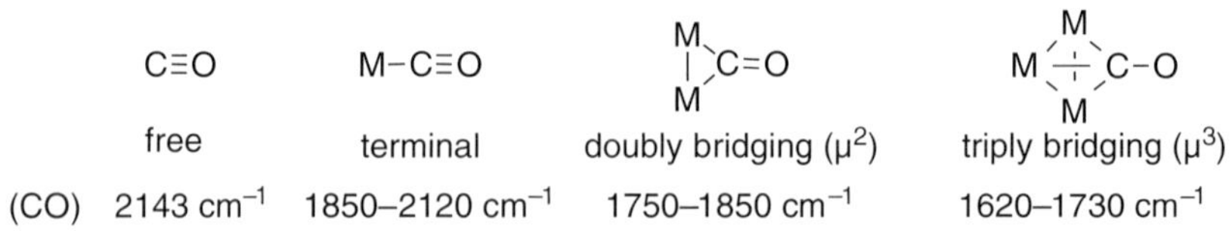

FIGURE 21 Modes of coordination of carbon monoxide to transition metals and IR frequencies of ν(CO) vibration modes.

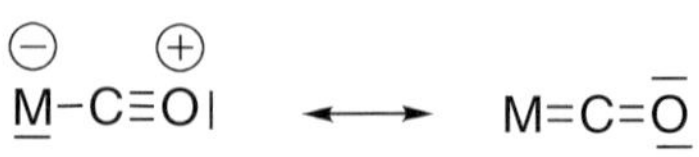

FIGURE 22 Resonance forms of the M–C–O bond.

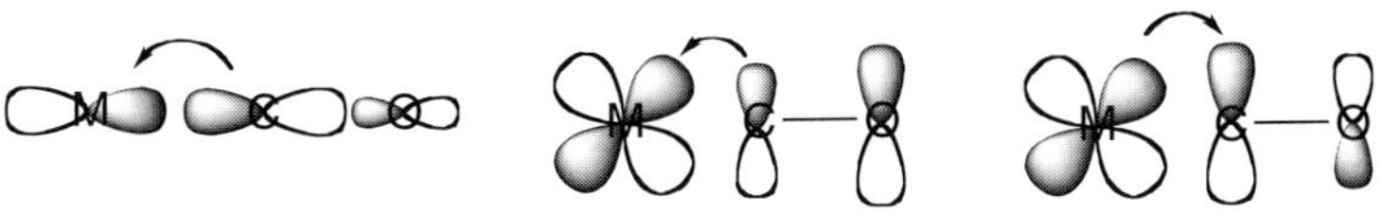

FIGURE 23 Bonding modes in M–CO.

The most informative technique to characterize metal carbonyls is IR spec troscopy. Carbon monoxide itself gives rise to a band at 2143 cm^{-1} assigned to the stretching mode of the C–O bond. Because π-bonding effects result in a weakening of the C–O bond, the ν(CO) vibration modes of metal carbonyls undergo a red shift of their position. Indeed the position of the ν(CO) bands is inversely correlated with the strength of the π-bonding between M and C. The position of the ν(CO) bands also depends on the bonding mode (see Fig. 21) and the σ-donor and π-acceptor properties of the other ligands surrounding the metal (s). Most importantly, the number and intensity of carbonyl stretching bands largely depend of the local symmetry. Therefore, the symmetry of the TMC complex is simply determined by counting the number of ν(CO) bands. Indeed the expected number of IR bands derives by means of group theory (Table 1).

Interestingly, ν(CO) bands can be up to 40 times more intense than saturated ketones [51]. This has been explained by the fact that stretching of the C–O bond lowers the energy of the π* antibonding orbital which acts as an acceptor of metal electron density (see above). This results in a transfer of the charge to the carbonyl groups.

Detection of biorecognition events by IR spectroscopy with TMC labels

In homogeneous media

Steroid hormonal receptor assay

To be used as a biological label, a chemical species must display the following features. It must be easy to introduce onto any kind of biological species by simple chemical means. Its presence must not disturb (or at least minimally disturb) the biological event under study; its detectability must be as high as possible in complex media. These characteristics were entirely fulfilled by radioisotopic labels. However, these species suffer from several drawbacks, such as high cost, health hazards, instability, and issues associated with their disposal. These limitations have prompted scientists to look for nonisotopic substitutes and given the features recalled above, TMC entities appeared as potentially attractive, especially with the simultaneous advent of Fourier transform IR apparatuses that tremendously increased the sensitivity of IR spectroscopy with respect to dispersive spectrometers.

TABLE 1 Infrared modes for some TMC complexes

Complex	Point group	Symmetry of IR-active CO stretching modes	Number of expected ν(CO) bands
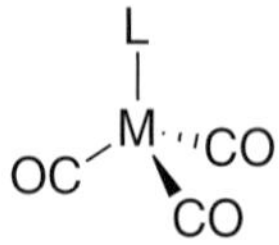	C_{3v}	$A_1 + E$	2
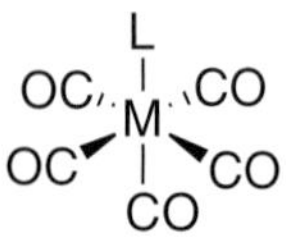	C_{4v}	$2\,A_1 + E$	3
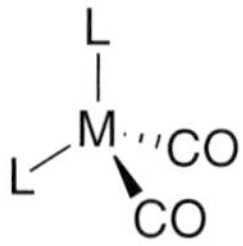	C_{2v}	$A_1 + B_1$	2
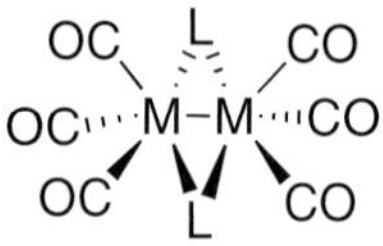	C_{2v}	$2\,A_1 + 2\,B_1 + B_2$	5

Estrogen and progesterone hormones (Fig. 24) are considered to play an important role in the development and progression of breast cancers.

They are associated to estrogen receptor (ER) and progesterone receptor (PR), respectively, both being cellular proteins binding the hormones with high affinity ($K_d = 0.05$–15 nM for the estradiol – ER association) and specificity. In normal breast cancer tissues, the level of ER and PR binding sites is low but for

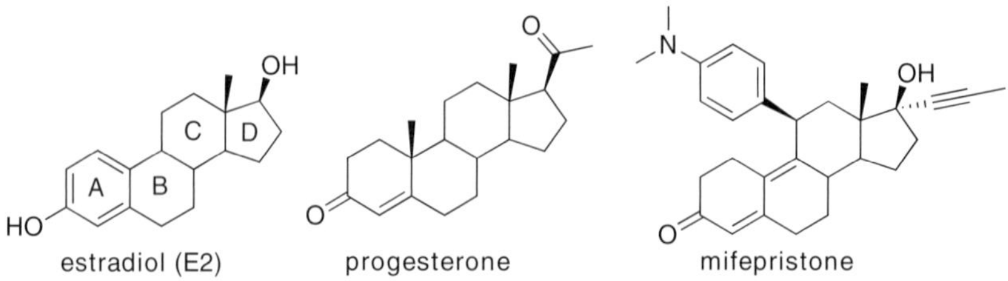

FIGURE 24 Structure of steroid hormones and mifepristone.

some of the primary breast cancer tumors, the level becomes measurable (40–600 fmol ER/mg protein in the cytosol of breast tissues), [52] indicating that the tumor will respond to hormonal therapy [53]. Indeed, the level of ER and PR in tumor cells is one of important criteria of decision in the choice of the therapeutic strategy for breast cancer treatment.

In the 1980s, measurement of ER and PR levels in clinical samples was exclusively carried out by radioligand assays involving hormones labeled by tritium isotope. To demonstrate the usefulness of metal carbonyl entities as labels for ER assay, a tricarbonyl chromium entity was introduced into the skeleton of estradiol (E2) by complexation of the A ring with $Cr(CO)_6$. For stability reasons, the phenolic group was converted to an ether group yielding E2-$Cr(CO)_3$. Alternatively, alkyne dicobalt hexacarbonyl and alkyne dimolybdenum tetracarbonyl dicyclopentadienyl entities were introduced at the 17β-position of estradiol. To assay the progesterone receptor, the alkyne group of mifepristone, a potent antagonist of the progesterone and glucocorticoic receptors was complexed with $Co_2(CO)_8$ to yield mifepristone-$Co_2(CO)_6$ (Fig. 25).

The relative binding affinity (RBA) of the TMC derivatives determined by a competitive ligand binding assay showed that despite the presence of bulky substituents on the skeleton of the hormones, the TMC derivatives were still able to bind to ER and PR (Table 2) and may be suitable to carry out ER and PR assay in biological samples.

The assay was carried out by incubating the TMC derivatives at 10 nM level with lamb uterine cytosol as a source of ER and PR and proteins were precipitated with protamine sulfate. The precipitate was collected by centrifugation, washed and lyophilized to yield a powder that was pressed into a pellet and analyzed by FT-IR in the transmission mode.

The IR spectrum of lamb uterine cytosol treated by estradiol shows a window of absorption around 2000 cm^{-1} (Fig. 26, (A)) whereas the IR spectrum of cytosol treated with E2-$Cr(CO)_3$ showed small by distinct band in this window that were indisputably assigned to the TMC label (Fig. 26, (B)). Furthermore, the IR spectrum of cytosol treated with E2-$Cr(CO)_3$ in the presence of a large excess of the nonsteroidal estrogen diethylstilbestrol (DES) lacked the 2 ν(CO)

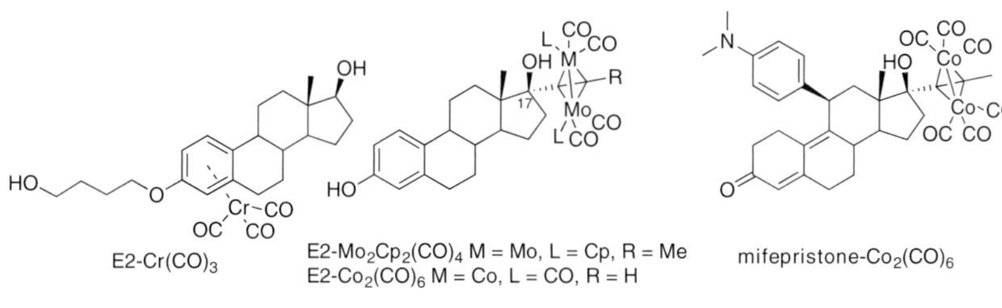

FIGURE 25 Structure of some TMC derivatives of estradiol and mifepristone.

TABLE 2 Relative binding affinity (RBA) values of TMC derivatives of estrogens and mifepristone (RBA(E2) = 100% and RBA (progesterone = 100%) and position of the detected ν(CO) bands in the IR spectrum of the protamine sulfate fraction of cytosol treated by the complexes

Molecule	RBA (%)	ν(CO) (cm^{-1})	Reference
E2-$Cr(CO)_3$	28	1957 (A_1) 1880 (E)	[54]–[56]
E2-$Co_2(CO)_6$	12	2049 (B_1) 2021 ($A_1 + B_1 + B_2$)	[57]
E2-$Mo_2Cp_2(CO)_4$	16	1999 1920 1893	[58]
mifepristone-$Co_2(CO)_6$	13	2090 (A_1) 2050 (B_1) 2025 ($A_1 + B_1 + B_2$)	[59],[60]

bands of the label, indicating that binding of E2-$Cr(CO)_3$ to ER was fully inhibited by DES. Similar results were obtained with the other estradiol derivatives and the mifepristone derivative (Table 2).

On similar lines, iron tricarbonyl derivatives of the flavanone naringenin (Fig. 27) were synthesized in view of detecting proteins involved in nodABC gene expression induction in *Rhizobium leguminosarium* when legume roots are formed.

The compound depicted in Figure 27 was shown as the most potent inducer of nodABC gene expression of the series [61]. It was therefore tentatively used to detect the regulator protein nodD in cell lysates of *R. leguminosarium* bacterial strains expressing or lacking nodD. Unfortunately, the characteristic ν(CO) bands of the TMC label were present in the IR spectrum of both strains, indicating a high level on nonspecific binding of naringenin-$Fe(CO)_3$.

Carbonyl metallo immunoAssay (CMIA) [62]

Immunoassays are widespread bioanalytical methods that are used to measure the concentration of substances in complex mixtures such as biological fluids (serum, plasma, and urine). They rely on the ability of an antibody to bind its antigen with high affinity and specificity. Most of the time, detection of this binding event requires the addition of a label that serves for quantification of the antigen–antibody complex concentration formed at equilibrium. Historically, the labels were radioisotopes such as ^{3}H, ^{14}C or ^{131}I and subsequently the immunoassay was called radioimmunoassay (RIA) [63]. The advent of immunoanalytical analysis tremendously improved the management of some

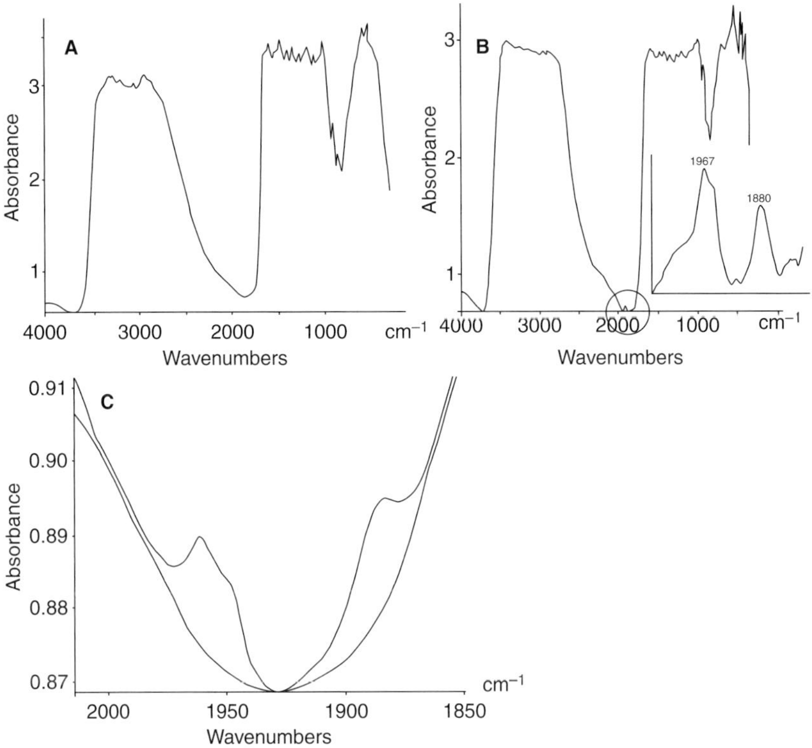

FIGURE 26 **IR spectra of lamb uterine cytosol treated by estradiol (upper left), 10 nM E2-Cr $(CO)_3$ (upper right, down, spectrum b) or 10 nM E2-Cr$(CO)_3$ + 1 μM DES (down, spectrum a), subsequent precipitation by protamine sulfate and lyophilization (pure protein pellet). Inset: expansion of the ν(CO) region of the spectrum and baseline correction. Pictures taken from [54,60] with permission.**

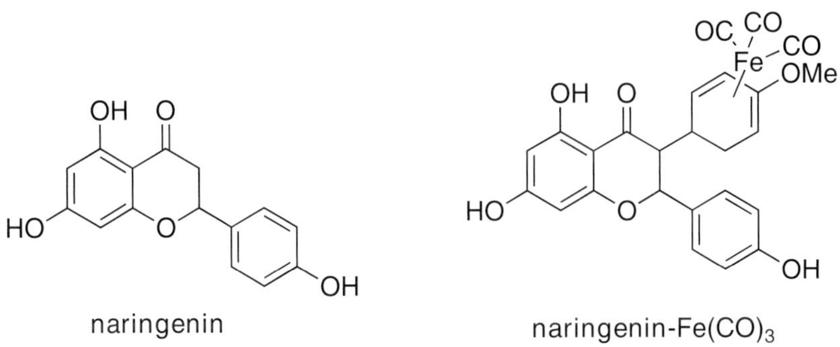

FIGURE 27 **Naringenin and one of its TMC derivatives.**

diseases, in particular endocrine disorders as they provide a simple and rapid means to measure the concentration of endocrine hormones that are present at low levels in blood (10–100 nmol/l). The assay principle is schematized in Figure 28. The sample containing the analyte Ag (acting as the antigen) is mixed with a fixed concentration of antibody Ab and a tracer Ag*. The tracer is a labeled molecule that shares some structural similarities with the antigen/analyte and will be recognized equally by the antibody. Both the analyte and the tracer compete for binding to the antibody, so that, at equilibrium, the concentration of antibody-tracer complex will be inversely related to the initial concentration of analyte in the sample (Fig. 28). The free fraction of tracer F is separated from the bound fraction of tracer B and B is measured by radioisotope counting.

For the same reasons recalled above, radioisotopic labels have been gradually replaced by nonisotopic species. By analogy with RIA, the term Carbonyl Metallo ImmunoAssay (CMIA) was coined to designate a competitive assay involving metal carbonyl labels and FT-IR spectroscopy for quantitative analysis of the labels. The question to address was whether and what conditions, FT-IR spectroscopy was sufficiently sensitive and reliable to achieve quantitative analysis in the concentration range required for immunoanalysis. IR spectroscopy being an absorption technique, Beer's law (1) applies for dilute solutions.

$$A = \varepsilon \times 1 \times c = \varepsilon \times 1 \times n/v = \varepsilon \times n \times 1/v. \tag{1}$$

Thus, for a given quantity n of compound, the absorbance A is proportional to the molar absorptivity ε which is an intrinsic feature of the molecule and to the optical pathlength l. But it is inversely proportional to the sample volume *v*. From this relationship, a light-pipe cell (Fig. 29) with a fill volume of 30 μl and a physical pathlength of 20 mm was designed [64].

With this cell in hand and an FT-IR spectrometer equipped with a liquid nitrogen cooled InSb detector, a linear relationship between the concentration of E2-$Co_2(CO)_6$ in CCl_4 and the absorbance of the B_1 vibration mode (see Table 2) was observed in the 10^{-8}–10^{-6} M range and the limit of detection was equal to 1 nM within a scan time of 3.5 min [65].

Metal carbonyl tracers of the three main anticonvulsants drugs, namely phenobarbital (PB), carbamazepine (CBZ), and diphenylhydantoin (DPH) were synthesized in the aim of implementing CMIA for therapeutic drug monitoring.

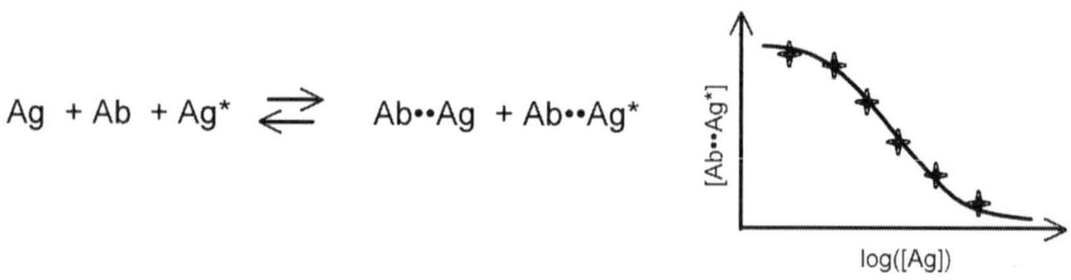

FIGURE 28 **Principle of competitive immunoassay and resulting standard curve.**

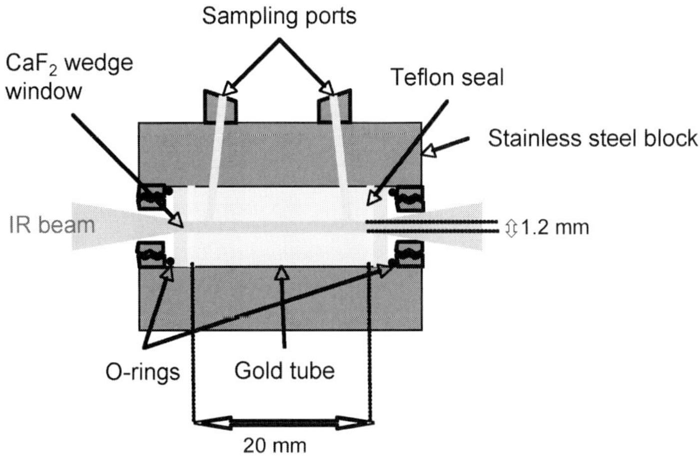

FIGURE 29 Schematic drawing of the IR light-pipe cell.

Some drugs, particularly anticonvulsants, have a narrow therapeutic index (Table 3), requiring their monitoring in blood of patients under treatment to avoid side effects due to overdoses. Polyclonal antibodies against the three drugs were produced in rabbits by immunization with suitable immunogens. Heterogeneous immunoassay schematized in Figure 30 was set up for each of these analytes, employing the corresponding tracer and antibody. In a typical assay, a series of tubes containing the antibody and the tracer (at fixed concentration) and standard concentrations of analyte was prepared. After incubation, the separation of the free and bound fractions of tracer was conveniently achieved by extraction with diisopropyl ether. The organic phases were evaporated under reduced pressure and redissolved in CCl_4 or $CHCl_3$ and injected in the light-pipe cell for IR measurement.

The percentage of bound tracer for each standard was calculated from the height of the analytical ν(CO) band and plotted as a function of analyte quantity. Standard curves were plotted from which was deduced the half-maximal inhibitory concentration IC_{50}, which gives an indication of the assay range for each analyte (Table 3).

Given the therapeutic index of each drug, 1–10 μl sample volume is sufficient to carry out an assay. In the same line, CMIA was applied to the endogeneous steroid hormone cortisol [70] and to the herbicides chlortoluron [71] and atrazine [72].

One of the most attractive features of TMC labels as compared to traditional labels such as fluorophores or enzymes is the possibility to simultaneously detect and quantify mixtures of different metal carbonyl species by IR spectroscopy thanks to their narrow absorption bands [73,74]. This is illustrated in Figure 31 depicting the superimposition of the IR spectra of PB-$Mn(CO)_3$, CBZ-$Co_2(CO)_6$, and DPH-$Cr(CO)_3$ and the IR spectrum of a mixture of the three tracers.

TABLE 3 Anticonvulsants, therapeutic indexes, TMC tracers, and IC_{50} values measured by CMIA

Analyte	Therapeutic index	Tracer	IC_{50} (nmol)	Reference
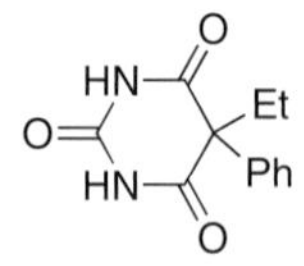 **phenobarbital (PB)**	43–172 nmol/ml	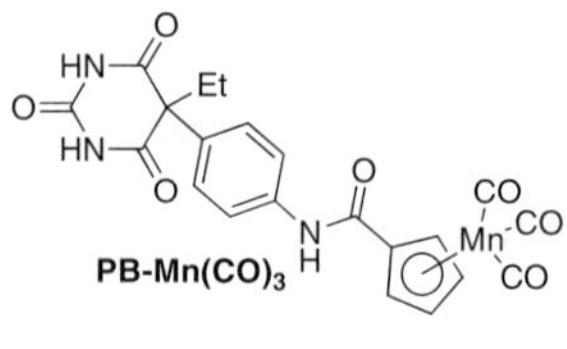 **PB-Mn(CO)$_3$**	0.2	[66]
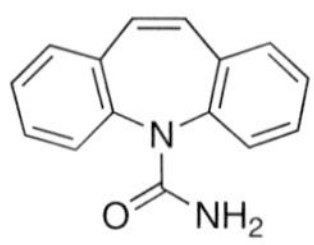 **carbamazepine (CBZ)**	17-51 nmol/ml	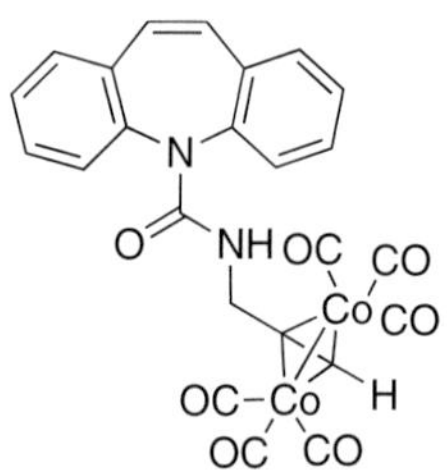 **CBZ-Co$_2$(CO)$_6$**	0.015	[67]
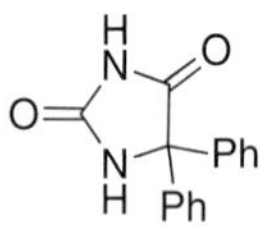 **phenytoin (DPH)**	20-99 nmol/ml	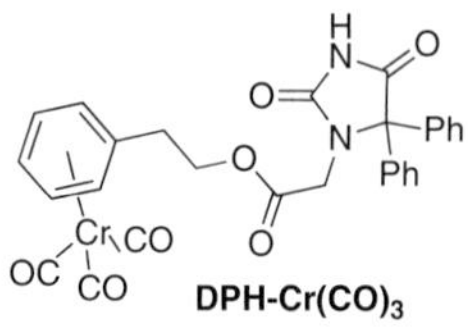 **DPH-Cr(CO)$_3$**	0.25	[68]
		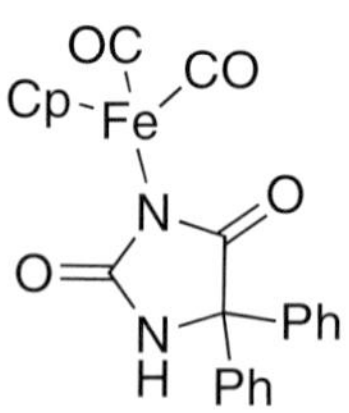 **DPH-Fe(CO)$_2$**	0.025	[69]

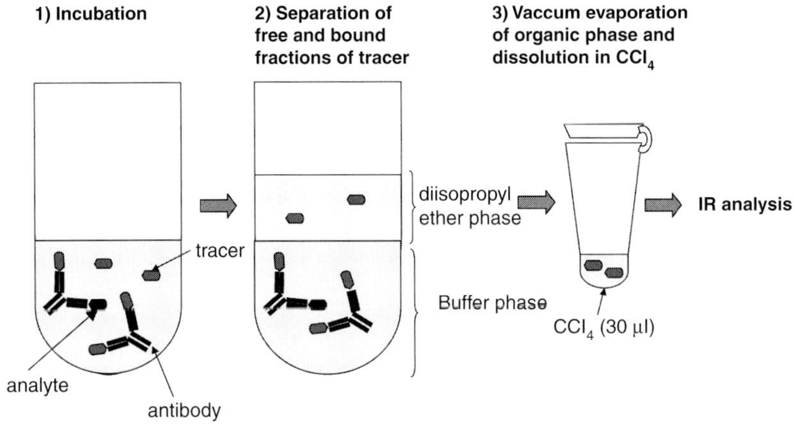

FIGURE 30 Principle of CMIA.

This feature was exploited to set up simultaneous multi-CMIA of two [76,77] and even three [75] anticonvulsant drugs involving the three TMC tracers. The analytical performances of the double and triple CMIA were identical to those of the mono-CMIA.

Mid-IR imaging

Osmium carbonyl derivatives of the fatty acid capric acid CA-$Os_3(CO)_{10}$ and phosphatidylcholine PC-$Os_3(CO)_{10}$ (Fig. 32) were synthesized and tested as mid-IR tracers for living cell imaging [78].

Oral mucosa cells soaked in aqueous solutions of these metal species were mapped with an IR microscope equipped with a 64 × 64 pixel PCT focal plane array detector at one of the tracer frequencies of absorption (2013 cm^{-1}). The resulting 2D-image (Fig. 7.13 (b)) was closely correlated with the optical image of the cells (Fig. 33 (a)), indicating that CA-$Os_3(CO)_{10}$ was a useful tracer for cell imaging.

Nevertheless, subcellular localization of the mid-IR tracers could only be achieved after cell fractioning. It resulted that the fatty acid like compound was found in the organelles and the membrane fractions while the phospholipids derivative was located in the cytoplasm and the membrane fractions.

Subcellular resolution with a TMC probe was very recently achieved by Raman microspectroscopy using [Mn(TMP)$(CO)_3$]Cl (TMP = tris(1-pyrazolyl)methane) as tracer on living cancer cells. This compound gives rise to a band at 1963 cm^{-1} in Raman spectroscopy owing to the asymmetric vibration mode of the Mn$(CO)_3$ entity. Images reconstituted from the integration of this band allowed to localize unambiguously the tracer in the cell nucleus and the nucleus membrane [79].

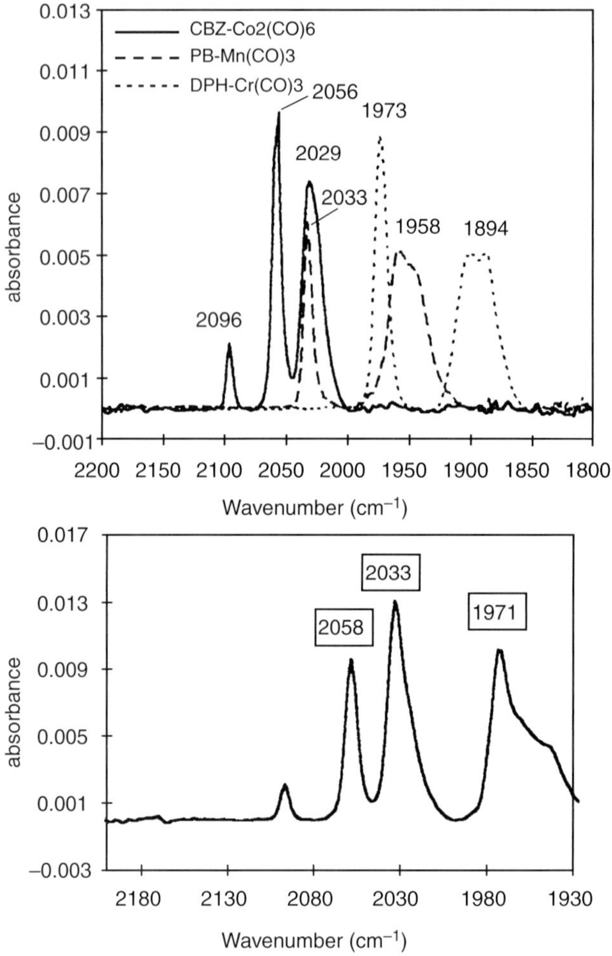

FIGURE 31 Left: superimposition of the IR spectrum of PB-Mn(CO)$_3$, CBZ-Co$_2$(CO)$_6$, and DPH-Cr(CO)$_3$; Right: IR spectrum of a mixture of the same compounds. Pictures taken from Ref. [75] with permission.

At the solid–gas interface

In the transmission mode

Most conventional immunoassays rely on the immobilization of the analytical reagent (antibody or antigen depending on the assay format) onto a solid phase in order to allow separation of bound and free fractions by simple washing steps. Antigen–antibody recognition steps are thus carried out at solid–liquid interfaces. Therefore, it was of great interest to examine whether CMIA could be transposed to a solid phase. The first attempt using classical polystyrene (PS) microtiter plates and transmission IR spectroscopy was unsuccessful because PS

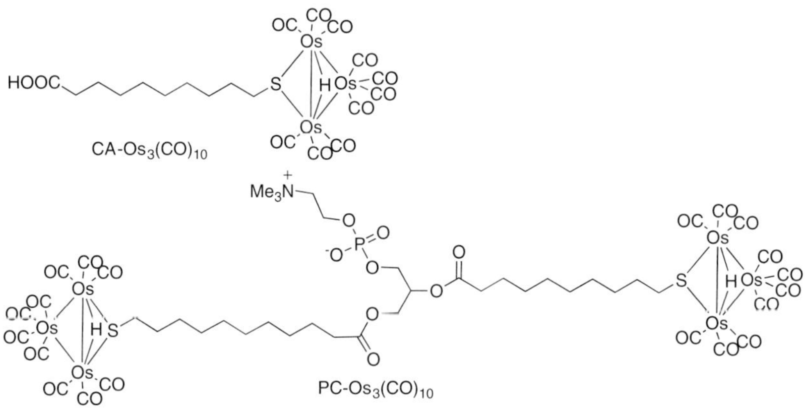

FIGURE 32 TMC derivatives of capric acid and phosphatidylcholine.

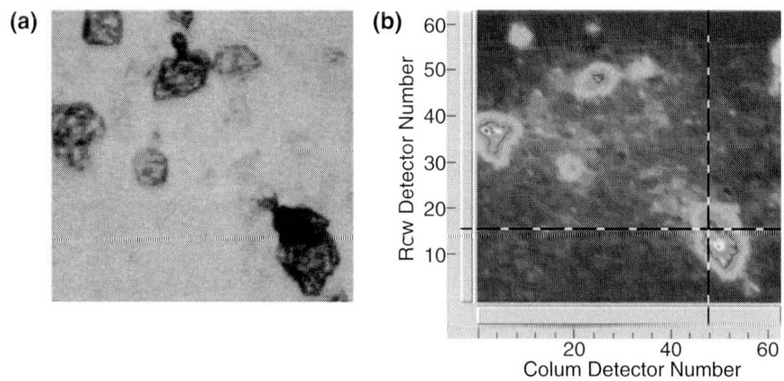

FIGURE 33 Normal optical image (a) and false-color IR image taken at 2013 cm^{-1} (b) of oral mucosa cells treated with CA-$Os_3(CO)_{10}$. Pictures taken from Ref. [78] with permission.

is almost completely opaque in the mid-IR region and the sensitivity too low [80]. Nitrocellulose (NC) appeared as a more promising solid phase because it is rather transparent in the 2000 cm^{-1} region and it displays high protein binding capacity (80–100 μg/cm^2) compared to PS (0.2 μg/cm^2). Simultaneously, the metal carbonyl entity was introduced onto the antibody (rather than onto the analyte analog) to increase the number of labels in the tracer. Furthermore, instead of directly conjugating the TMC labels to the antibody, which may result in a loss of immunoreactivity, the labels were attached to a synthetic polymer called PAMAM dendrimer (Fig. 34) by its terminal primary amine functions and the labeled dendrimer was attached to the antibody via its carbohydrate residues (Fig. 34).

The IR solid phase detection of antigen–antibody recognition was demonstrated by coating various amounts of rabbit IgG onto NC membranes then

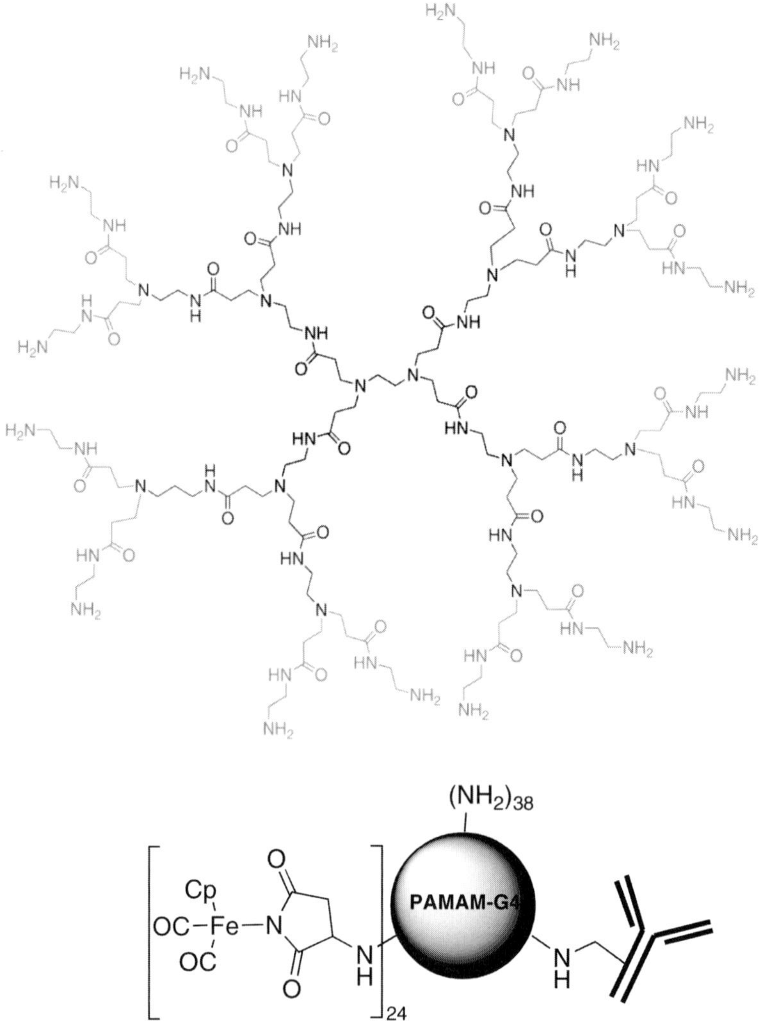

FIGURE 34 Structure of generation 2 PAMAM dendrimer and Ab-PAMAM-Fe(CO)$_2$ tracer.

expose them to anti-rabbit IgG labeled by the Fe(CO)$_2$ probes. The ν(CO) bands of the TMC label were observed by IR analysis of the membranes. Moreover, the intensity of the signals could be related to the quantity of spotted rabbit IgG in the range between 0.7 and 270 pmol [81].

In the specular reflection mode (IRRAS)

Infrared reflection–absorption spectroscopy (IRRAS or RAIRS) is a specially useful spectral technique to study ultrathin organic films at the surface of reflective materials. It not only provides information on the molecular groups of the thin film but also about their orientation relative to the metal surface (see, section, detection of biorecognition events by IR spectroscopy with TMC labels). It is particularly well suited to the analysis of self-assembled monolayers (SAMs) chemisorbed onto noble metals such as gold. These SAMs are readily obtained by exposure of metal surfaces to solutions of thiols or disulfides [82]. In a preliminary step, a disulfide derivative carrying two $Mn(CO)_3$ units was synthesized and chemisorbed on planar gold chips (Fig. 35).

The IRRAS spectrum of the gold chip displayed two intense ν(CO) bands at 2041 and 1968 cm^{-1}, confirming that IRRAS was an adequate technique to detect chemisorbed TMC species at the surface of gold [83].

Next, the interfacial high affinity recognition between biotin derivatives chemisorbed on gold substrate and avidin in solution was detected *ex situ* by IRRAS thanks to the labeling of avidin with a dicobalt hexacarbonyl label (Av-$Co_2(CO)_6$, Fig. 36). The presence of a set of three ν(CO) bands on the IR spectra

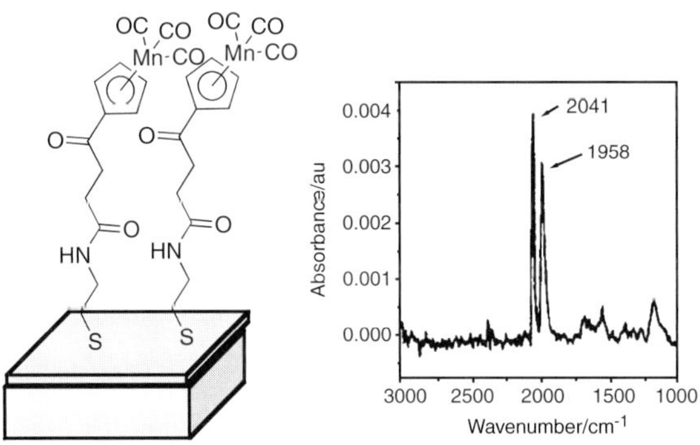

FIGURE 35 TMC entity chemisorbed on a planar gold-coated chip and IRRAS analysis of the surface (spectrum taken from Ref. [83] with permission).

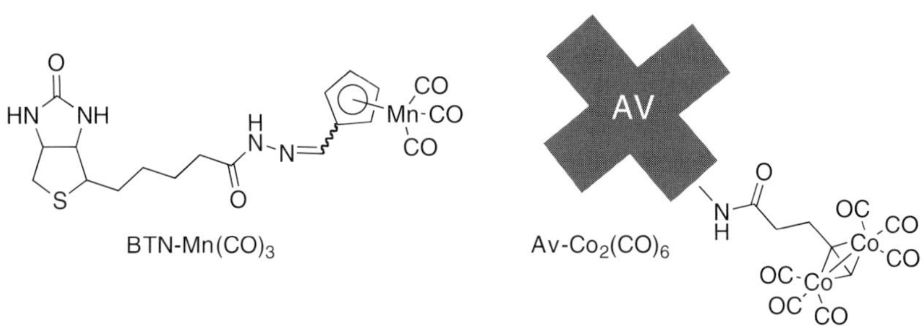

FIGURE 36 Structure of the TMC tracers of biotin BTN-$Mn(CO)_3$ and avidin Av-$Co_2(CO)_6$.

provided undisputable evidence that avidin was bound to gold by affinity for biotin [33,84, 85]. Furthermore, capture of a TMC derivative of biotin (BTN-Mn $(CO)_3$, Fig. 36) by the avidin layer was also detected *ex situ* by IRRAS thanks to the two ν(CO) bands of the $Mn(CO)_3$ entity.

Exposure of a series of gold chips coated with biotin to solutions of Av-$Co_2(CO)_6$ at various concentrations followed by plotting of the integrated area of the ν(CO) bands versus the concentration of Av-$Co_2(CO)_6$ yielded the curve shown in Figure 37, the form of which being reminiscent of a Langmuir isotherm [33]. Curve-fitting of the data according to the Langmuir isotherm equation

$$A = A_{\max} \times \frac{K \times c}{1 + K \times c} \tag{2}$$

yielded a equilibrium constant K of $3.2 \pm 0.9 \times 10^7\ \mathrm{M}^{-1}$.

In a similar fashion, interfacial selective recognition between an antigen (rabbit IgG) immobilized on gold substrates and its antibody (goat antirabbit IgG) labeled with the same TMC label in the presence of nonbinding proteins was detected *ex situ* by PM-IRRAS. A Langmuir isotherm curve was also obtained by plotting the integrated area of the ν(CO) bands versus the concentration of labeled antibody which yielded a Langmuir equilibrium constant K of $1.3 \pm 0.12 \times 10^7\ \mathrm{M}^{-1}$[39].

On the basis of these last results, an IR optical immunosensor was set up for the detection of the herbicide atrazine (ATZ). A derivative of ATZ (OVA-ATZ) was chemisorbed to gold substrates that were further exposed to solutions containing a fixed quantity of anti-ATZ and a variable quantity of ATZ. The presence of ATZ in solution inhibited the binding of anti-ATZ to the

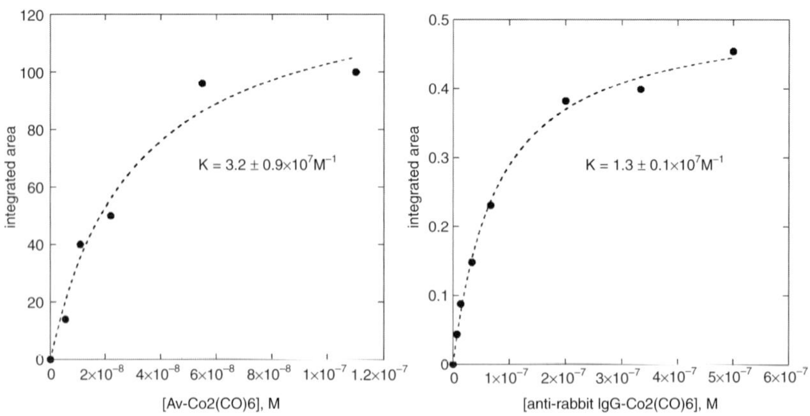

FIGURE 37 Left: binding curve of Av-$Co_2(CO)_6$ to immobilized biotin. Right: binding curve of antirabbit IgG-$Co_2(CO)_6$ to immobilized rabbit IgG.

immobilized ATZ derivative, which resulted in lesser antibody on the surface as the concentration of ATZ in solution increased. The amount of bound anti-ATZ antibody (of the rabbit IgG type) was revealed by exposure to antirabbit IgG–$Co_2(CO)_6$ followed by detection of the TMC label by PM-IRRAS (Fig. 38, left). A standard curve was established by plotting the integrated area of the ν(CO) bands as a function of atrazine concentration (Fig. 38, right) characterized by a IC_{50} of 11 ± 2 nM [86].

Very recently, a metal carbonyl probe was employed to detect *ex situ* by PM-IRRAS the hybridization between an oligonucleotide chemisorbed on gold substrate and its complementary oligonucleotide in solution. Conversely to the previous examples, the TMC tag is not chemically attached to the oligonucleotide in solution but to a naphthalene diimide entity (Fig. 39) that binds specifically to double strand DNA by threading intercalation [87].

In the diffuse reflectance mode (DRIFTS)

The molecular recognition between avidin covalently bound to acrylic beads and biotin labeled by a $Fe(CO)_2$ tag via generation 5 polypropylenimine

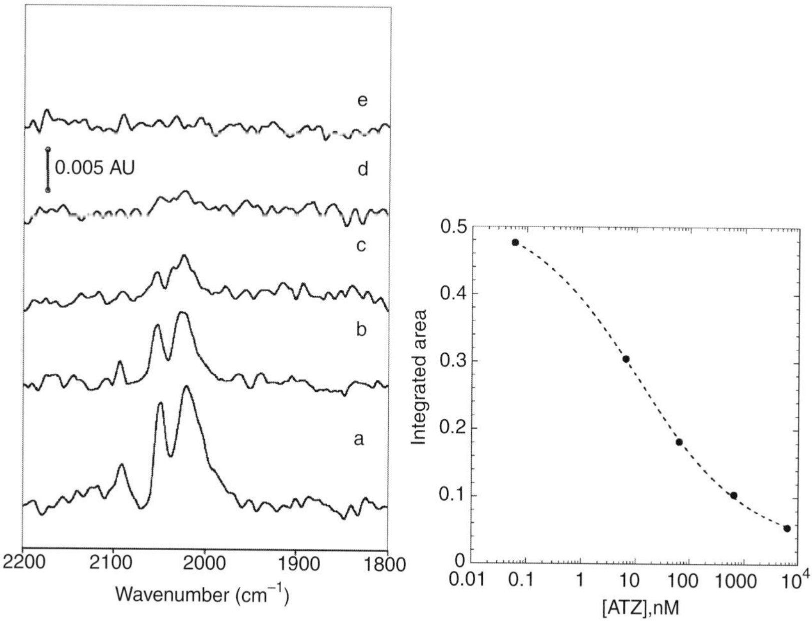

FIGURE 38 Left: PM-IRRAS of sensor chips coated with OVA-ATZ. Spectra recorded after sequential exposure to mixtures of anti-ATZ antibody and atrazine (0 nM (a), 6.4 nM (b)), 64 nM (c), 640 nM (d), and 6400 nM (e)) and antirabbit IgG-$Co_2(CO)_6$; expansion in the ν(CO) spectral range. Right: standard curve for atrazine established by the IR optical immunosensor.

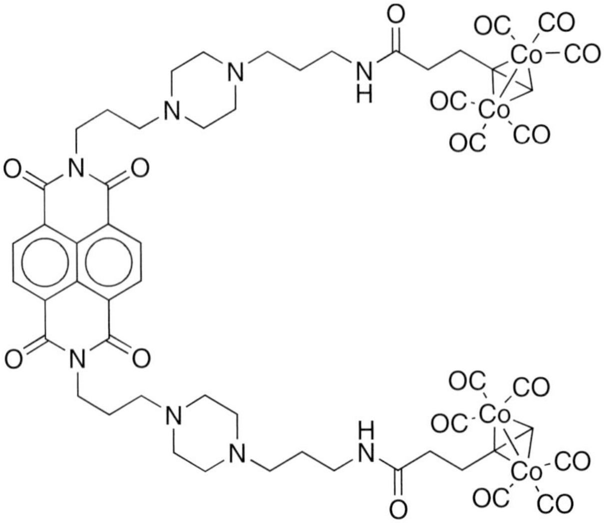

FIGURE 39 **Naphthalene diimide threading intercalator labeled by a TMC probe.**

dendrimer was qualitatively detected *ex situ* by diffuse reflectance Fourier transform IR spectroscopy (DRIFTS). Although acrylic resin beads are a highly absorbing material, it displayed a window of absorption in the 2000 cm^{-1} region where the two expected ν(CO) bands of the $Fe(CO)_2$ entity were observed after treatment of the beads with a solution of biotin tracer [88].

Molecular sensing by IR spectroscopy

For all the results reviewed above, the spectral properties of the TMC tracer once bound to its biological target were undistinguishable from those of the free tracer. In other words, the biorecognition event had no influence whatsoever on the IR response of the label, so that a separation step between the bound and free fractions of TMC tracer was necessarily included in the analytical procedure. Conversely, a molecular *bioprobe* is a functional molecule that, upon recognition and binding of its biological target, sees its spectral properties change in accordance and can thus sense the binding event directly. More generally, a molecular sensor is a device for which the binding of an analyte provides a response owing to the presence of a signal transduction component [89].

Several investigators have explored the ability of TMC molecules to act as molecular sensors, given the fact that IR signals resulting from the vibration of the carbonyl ligands have long been known to be solvent sensitive [51] and that high affinity molecular associations involve an ensemble of noncovalent interactions (Van der Waals, π-stacking, H-bonds etc.) that contribute to the tight binding of the molecules to one another and place both partners in an environment different from their "free" state.

In homogeneous media (transmission mode)

As a starting point, Anson and Stephenson rationalized the solvent influence on the symmetric (A_1) and asymmetric (E) vibration modes of $Fe(CO)_3$ units. They confirmed that the ν(CO) bands of (η^4-cyclohexadiene)$Fe(CO)_3$ complexes were sensitive to the solvent polarity both in their position (solvent-induced shifts up to 38 cm^{-1}) and width [90]. More specific interactions were transduced by IR spectroscopy of TMC probes. For example, the two ν(CO) bands of (η^6-benzene)$Cr(CO)_3$ in cyclohexane underwent a significant negative shift upon addition of 1,2-dimethoxybenzene as a result of π-stacking interaction between the arene groups (Fig. 40). Spectral analysis and decomposition by a least-squares fitting program allowed to identify component bands assigned to "free" and "bound" probe states [91].

In the same line, metal carbonyl complexes including ionizable groups (COOH, OH, NH_2) were shown as IR-active pH probes. The position of the ν(CO) bands was related to the pH of the solution in a certain range according to the pK_a of the ionizable group. pH-induced shifts up to 26 cm^{-1} were observed [92].

More sophisticated IR-active molecular probes were designed for the detection of alkali metal cations 15-crown-5/18-crown-6 ethers were combined with an arene chromium tricarbonyl entity (Fig. 41) to provide probes that respond to Na^+ and K^+ by positive shift of the ν(CO) bands in a concentration-dependent fashion. This shift was attributed to reduced back donation from the metal to the π* antibonding orbitals of the CO ligands [93].

Peris *et al.* designed two TMC molecular probes for H-bond donors (Fig. 42). Both complexes include an uncoordinated N atom that served as H-bond acceptor. Association between several H-donors and the TMC

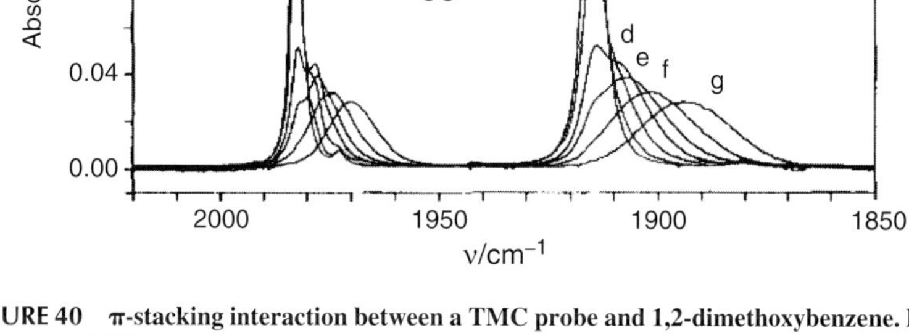

FIGURE 40 **π-stacking interaction between a TMC probe and 1,2-dimethoxybenzene. FTIR spectra of TMC probe in cyclohexane (a), an in 39 mM (b), 79 mM (c), 392 mM (d), 785 mM (e), 1.57 M (f), and 3.92 M (g) solutions of 1,2-dimethoxybenzene in cyclohexane (figure taken from [91] with permission).**

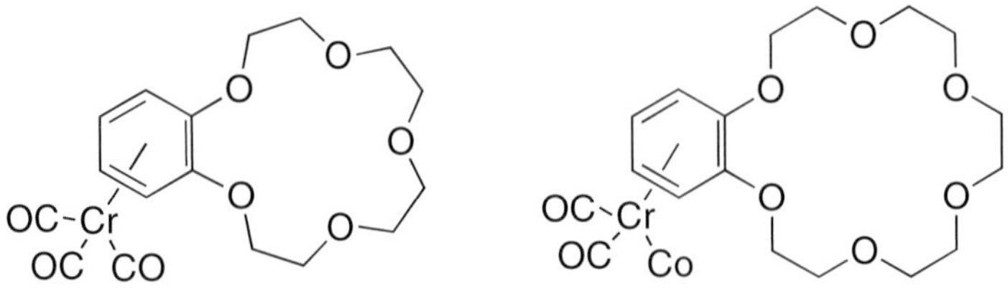

FIGURE 41 TMC molecular sensors for alkali metal cations.

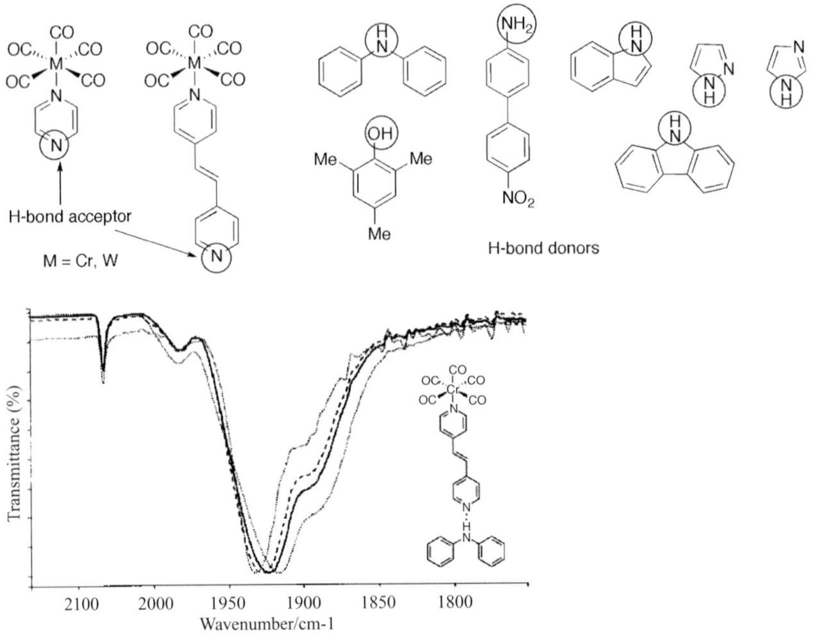

FIGURE 42 TMC molecular probes for H-bond and H-bond donors. FTIR spectra of the 1,2-dipyridyl-4-ethene tungsten complex with different amounts of diphenylamine (figure taken from [94] with permission).

probes resulted in a positive shift of the most intense E vibration mode of the $M(CO)_5$ entity up to $20\,cm^{-1}$ (Fig. 42). Binding constants were deduced from $\Delta\nu$ versus H-donor concentration plots and ranged between 0.08 and $1.1\,mol^{-1}$ [94].

More recently, TMC probes (Fig. 43) were designed for selective molecular recognition of acetate anion in ethanol. Addition of AcO^- to a solution of the probes induced a negative shift of the 2 $\nu(CO)$ vibration modes of the $Cr(CO)_3$ unit of up to $10\,cm^{-1}$. The association constant derived from the binding isotherm was found equal to $6.8 \times 10^4\,M^{-1}$ [95].

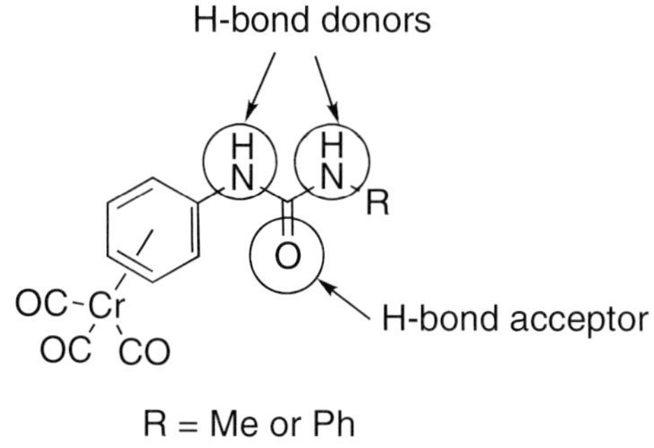

FIGURE 43 TMC probes for acetate anion sensing.

In conclusion, all the individual forms of interaction typically encountered in high affinity biomolecular associations have successfully been probed by suitable TMC complexes.

At the solid–liquid interface (ATR)

For evident convenience reasons and to allow their reusability, molecular sensors should be best attached to a solid support. Therefore, Creaser and Stephenson have reported the chemisorption of (η^4-cyclohexadiene)Fe(CO)$_3$ species to silica gel and fumed silica particles via organosilane chemistry (Fig. 44).

These immobilized sensors were successfully used to probe the polarity of solvents environment by ATR-IR giving similar responses as the probe in solution [96]. Two immobilized TMC probes were combined to assay dodecane in cyclohexanol in the range between 0% and 5% [97]. A very convenient setup called optical optrode (Fig. 45) was proposed for pH and solvent-environment remote sensing based on ν(CO) shifts of the immobilized probe [98]. The optical optrode is based on a ATR accessory including a ZnSe reflectance element coupled to a bifurcated optical fiber bundle made of chalcogenide. The device is interfaced to the FT-IR spectrometer through a launch system. The ZnSe ATR element is coated with the TMC probe immobilized on 7 nm fumed silica

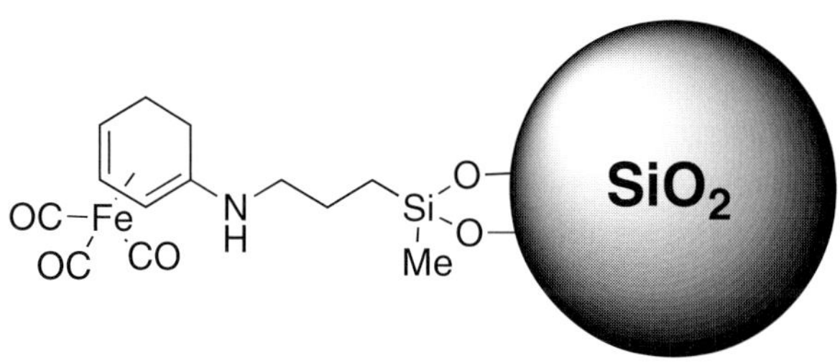

FIGURE 44 TMC probe attached to silica.

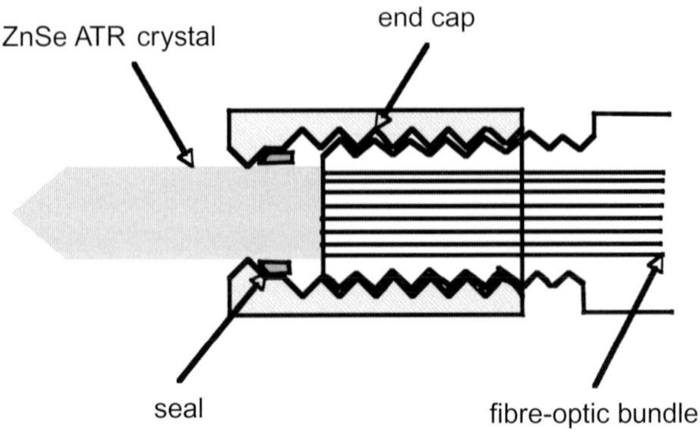

FIGURE 45 Diagram of the ATR optical optrode.

particles by dip coating. As observed by transmission measurements in solution, the probe was able to sense the solvent environment with a similar efficiency and also the pH between 10 and 12 thanks to the amine function of the silyl linker.

Non-TMC probes for molecular sensing and biomolecular interaction studies

Within the last 10 years, two other molecular groups, namely the cyano/nitrile group (CN) and the azido group (N_3) were proposed as vibrational probes for molecular biosensing and biorecognition event studies (see Ref. [99] for a recent review on the cyano probes). They share several similarities with the TMC entities as they give rise to intense vibration bands in a spectral region relatively free of any other absorption (2200–2260 cm^{-1} for ν(CN); 2080–2135 cm^{-1} for ν_{as}(N=N=N) and 2140–2175 cm^{-1} for ν(CN) in SCN). Moreover, these vibration modes have been shown to be sensitive to the environment, which can provide useful structural information at the local scale, that is, at close proximity of the vibrational probe. But, conversely to the TMC groups, they are both of small size and should not be prone to alter the global nor the local structure of the biomolecule they are attached to, particularly in the case of proteins and DNA.

Cyano probes

Cyano groups were historically the first vibrational probes used to sense local environments. Indeed, cyanide anion's ability to coordinate to Fe^{3+} in hemes of several hemoproteins was evidenced by IR spectroscopy that also provided useful information on how the cyanide was actually coordinated to the metal and characteristics of the ligand binding site [100]. Conversely to CO (another strong ligand of hemes) that can only be used to probe the environment around the metals they are coordinated to, the cyano/nitrile group appeared as a much more versatile label as it can be attached to carbon and sulfur atoms and furthermore, many compounds with biological activity contain a CN substituent. Indeed a two-step strategy was developed to chemically introduce a nitrile group to proteins' accessible cysteine residues based on thiol–disulfide exchange with 5,5′-dithio-bis(2-nitrobenzoic acid) (DTNB) followed by reaction with potassium cyanide (Fig. 46) [101].

Alternatively, unnatural aminoacids derived from Phe and Trp and including a nitrile group (Fig. 47) have been included in polypeptide sequences in a site-specific manner by peptidic synthesis [102–105] or genetic incorporation (amber codon suppression technology) [106] as a more general approach. Using simple IR spectral measurements, it has been shown that the vibration frequency of the CN group in Phe_{CN} (and to a lesser extent in Trp_{CN}) was sensitive to the polarity of solvent environment with a positive shift of 10 cm^{-1} observed when THF was replaced by H_2O [102]. Consequently, the interaction of several polypeptides including a cyanylated aminoacid with proteins [102] and model membranes [103,105,107] was successfully probed by IR spectroscopy of the ν(CN) band.

The sensitivity of the ν(CN) vibration mode was also exploited to probe conformational changes as a function of temperature (from α-helix to random coil) in peptide models including a cyanylated cysteine residue [108]. Figure 48 depicts the IR spectrum in the CN stretching region of one of the model helical peptides, illustrating the effect of temperature on the position and shape of the ν(CN) band. Both a shift to the red and band narrowing were noticed, indicating that changes in peptide conformation could be probed by IR spectroscopy of the nitrile group chemically attached to cysteine.

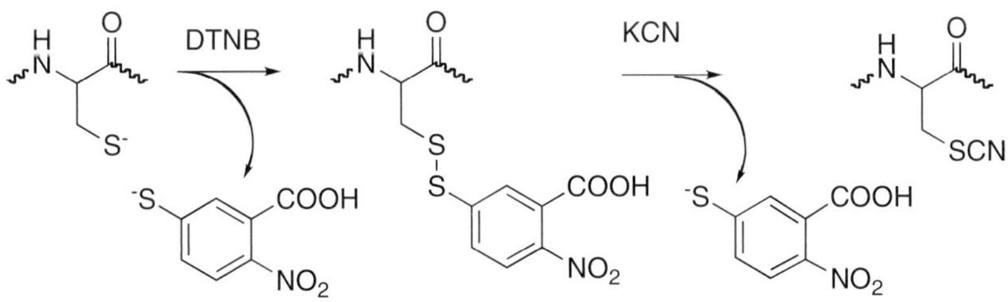

FIGURE 46 Cysteine cyanylation.

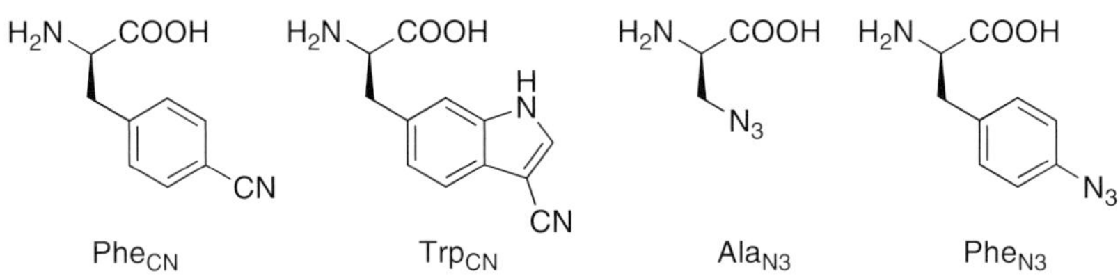

FIGURE 47 Unnatural amino acids carrying a nitrile or an azido group.

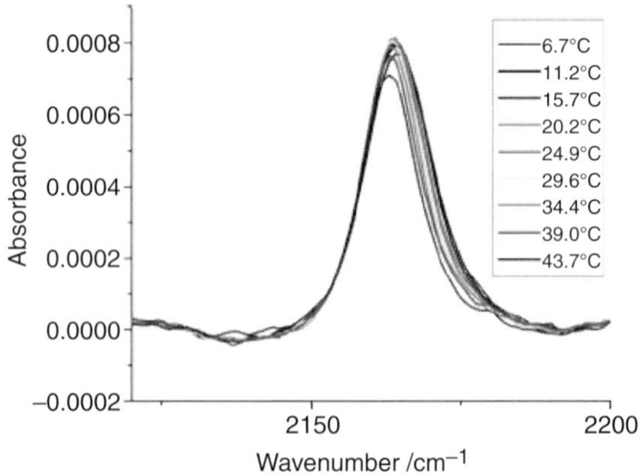

FIGURE 48 Temperature dependence of IR absorption spectra in the CN stretching region for Ac-AC*AAK-(AAAAK)$_4$-NH$_2$; C* denotes cyanylated cysteine. Figure taken from Ref. [108] with permission.

The change of polarity at the heme cofactor of myoglobin upon coordination of water, NO, CO, and O_2 to the central iron atom was probed spectroscopically by replacing the distal His64 residue (Fig. 49) by Phe_{CN} [106].

ATR-IR using polarized light source was used to study the reversible interaction between the inhalational anesthetic halothane and an anesthetic-binding model ion-channel protein including Phe_{CN} in its sequence deposited as a Langmuir–Blodgett film at the surface of a Ge ATR crystal. Formation of the halothane-protein complex led to a blue shit of the ν(CN) band by 6 cm^{-1} [104]. Polarized ATR-IR was employed to shed light on the binding of the well-studied membrane-binding peptide mastoparan x to a phospholipid bilayer. Seven analogues of this peptide including Phe_{CN} at various positions along the sequence were synthesized. Information on the hydration state of specific sites of

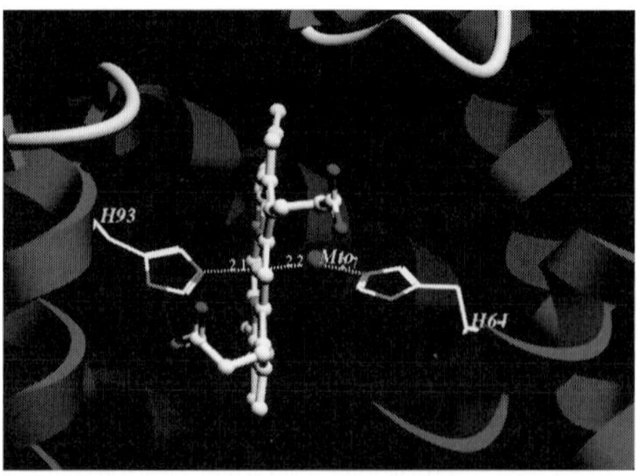

FIGURE 49 Heme pocket in myoglobin depicted the proximal His93 residue, the distal His64 residue and a water molecule coordinated to the iron atom (PDB file 1VXH).

mastoparan x upon binding to the membrane model was deduced from the position of the ν(CN) band. Spatial orientation of individual side chains was uncovered from polarization measurements [103]. Boxer *et al.* made use of the vibrational Stark effect on CN molecular groups to study electric fields in proteins. IDD743 (Fig. 50), an inhibitor of human aldose reductase (hALR2) containing a nitrile substituent was used as a local reporter to probe the electrostatic environment within the active site of hALR2. Stark shifts were used to measure changes in electric fields caused by mutations in the enzyme's active site [109].

2D IR spectroscopy is a relatively new technique that provides information about structural dynamics as a result of the coupling between a vibrational mode and its surroundings [110]. A very interesting work in this field dealt with the study of the interaction between HIV-1 reverse transcriptase and the nonnucleoside inhibitor TMC278 that contains two cyano groups (Fig. 50). 2D IR experiments not only shed light on the dynamics of the interaction between the enzyme and its inhibitor but also indicated that both CN groups are located in different protein environments [111].

Krummel and Zanni showed that IR spectroscopy could be used to probe the distance and the angle between coupled nitrile groups carried by deoxyuridine (5-CN dU, Fig. 50) and brought to close distance by DNA hybridization [112].

Azido probes

Comparatively less work has been published so far on the use of azido groups as IR probes of environment, although the spectral features of the ν_{as}(N=N=N) vibration mode are even more meaningful than those of the ν(CN) mode. Indeed the dipole strength of the ν_{as}(N=N=N) mode is found to be up to 19 times larger than that of the ν(CN) mode. The ν_{as}(N=N=N) vibration is also sensitive to solvent polarity as it is blue shifted by 14 cm^{-1} upon replacing water by DMSO (Fig. 51) [113].

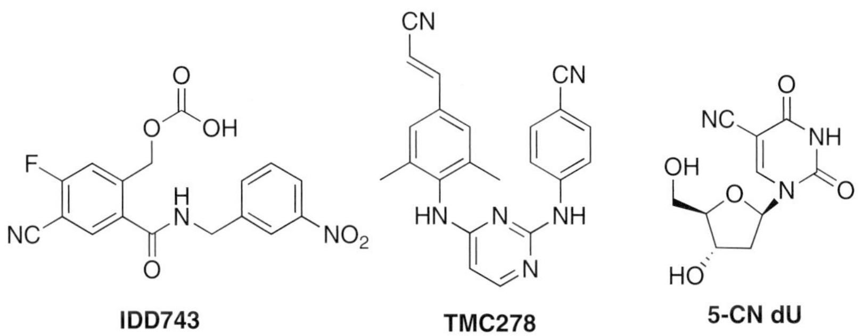

FIGURE 50 Some nitrile probes.

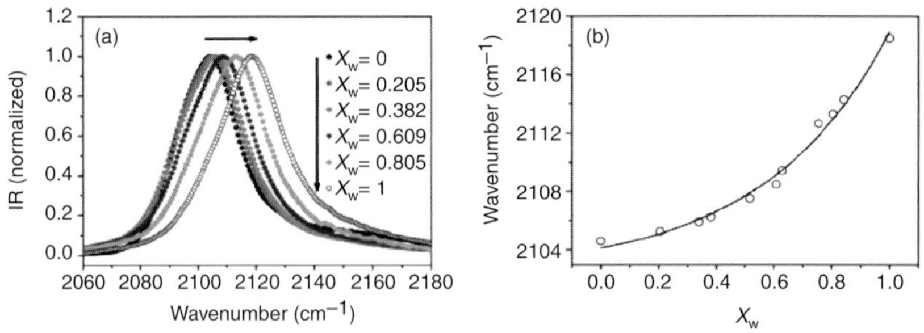

FIGURE 51 (a) Normalized IR spectra of Ac-Ala$_{N3}$-NHMe in DMSO/H_2O mixtures for varying water mole fraction X_w. (b) Plot of the $\nu(N_3)$ wavenumber versus X_w; the solid curve is a fitted exponential function. Figures taken from Ref. [113] with permission.

As an example, β-azidoalanine Ala$_{N3}$ (Fig. 47) was incorporated by peptide synthesis in the Aβ(16-22) peptide of the Alzheimer's disease amyloid β-protein at position Ala21. The peptide aggregate displayed an azido band at 2104 cm^{-1} in water, that is, at the same position as the peptide monomer in DMSO. This indicated that the azido group was in a less polar environment in its aggregated form [113].

Using the amber codon suppression technology, Phe$_{N3}$ was incorporated at different target sites into rhodopsin, a G-protein coupled receptor involved in the first events of light perception and three of the mutant proteins were analyzed by IR spectroscopy. Upon light irradiation, this receptor underwent a change of the position of the azido group in two mutants out of three, as observed by the shift of the ν(N_3) band in FT-IR difference spectroscopy. Simultaneously, the position of the ν(N_3) of the free Phe$_{N3}$ was seen to undergo a downshift by 13 cm^{-1} by switching form water to isopropanol. Thus, it was concluded that the activation of rhodopsin by light irradiation led to significant structural changes. Residue 250 (valine in the native structure) went from a very polar to a less polar environment, while residue 227 (valine in the native structure) translocated to a more polar environment. No change of the environment around residue 102 (tryptophane in the native sequence) was noticed as this residue is located in the extracellular domain of rhodopsin [114].

Conclusion

A large variety of high affinity association processes between two biological partners can be efficiently transduced, thanks to IR spectroscopic probes either present or intentionally introduced on one of the two partners. In some cases, the IR probes operate as molecular sensors, that is, their signal is affected by the association. While most of the published work involved transition metal carbonyl probes, the use of other chemical species with close spectral features (nitrile and azide groups) has been recently reported that shed light on various important biological processes.

REFERENCES

[1] Mirabella F. MJ, Harrick NJ. *Appl. Spectrosc. Rev.* 1985;21:45–178.

[2] Lefèvre G. *Adv. Colloid Interface Sci.* 2004;107:109–124.

[3] Griffiths PR, de Haseth J. *Fourier Transform Infrared Spectroscopy.* New York: Wiley-Interscience; 1986.

[4] Mirabella FM. In: F. M. Mirabella (Ed.), *Internal Reflection Spectroscopy.* New York: Marcel Dekker; 1993 pp. 17–52.

[5] Schmitt J, Nivens D, White DC, Flemming H-C. *Wat. Sci. Tech.* 1995;32:149–155.

[6] Sperline RP, Jeon JS, Raghavan S. *Appl. Spectrosc.* 1995;49:1178–1182.

[7] Devouge S, Conti J, Goldsztein A, Gosselin E, Brans A, Voué M, De Coninck J, Homblé F, Goormaghtigh E, Marchand-Brynaert J. *J. Colloid Interface Sci.* 2009;332:408–415.

[8] Goldsztein A, Aamouche A, Homblé F, Voué M, Conti J, De Coninck J, Devouge S, Marchand-Brynaert J, Goormaghtigh E. *Biosens. Bioelectron.* 2009;24:1831–1836.

[9] Onodera K, Hirano-Iwata A, Miyamoto K, Kimura Y, Kataoka M, Shinohara Y, Niwano M. *Langmuir* 2007;23:12287–12292.

[10] Hirano A, Onodera K, Miyamoto K, Kimura Y, Kataoka M, Shinohara Y, Niwano M. *Sens. Lett.* 2009;6:613–617.

[11] Susi H, Byler M. *Methods Enzymol.* 1986;130:290–311.

[12] Buijs J, Norde W, Lichtenbelt J. WT. *Langmuir* 1996;12:1605–1613.

[13] Reiter G, Hassler N, Weber V, Falkenhagen D, Fingeli UP. *Biochim. Biophys. Acta* 2004;1699:253–261.

[14] Reiter G, Siam D, Falkenhagen D, Gollneritsch W, Baurecht D. *Langmuir.*, 2002;18:5761–5771.

[15] Gosselin E, Gorez M, Voué M, Denis O, Conti J, Popovic N, Van Cauwenberge A, Noel E, De Coninck J. *Biosens. Bioelectron.* 2009;24:2554–2558.

[16] Hirano-Iwata A, Yamaguchi R, Miyamoto K, Kimura Y, Niwano M. *J. Appl. Physics* 2009;105:102039.

[17] Liao W, Wei F, Liu D, Qian MX, Yuan G, Zhao XS. *Sens. Actuators B* 2006;114:445–450.

[18] Liley M, Keller TA, Duschl C, Vogel H. *Langmuir* 1997;13.4190–4192.

[19] Johnson BW, Bauhofer J, Doblhofer K, Pettinger B. *Electrochim. Acta* 1992;37:2321–2329.

[20] Osawa M. *Bull. Chem. Soc. Jpn.* 1997;70:2861–2880.

[21] Nishikawa Y, Nagasawa T, Fujiwara K, Osawa M. *Vib. Spectrosc.* 1993;6:43–53.

[22] Haynes C, McFarland A, Smith M, Hulteen J, Van Duyne R. *J. Phys. Chem. B* 2002;106:1898–1902.

[23] Ataka K, Heberle J. *J. Am. Chem. Soc.* 2004;126:9445–9557.

[24] Seelenbinder J, Brown C, Pivarnik P, Rand A. *Anal. Chem.* 1999;71:1963–1966.

[25] Aroca R, Bujalski R. *Vib. Spectrosc.* 1999;19:11–21.

[26] Holman H. -YN, Perry DL, Hunter-Cevera JC. *J. Microbiol. Methods* 1998;34:59–71.

[27] Busalmen JP, Bernà A, Feliu JM. *Langmuir* 2007;23:6459–6466.

[28] Busalmen JP, Esteve-Nuñez A, Berná A, Feliu JM. *Bioelectrochem.* 2010;78:25–29.

[29] Nuzzo RG, Dubois LH, Allara DL. *J. Am. Chem. Soc.* 1990;112:558–569.

[30] Chaki NK, Vijayamohanan K. *Biosens. Bioelectron.* 2002;17:1–12.

[31] Boujday S, Bantegnie A, Briand E, Marnet P-G, Salmain M, Pradier C-M. *J. Phys. Chem. B* 2008;12:6708–6715.

[32] Briand E, Gu C, Salmain M, Herry J-M, Pradier C-M. *Surf. Sci.* 2007;601:3850–3855.

[33] Pradier C-M, Salmain M, Zheng L, Jaouen G. *Surf. Sci.* 2002;502-503:193–202.

[34] Arafat A, Daous MA. *Sens. Actuators B.*, 2011. doi: 10.1016/jsnb.2010.12.013.

[35] Porter MD, Bright TB, Allara DL, Chidsey CED. *J. Am. Chem. Soc.* 1987;109:3559.

[36] Sayed MA, Uno H, Harada K, Tanaka K, Kim Y-H, Nakaoki Y, Okumura K, Tero R, Urisu T. *Chem. Phys. Lett.* 2008;466:235–239.

[37] Enders D, Rupp S, Küller A, Pucci A. *Surf. Sci.* 2006;600:L305–L308.

[38] Briand E, Salmain M, Compère C, Pradier C-M. *Colloids Surf. B: Biointerfaces* 2006;53:215–224.

[39] Briand E, Salmain M, Compere C, Pradier C-M. *Biosens. Bioelectron.* 2007;22:2884–2890.

[40] Tielens F, Humblot V, Pradier CM, Calatayud M, Illas F. *Langmuir* 2009;25:9980–9985.

[41] Thebault, P.; Boujday, S.; Senechal, H.; Pradier, C. M. *J. Phys. Chem. B, 114*, 10612–10619.

[42] Boujday S, Gu C, Girardot M, Salmain M, Pradier CM. *Talanta* 2009;78:165–170.

[43] Boujday S, Nasri S, Salmain M, Pradier C-M. *Biosens. Bioelectron.* 2010;26:1750–1754.

[44] Brewer SH, Anthireya SJ, Lappi SE, Drapcho DL, Franzen S. *Langmuir* 2002;18:4460–4464.

[45] Moses S, Brewer SH, Lowe LB, Lappi SE, Gilvey L. BG, Sauthier M, Tenent RC, Feldheim DL, Franzen S. *Langmuir* 2004;20:11134–11140.

[46] Tokareva I, Hutter E. *J. Am. Chem. Soc.* 2004;126:15784–15789.

[47] Mateo-Marti E, Pradier CM. In: V. Somerset (Ed.), *Intelligent and Biosensors.* Croatia: Intech; 2010 pp. 323–344.

[48] Mateo-Marti E, Briones C, Roman E, Briand E, Pradier CM, Martin-Gago JA. *Langmuir* 2005;21:9510–9517.

[49] Mateo-Marti E, Briones C, Pradier CM, Martin-Gago JA. *Biosens. Bioelectron.* 2007;22:1926–1932.

[50] Elschenbroich C, Salzer A. *Organometallics. A Concise Introduction, 2nd edn. Weinheim: VCH; 1992.*

[51] Kettle S. FA, Paul I. *Adv. Organomet. Chem.* 1972;10:199–235.

[52] Hähnel R, Twaddle E. *Cancer Res.* 1973;33:559–566.

[53] Habel LA, Stanford JL. *Epidemiologic Rev.* 1993;15:209–219.

[54] Jaouen G, Vessières A, Top S, Ismail AA, Butler IS. *J. Am. Chem. Soc.* 1985;107:4778–4780.

[55] Vessières A, Jaouen G. *Pure Appl. Chem.* 1985;57:1865–1874.

[56] Vessieres A, Top S, Ismail AA, Butler IS, Louer M, Jaouen G. *Biochemistry* 1988;27:6659–6666.

[57] Butler IS, Vessières A, Jaouen G. *Comments Inorg. Chem.* 1989;8:269–286.

[58] Jaouen G, Vessières A, Top S, Savignac M, Ismail AA, Butler IS. *Organometallics* 1987;6:1985–1987.

[59] Vessières A, Tondu S, Jaouen G, Top S, Ismail AA, Teutsch G, Moguilewski M. *Inorg. Chem.* 1987;27:1850–1852.

[60] Jaouen G, Vessières A. *Pure Appl. Chem.* 1989;61:565–572.

[61] Anson CE, Creaser CS, Downie JA, Egyed O, Malkov AV, Mojovic L, Stephenson GR, Turner AT, Wilson KE. *Bioorg. Med. Chem. Lett.* 1998;8:3549–3554.

[62] Vessières A, Salmain M, Brossier P, Jaouen G. *J. Pharm. Biomed. Anal.* 1999;21:625–633.

[63] Yalow RS, Berson SA. *J. Clin. Invest.* 1960;39:1157–1175.

[64] Vessières A, Jaouen G, Salmain M, Butler IS. *Appl. Spectrosc.* 1990;44:1092–1094.

[65] Salmain M, Vessieres A, Jaouen G, Butler IS. *Anal. Chem.* 1991;63:2323–2329.

[66] Salmain M, Vessières A, Brossier P, Butler IS, Jaouen G. *J. Immunol. Methods* 1992;148:65–75.

[67] Varenne A, Vessieres A, Salmain M, Brossier P, Jaouen G. *J. Immunol. Methods* 1995;186:195–204.

[68] Varenne A, Vessieres A, Brossier P, Jaouen G. *Res. Commun. Chem. Pathol. Pharmacol* 1994;84:81–92.

[69] Vessieres A, Kowalski K, Zakrzewski J, Stepien A, Grabowski M, Jaouen G. *Bioconjugate Chem.* 1999;10:379–385.
[70] Philomin V, Vessières A, Jaouen G. *J. Immunol. Methods* 1994;171:201–210.
[71] Vessieres A, Fischer-Durand N, Le Bideau F, Janvier P, Heldt JM, Ben Rejeb S, Jaouen G. *Appl. Organomet. Chem.* 2002;16:669–674.
[72] Fischer-Durand N, Vessieres A, Heldt JM, le Bideau F, Jaouen G. *J. Organomet. Chem.* 2003;668:59–66.
[73] Salmain M, Vessières A, Brossier P, Jaouen G. *Anal. Biochem.* 1993;208:117–120.
[74] Salmain M, Varenne A, Vessieres A, Jaouen G. *Appl. Spectrosc.* 1998;52:1383–1390.
[75] Salmain M, Vessières A, Varenne A, Brossier P, Jaouen G. *J. Organomet. Chem.* 1999;589:92–97.
[76] Varenne A, Vessières A, Salmain M, Brossier P, Jaouen G. *Immunoanal. Biol. Spéc.* 1994;9:205–214.
[77] Varenne A, Vessieres A, Salmain M, Durand S, Brossier P, Jaouen G. *Anal. Biochem.* 1996;242:172–179.
[78] Kong KV, Chew W, Lim L. HK, Fan WY, Leong WK. *Bioconjugate Chem.* 2007;18:1370–1374.
[79] Meister K, Niesel J, Schatzschneider U, Metzler-Nolte N, Schmidt DA, Havenith M. *Angew. Chem. Intern. Ed.* 2010;49:3310–3312.
[80] Varenne A, Salmain M, Brisson C, Jaouen G. *Bioconjugate Chem.* 1992;3:471–476.
[81] Fischer-Durand N, Salmain M, Rudolf B, Vessieres A, Zakrzewski J, Jaouen G. *Chembiochem* 2004;5:519–525.
[82] Ulman A. *Chem. Rev.* 1996;96:1533–1554.
[83] Chavigny C, Le Bideau F, Pradier CM, Palacin S, Brossier P, Jaouen G. *Chem. Commun.*, 1998;1727–1728.
[84] Yam C-M, Pradier C-M, Salmain M, Fischer-Durand N, Jaouen G. *J. Colloid Interface Sci.* 2002;245:204–207.
[85] Pradier C-M, Salmain M, Zheng L, Méthivier C. *Surf. Interface Anal.* 2002;34:67–71.
[86] Salmain M, Fischer-Durand N, Pradier CM. *Anal. Biochem.* 2008;373:61–70.
[87] Ohtsuka K, Komizo K, Takenaka S. *J. Organomet. Chem.* 2010;695:1281–1286.
[88] Rudolf B, Zakrzewski J, Celichowski G, Salmain M, Vessieres A, Jaouen G. *Appl. Organomet. Chem.* 2004;18:105–110.
[89] Stephenson GR. In: G. Jaouen (Ed.), *Bioorganometallics.* Weinheim: Wiley-VCH; 2006pp. 215–262.
[90] Creaser CS, Fey MA, Stephenson GR. *Spectrochim. Acta A* 1994;50:1295–1299.
[91] Anson CE, Creaser CS, Stephenson GR. *Spectrochim. Acta A* 1996;52:1183–1191.
[92] Anson CE, Baldwin TJ, Creaser CS, Fey MA, Stephenson GR. *Organometallics* 1996;15:1451–1456.
[93] Anson CE, Creaser CS, Stephenson GR. *J. Chem. Soc., Chem. Comm* 1994;2175–2176.
[94] Peris E, Mata JA, Moliner V. *J. Chem. Soc., Dalton Trans* 1999;3893–3898.
[95] Tian XH, Song L, Yuan YZ, Lin JP. *Organometallics* 2010;29:509–511.
[96] Creaser CS, Hutchinson WE, Stephenson RG. *Appl. Spectrosc.* 2000;54:1624–1628.
[97] Creaser CS, Hutchinson WE, Stephenson GR. *Analyst* 2001;126:647–651.
[98] Creaser CS, Hutchinson WE, Stephenson GR. *Sens. Actuators B: Chem.* 2002;82:150–157.
[99] Lindquist BA, Furse KE, Corcelli SA. *Phys. Chem. Chem. Phys.* 2009;11:8119–8132.
[100] Yoshikawa S, O'Keefe DH, Caughey WS. *J. Biol. Chem.* 1985;280:3518–3528.
[101] Fafarman AT, Webb LJ, Chuang JI, Boxer SG. *J. Am. Chem. Soc.* 2006;128:13356–13357.

[102] Getahun Z, Huang C-Y, Wang T, De Len B, DeGrado WF, Gai F. *J. Am. Chem. Soc.* 2002;125:405–411.
[103] Tucker MJ, Getahun Z, Nanda V, DeGrado WF, Gai F. *J. Am. Chem. Soc.* 2004;126:5078–5079.
[104] Liu J, Strzalka J, Tronin A, Johansson JS, Blasie JK. *Biophys. J.* 2009;96:4176–4187.
[105] Waegele MM, Tucker MJ, Gai F. *Chem. Phys. Lett.* 2009;478:249–253.
[106] Schultz KC, Supekova L, Ryu Y, Xie J, Perera R, Schultz PG. *J. Am. Chem. Soc.* 2006;128:13984–13985.
[107] McMahon HA, Alfieri KN, Clark K. AA, Londergan CH. *J. Phys. Chem. Lett.* 2010;1:850–855.
[108] Edelstein L, Stetz MA, McMahon HA, Londergan CH. *J. Phys. Chem. B* 2010;114:4931–4936.
[109] Suydam IT, Snow CD, Pande VS, Boxer SG. *Science* 2006;313:200–204.
[110] Kim YS, Hochstrasser RM. *J. Phys. Chem. B* 2009;113:8231–8251.
[111] Fang C, Bauman JD, Das K, Remorino A, Arnold E, Hochstrasser RM. *Proc. Natl. Acad. Sci. USA* 2008;105:1472–1477.
[112] Krummel AT, Zanni MT. *J. Phys. Chem. B* 2008;112:1336–1338.
[113] Oh K-I, Lee J-H, Joo C, Han H, Cho M. *J. Phys. Chem. B* 2008;112:10352–10357.
[114] Ye S, Huber T, Vogel R, Sakmar TP. *Nat. Chem. Biol.* 2009;5:397–399.

Chapter 8

Advanced infrared glasses for biochemical sensing

Pierre Lucas[1] and Bruno Bureau[2]
[1]*Department of Materials Science and Engineering, University of Arizona, 4715 E. Fort Lowell Road, Tucson, AZ 85712, USA*
[2]*UMR CNRS 6226 Sciences Chimiques, Groupe Verres et Céramiques, Université de Rennes I, Campus de Beaulieu, 35042 Rennes cedex, France*

INTRODUCTION

Overview

One of the main advantage or IR spectroscopy is the noninvasive and nondestructive nature of its probing mechanism. This is particularly relevant for the characterization of live microorganisms which can then be continuously monitored with minimal disturbance to their metabolism. The spectral range of IR spectroscopy allows one to probe the vibrational fingerprint of biomolecules and

Biointerface Characterization by Advanced IR Spectroscopy. DOI: 10.1016/B978-0-444-53558-0.00008-4

to collect spectral information of high specificity with no observable damage to the target molecule. This technique has therefore been applied to the characterization and identification of many microorganisms including human and microbial cells [1–4]. However, with a conventional infrared spectrometer, the measurement is typically performed in transmission and the sample has to be extracted and introduced inside the spectrometer between the light source and detector (Fig. 1(a)). This might not be a possible or desirable approach for the characterization of many samples such as live skin tissues or cell cultures grown on substrates that are not transparent to infrared light. In these cases, remote FTIR methods are required in order to bring the light signal to the sample instead of bringing the sample into the analytical instrument. One of these methods is fiber evanescent wave spectroscopy (FEWS) which makes use of infrared transparent fibers probes to guide the light source to the sample and collect the resulting signal onto a detector (Fig. 1(b)). This technique has recently gained much interest, thanks to the development of high-quality infrared transparent fibers [5–7] and has found applications in a wide range of fields, including biomedical sensing [8], cancer diagnosis [9,10], *in vivo* monitoring of cells

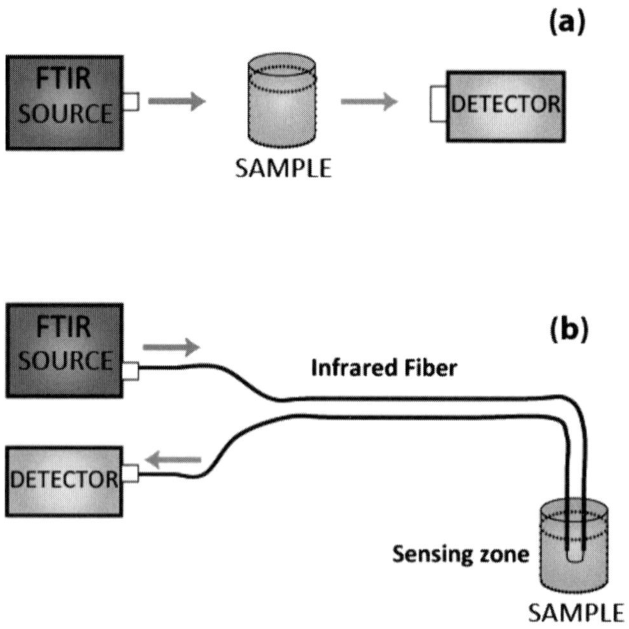

FIGURE 1 (a) Conventional FTIR set up allowing measurements of vibrational signals in transmission. (b) Fiber evanescent wave spectroscopy setup allowing remote IR sensing with the use of an infrared transparent fiber.

and bacteria [11,12] as well as environmental and industrial chemical monitoring [13–15]. Mainly two types of materials are currently used for the design of high-quality IR optical fibers, chalcogenides such as TeAsSe glass [16] and silver halides such as AgClBr polycrystals [17]. Both materials have advantages and drawbacks. Chalcogenides are easy to shape into complex optical components and have excellent chemical resistance to biological materials [12,18,19]; however, they have fairly poor mechanical resistance. On the other end, silver halides posses outstanding mechanical properties but suffer from poor chemical resistance to biological fluids [20,21]. This chapter intends to review the use of FEWS for the characterization of biological materials; hence, we will focus on the use of chalcogenide fibers and their application in biochemical sensing. An extensive review of the use and applications of infrared fiber sensors can be found in Refs. [5–7].

Development and historical background of FEWS

The development of optical grade infrared glass fibers dates back to the early 1980s. Both chalcogenide [22–24] and silver halide fibers [25] were produced for various optical applications such as radiometry. Optical attenuations below 1 dB/m could be obtained but these were limited to fairly small spectral ranges [22], therefore, limiting their use for remote spectroscopy.

The early development of FEWS occurred in the early 1990s and grew rapidly to a mature advanced technique throughout this decade. Early demonstrations of FEWS involved monitoring of chemical reactions [26] as well as quantitative analysis of chemical substances [20,27]. These were quickly followed by the development of surface functionalized fibers aimed at improving the selectivity of the sensors [28,29].

Meanwhile, a theoretical analysis of light propagation in FEWS was performed by Katzir [30,31] who modeled the effect of instrumental parameters such as fiber diameter, launching conditions, etc., on the absorption coefficient. It was shown that the evanescent wave intensity increases with the decreasing fiber diameter, therefore, providing greater detection sensitivity in fibers with a smaller diameter. This led to the development of fibers with tapered geometry aimed at improving the detection limit of FEWS sensors [32].

Following these developments and the design of optimized FEWS instrumentation, the technique was increasingly applied to biomedical applications such as blood monitoring [8,33] or cancer diagnostic [9,34]. Indeed, the possibility of remote detection using flexible fibers provided a convenient approach for medical analysis such as urology [35] or monitoring of skin tissues [36]. It was shown that significant spectral differences could be observed between malignant and healthy bladder and liver tissues therefore demonstrating the potential of this technique for *in vivo* monitoring using endoscopic probes.

However, while infrared spectral analysis provides outstanding detection selectivity due to the highly specific vibrational signatures of molecules, the detection sensitivity of FTIR is not optimal in comparison to other analytical techniques. This limitation lead to the development of FEWS coating aimed at enriching the evanescent wave region with specific analytes for improved detection resolution [37, 38]. This type of coating was shown to provide detection limit down to the ppb level and appeared to function reversibly to be regenerated upon removal of the analyte. This type of sensitized fibers therefore combined high selectivity with high sensitivity.

More recently, remote fiber sensors have found applications in environmental monitoring of both water supplies and atmospheric pollutants [14, 15]. It was shown that these fiber sensors could effectively monitor the presence of chlorinated hydrocarbons in water wells, *in situ* and in real time at concentrations down to 1 ppm. Additionally, similar fibers were used as gas sensors to monitor the concentration of CO_2 gas held in geological storage in an effort to reduce greenhouse effects.

Finally, chalcogenide fibers have recently been applied to the monitoring and study of live microorganisms such as human cell cultures and bacteria [11,12]. It was shown that these fibers have good chemical resistance toward biological materials and aqueous environments [18,19] and could be used to monitor the dynamics of biofilms or the metabolism of live human lung cells. Chalcogenide infrared fibers indeed constitute a very promising technique for the study of biological systems as will be shown in the following.

INSTRUMENT DESCRIPTION

FEWS principle

The detection principle associated with the present technique is based on attenuated total reflection. When a light wave propagates in a material of higher refractive index than the surrounding air (such as a glass fiber), the light is trapped within the material through total internal reflection. The light wave can then propagate through a fiber without energy loss other than that associated with intrinsic absorptions due to impurities (Fig. 2). However, during total internal reflection, a part of the electric field of the light is evanescent above the fiber surface and can couple with an absorbing medium in contact with the fiber (Fig. 2). If the light wavelength is resonant with a vibrational mode of the medium, energy transfer occurs and the light intensity decreases as its energy is dissipated. This absorption mechanism is identical to that of conventional ATR crystals, but occurs with greater frequency in a fiber.

When a broadband IR source is coupled into the fiber (Fig. 1) the resulting absorption spectra is essentially identical to that of a conventional transmission measurement except for a slight wavelength dependence of the peak

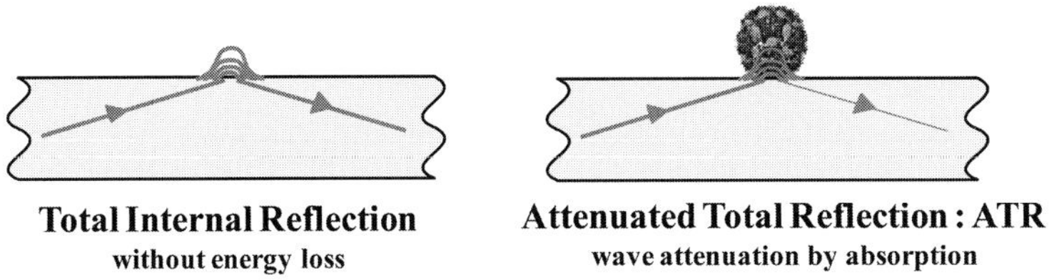

FIGURE 2 Principle of evanescent wave absorption of a lightwave confined in an optical fiber. The electric field evanescent upon total internal reflection can couple with an absorbing molecule such as a microorganism. The wave intensity decreases as its energy is dissipated in the absorbing medium.

intensity due to a lower penetration depth of the evanescent wave at longer wavelengths. However, the peak position is not affected and normal spectral analysis tools can be used to interpret these spectra. The experimental set up for FEWS makes use of a standard FTIR spectrometer where the broadband source from a glow-bar is coupled into one end of the fiber using gold-coated concave mirrors and the resulting light signal is collected on a mercury cadmium telluride (MCT) detector on the other end of the fiber. High sensitivity detectors such as MCT are required due to the low total light intensity propagating through the fiber. Fiber coupling attachments are now commercially available from several companies.

Fiber sensor design

Based on the absorption mechanism described above it is expected that an increasing number of total internal reflections will lead to greater absorption and higher detection sensitivity. Following simple geometrical optic considerations, it can be shown that a decrease in fiber diameter leads to an increase in the number of absorption events (Fig. 3).

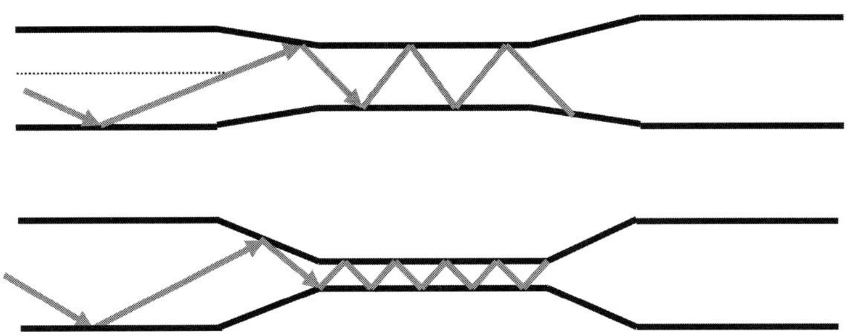

FIGURE 3 Correlation between the fiber diameter and detection sensitivity based on geometric optics. A smaller fiber diameter leads to an increasing number of absorption events along a given section of fiber.

The number of reflections over length L is given by equation (1),

$$N(\theta, d, L) = L * \frac{\tan(90 - \theta)}{d}. \tag{1}$$

where d is the fiber diameter and θ is the angle of incidence of a light ray on the fiber surface.

Assuming identical launching condition in the fiber, and identical taper reduction gradient, the sensitivity should have an inverse dependence on the diameter. The use of fiber taper for increasing detection sensitivity is therefore widely applied in FEWS technology. Figure 4(a) shows the image of a tapered Ge–Se chalcogenide fiber whose diameter was reduced from 300 to 100 μm. Tapers can be produced by simply reheating the glass above the glass transition and applying a tensile stress [39] but also through etching [32] or using photo-induced fluidity [40]. The production of tapers have several advantages including limiting the signal collection to a small section of the fiber called the sensing zone therefore enabling probing of very small spatial areas. However, the main advantage is the considerable increase in detection sensitivity as described in Figure 4.

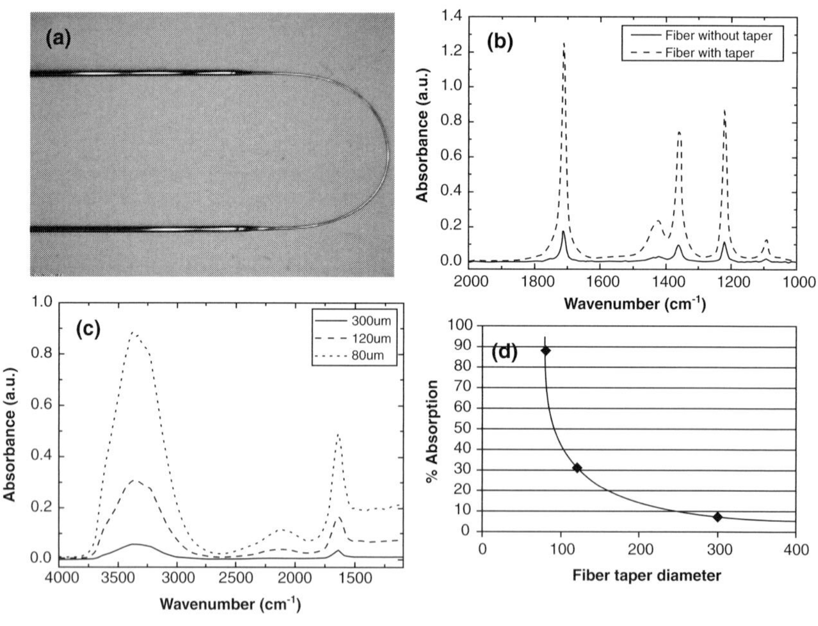

FIGURE 4 **(a) Microscope image of the tapered fiber-sensing zone. The original fiber diameter is 300 μm and the taper length is less than 2 cm. (b) Increase in sensitivity of the detection of acetone after reduction of the fiber diameter from 300 to 120 μm. (c) Increase in sensitivity of the detection of water for decreasing fiber diameters. (d) Rapid increase of the main water absorption peak at 3300 cm^{-1} with the decreasing fiber diameter.**

The detection sensitivity can be increased by several orders of magnitude through tapering and shows a rapid increase with decreasing fiber diameter as expected from equation (1) (see Fig. 4(d)). The fiber diameter reduction is only limited by the decreasing mechanical resistance of the fiber at diameters below 50 μm which can become problematic for handling the sensor.

It should also be noted that the tapering process has a significant effect on the extent of the penetration depth of the evanescent wave above the fiber surface. It can be shown that the electric field E of the propagating wave decreases exponentially above the fiber surface according to equation (2). The penetration depth d_p is then defined as the depth at which E drops to $1/e$ of its original value at the fiber surface:

$$E = E_0 \exp(-z\gamma). \tag{2}$$

Conventional principles of total internal reflection show that d_p is a function of both the refractive index of the glass n_2 and the index of the surrounding medium n_1 as well as the angle of incidence θ_i of light rays on the fiber surface according to equation (3) [41].

$$d_p = \frac{\lambda}{2\pi\sqrt{n_2^2 \sin^2\theta_i - n_1^2}}. \tag{3}$$

Considering the high index of chalcogenide glasses ($n_2 = 2.8$), the corresponding lower index of a surrounding aqueous solution ($n_1 = 1.3$) and the index of the air ($n = 1$), it is predicted from Snell's law that the critical angle for light propagation is $\approx 69°$ due to the light coupling limitations in the fiber; therefore, light rays with a range of angle $69° \leq \theta_i < 90°$ can propagate in the fiber. Figure 3 shows qualitatively that the incident angle θ_i decreases with the decreasing fiber diameter therefore increasing the penetration depth d_p and consequently the intensity of the detection signal. It is then clear that the fiber tapering has two beneficial effects on the detection sensitivity, one due to the increasing number of attenuated reflections and the second due to an increase in penetration depth of the evanescent wave.

Additionally, equation (3) indicates that the penetration depth is directly proportional to the wavelength of the incident light. It is therefore expected that the detection intensity should increase at longer wavelength, and this effect is shown to be responsible for the wavelength dependence of ATR and FEWS spectra in comparison to standard transmission spectra as mentioned in the FEWS principle section. The penetration depth d_p can then be estimated for the mid-IR range using equation (3) and is shown to be lower than 1 μm as displayed in Figure. 5. The signal collection can therefore be restricted to a small spatial region near the fiber surface and this characteristic offers attractive potentials for the study of biointerfaces such as cell membranes since the main signal can be limited to the region of interest without significant interference

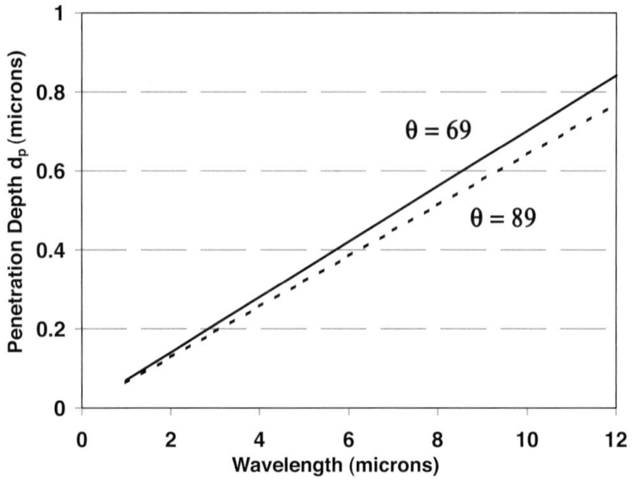

FIGURE 5 Penetration depth of an evanescent wave as a function of wavelength predicted for a chalcogenide glass fiber of index n = 2.8 immersed in an aqueous solution of index n = 1.3.

from other components of the cell [11]. Figure 5 also shows the range of penetration depth probed by light rays of different incident angles. Again, lower incident angles resulting from fiber tapering lead to higher absorption and improved detection signal.

IR materials for FEWS

There are three main requirements for a good infrared fiber material aimed at biosensing. (i) Good rheological properties that allows shaping of the glass into a fiber without risk of recrystallization, (ii) a wide transparency window in the infrared region that allows for collection of relevant spectral signatures, and (iii) a good chemical stability toward biological fluids and aqueous environments to maintain the integrity of the optical element during signal collection.

Rheological properties

Many crystalline infrared materials such as Ge, ZnSe, or CaF_2 are available for designing optical systems. However, one of the major benefits of a glass versus a crystal is the ability to shape complex optical elements by simply reheating the material above it softening point. This characteristic enables the production of molded lenses at relatively low cost in comparison to single crystals lenses and also allows the production of unique shapes such as optical fibers that would not

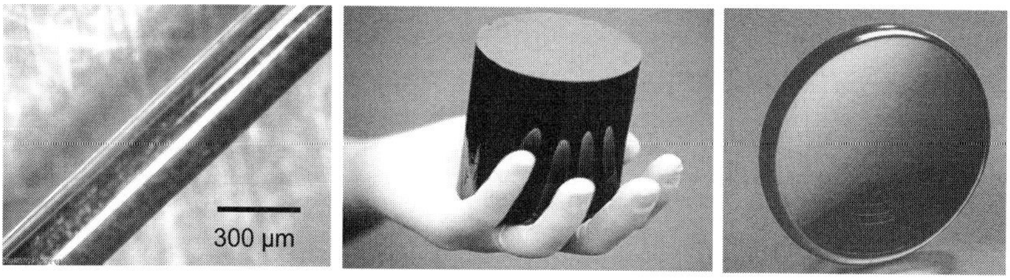

FIGURE 6 Examples of optical elements made of chalcogenide glass. The outstanding rheological properties permit to produce large optics or to shape the glass into fiber and molded lenses.

be available from a crystalline material. In that respect chalcogenide glasses offer a unique advantage due to their excellent resistance to crystallization which allows shaping of large optics such as lenses or complex optics such as fibers without risk of crystallite formation. Indeed, the presence of crystallites in the glass matrix is very detrimental to the optical properties as they act as scattering centers and would effectively ruin the transparency. Figure 6 shows several examples of optical elements produced from chalcogenide glass. Large blocks of glass can be safely prepared without crystallization, or fibers with diameter of a couple hundred μm can be drawn in a dedicated fibering tower. These materials are now becoming widely commercially available for applications ranging from night vision cameras to driving assistance in the automobile industry [42]. Chalcogenide fibers are now also becoming commercially available [43] due to their potential applications for infrared laser delivery such as CO_2 laser at 10.6 μm used for surgery or their use for defense application due to their transparency in the strategic atmospheric windows at 3–5 and 8–12 μm.

Optical window

While the vast majority of glassy materials are made out of silicate and have an outstanding chemical and mechanical resistance, oxide-based glasses are not transparent in the mid-infrared spectral region (Fig. 7). Fluoride glasses such as ZBLAN [44] have an optical window extending further out in the mid-IR up to ~7 μm but this spectral range does not cover the most important "signature region" of biological molecules between 6 and 12 μm. Hence, chalcogenide glasses emerge as the ideal materials choice for designing optical fibers in the long wavelength range.

The wide optical window results from the large mass of the components elements in comparison to oxide and fluorides. The multiphonon cutoff is a direct function of the mass of the elements and can be tuned to fit specific applications. Sulfide glasses cutoff around 10 μm and would be inappropriate for many bioapplications, while selenides extend to 16 μm and are suitable

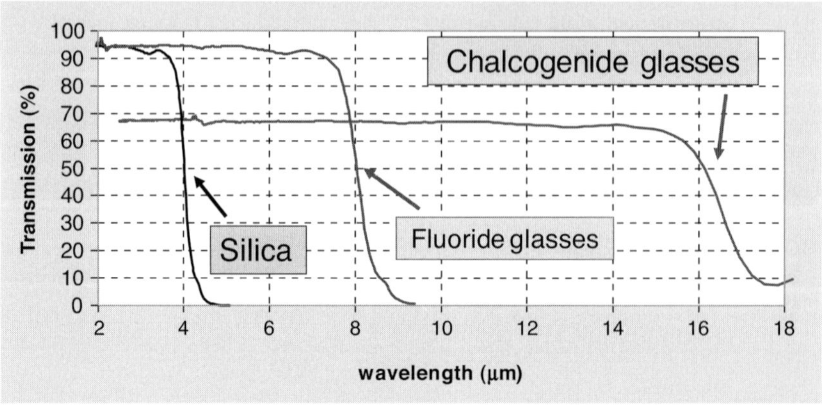

FIGURE 7 Transparency window of glasses based on oxide (silica), fluoride and chalcogenide elements. The heavier the component element, the wider the optical window in the infrared.

candidates [12]. Telluride glasses have recently been the subject of much research and are shown to extend down to more than 22 µm [45–47]. They have excellent chemical resistance and are therefore adequate for biosensing and have also found application in sensing long IR molecular vibrations such as CO_2 at 16 µm for space exploration [47]. It should also be noted that these glasses are amorphous semiconductors with bandgap ranging from 1 to 2 eV; hence, most chalcogenides are opaque to visible light as shown in Figure 6.

In order to maintain high transparency, it is essential that the glasses be synthesized in very high purity [48, 49]. In particular, all traces of low-mass impurities such as carbon, oxygen, and water must be carefully removed as they would generate absorption within the optical window which would alter the optical quality. This is a particularly significant problem in glass fibers where the optical path is very long and where even traces amount of impurities would generate unacceptable losses and render the fiber opaque in the spectral region of interest. Hence, extreme care must be taken during synthesis of these glasses to remove all contaminants through multiple purification steps. The full synthesis must also be performed under high vacuum or controlled atmosphere in order to avoid contact with oxygen. When these purification steps are followed, fiber losses as low as 0.05 dB/m can be reached [48, 49].

Chemical stability

Chalcogenide glasses are composed of divalent atoms such as S, Se, or Te that generate structures based on chains of atoms. In order to improve the thermal

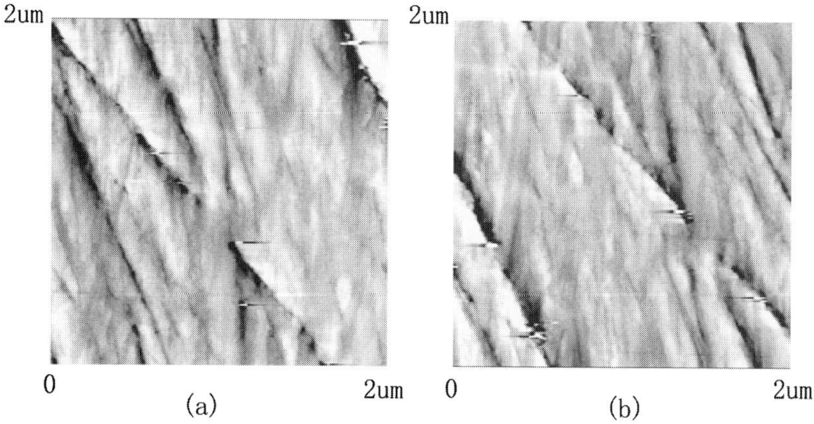

FIGURE 8 AFM topography of the surface for $Ge_{10}As_{15}Te_{75}$ glass before (a) and after (b) being soaked in tap water for 120 h.

and mechanical stability of the glass, these chains are typically cross-linked with higher valence atoms such as Ge, As, or Sb. The presence of elements such as As or Sb has raised some concern of toxicity for sensitive applications such as biomedical sensing. However, it was shown that these glasses are chemically very stable and can be safely used in aqueous environment or in contact with live microorganisms granted they have not been previously oxidized [12, 18, 19]. While a chalcogenide glass can slowly oxidized upon extended exposure to air, the oxidation is limited to a very thin surface layer which progress at a rate of about 20 Å per year [18]. In addition, this oxide layer is very soluble and can be readily removed by rinsing with water. Fresh glasses on the other end are fully insoluble and do not degrade or dissolve in the aqueous solution. Figure 8 shows an atomic force microscopy (AFM) image of the surface of a Ge–As–Te glass before and after immersion in water for 120 h. The polishing scratches on the glass surface remain unchanged after immersion therefore indicating that no substantial dissolution occurs. Similar tests performed on Te–As–Se glasses have shown the same behavior. The quantity of As that leached out of a Te–As–Se fiber immersed in water for 50 h was measured by inductively coupled plasma mass spectroscopy and shown to be negligible, below 0.05 μM/l. This concentration did not evolve measurably with time therefore confirming the glass stability [18]. These results were also confirmed by cyclic voltammetry, which showed that no measurable oxidation process occurred during immersion in the aqueous solution.

For extended storage, it is best to conserve chalcogenide fibers away from oxygen in order to prevent surface oxidation. An X-ray photoelectron study showed that the initial oxide layer develops within a few months; however, this layer has no visible effect on the optical quality of the fiber. Figure 9 shows the

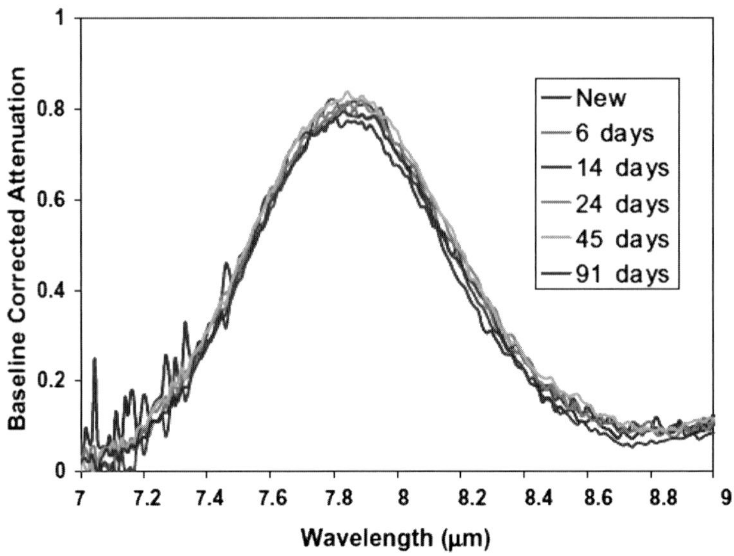

FIGURE 9 **Evolution of the absorption peak associated with the As–O vibration as a function of time during exposure to air for up to three months.**

evolution of the attenuation peak associated with the As–O vibrational mode in a Te–As–Se fiber. It is shown that the peak does not evolve significantly with time. This indicates that the absorption results mainly from intrinsic oxygen impurities in the glass and that the surface oxidation has no detrimental effect on the optical quality of the fiber.

Chaclogenide fibers are also stable in contact with biological material and are biocompatible with live microorganisms. Figure 10(a) shows a fluorescence micrograph of a layer of human lung cells coated on a Te–As–Se fiber and stained with two different dyes revealing dead and viable cells. A confluent monolayer of live cells is visible while few dead cells are present. Additionally, a viability tests performed with a colorimetric assay after 24 h of incubation shows that washed fibers have no toxic effect on the lung cells while the oxidized fiber and the soaking solution do (Fig. 10(b)). The oxide layer on the oxidized fiber surface dissolves readily and releases As, whereas the glass is stable against dissolution.

In conclusion, chalcogenide fibers are shown to have sufficient chemical stability for biochemical sensing although care must be taken to avoid the use of oxidized fibers in order to prevent the release of toxic elements. Regardless, several glass compositions devoid of toxic element are available to produce high quality optical fibers [49].

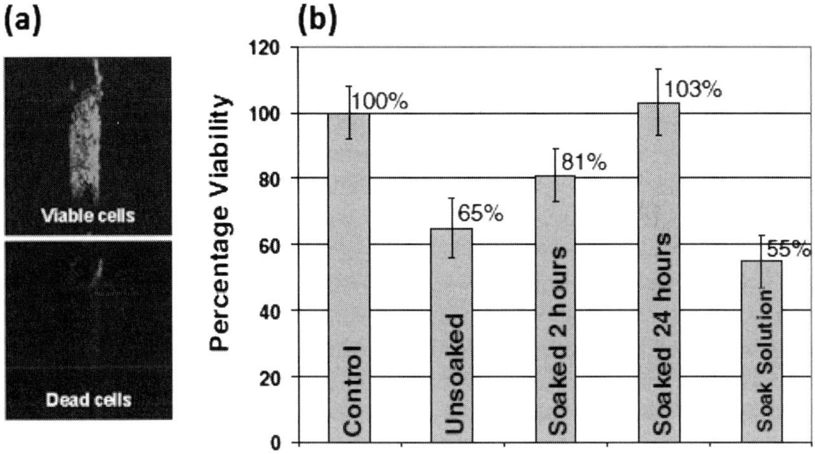

FIGURE 10 (a) Fluorescence micrographs of a lung cell monolayer attached to a chalcogenide fiber. Viable cells stained with green SYTO 10 dye are shown in the top-right micrograph, and nonviable cells stained with Dead Red dye are shown in the bottom-right micrograph. Each image shows the same field. (b) Viability of lung cells exposed to oxidized Te–As–Se fibers, and after soaking for 2 and 24 h in solution, and to the soaking solution.

APPLICATION OF FEWS TO BIOSENSING

Hydrophobic fiber surface for sensing in aqueous environments

An added advantage of chalcogenide fibers is that the glass surface displays a hydrophobic character that promotes stronger interactions with organic molecules than with water. This provides a notable benefit for IR sensing due to the strong water signal that often overlap with the signature of target molecules. The water signal can indeed mask some important spectral features of biological molecules, in particular near the amide region. Measurements performed in transmission do not permit to separate the two signals; however, FEWS measurements allow us to minimize the effect of water by maximizing the surface interactions with organic materials and increasing their concentration within the evanescent wave region. This results in an overall stronger signal of the target molecule. Figure 11 illustrates this phenomenon for a mixture of ethanol and water. A comparison of the water (1600 cm^{-1}) to ethanol (1050 cm^{-1}) peak ratio clearly shows an intensity enhancement of the organic peak in the FEWS spectra relative to the transmission spectra. In fact, measurements of the spectral enhancement performed with organic solvents ranging from polar to highly nonpolar show a linear correlation between the spectral enhancement and the dielectric constant of the solvent. The origin of the enhancement is therefore clearly associated with the surface hydrophobicity of the glass fibers.

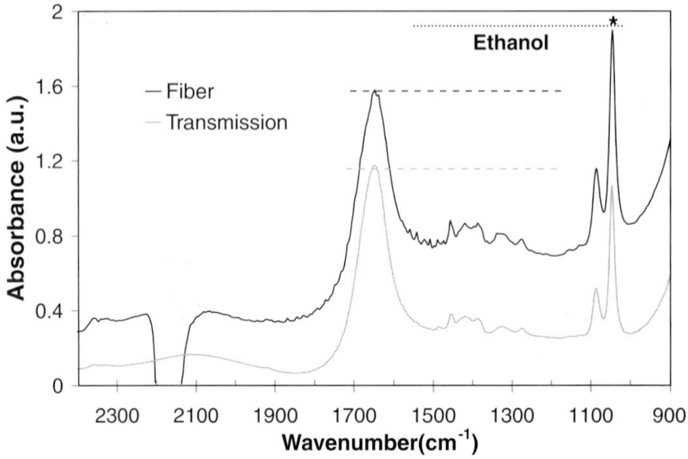

FIGURE 11 **Comparison of the peak intensity between water and ethanol when collected in transmission versus FEWS. The ethanol peak is more intense in FEWS.**

A series of contact angle measurements between drops of solvent and a glass disc also confirm the direct correlation between the strength of the solvent–glass interaction and the spectral enhancement (Fig. 12) [50]. The fiber surface selectively attracts the nonpolar solvent molecules in the vicinity of the glass surface where the evanescent wave is the most intense. This results in a net increase of the target molecule signal.

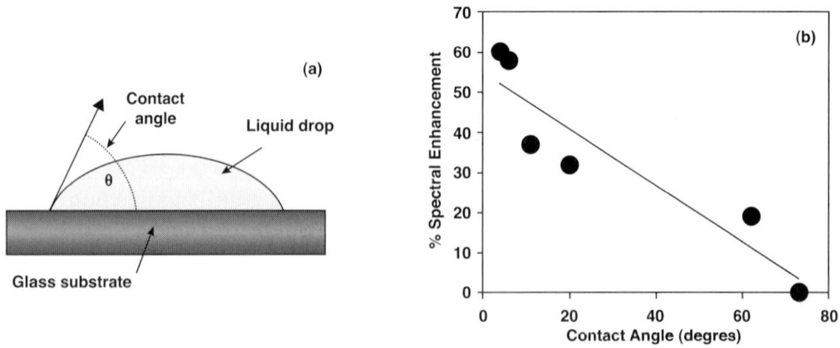

FIGURE 12 **(a) Contact angle measurements as a mean of quantifying the strength of the interaction between the solvent and the glass surface. (b) Correlation between the contact angle and the specral enhancement.**

Monitoring of live cells

The selective sensitivity of chalcogenide fiber sensors toward organic material versus water makes them interesting candidate for monitoring microorganisms in aqueous environment. Figure 13 shows two optical micrographs of human lung cells coated on the surface of a clean Te–As–Se glass fiber. The A549 type II alveolar epithelial cells used in this study were an anchorage-dependent immortalized lung cell line that form strong attachments to surfaces. The strong interaction of the cell with the sensing zone of the TAS fiber is very beneficial for the FEWS experiment as it increases the cell signal intensity.

Figure 14 shows the spectrum of a monolayer of epithelial cells collected by FEWS. The spectrum underlines the validity of the technique for studying biological compounds. The common spectral features expected for biological samples are assigned as follow: O–H elongation of water (A), amide A and B of proteins (B), CH_2 and CH_3 of lipids (C), amide I and II of proteins (D), fatty acids and amino acids (E), phospholipids and nucleic acids (F), and polysaccharides (G). The ability to observe these spectral features offers the possibility of monitoring the cell integrity as a function of time and environments. The response of the live cells to different toxins can then be detected through changes in specific spectral features [51, 52]. This allows not only detecting the presence of a toxin but also provides information on the way the toxin affects the cell metabolism and therefore can help identifying its nature. Toxin concentrations in the μM range can be detected that way [11]. Such low concentrations are below the detection limit of a conventional FTIR measurement; hence, the cell layer also serves as a mean of sensitizing the fiber sensor.

Figure 15 shows an example of cell response upon exposure to a surfactant. The cells were exposed to 1 mM of Triton X-100 and their signal was monitored for up to 4 h. The cell spectra show a rapid decay of two absorption peaks near 2920 and 2850 cm^{-1} and the response is complete in less than 20 min. These two

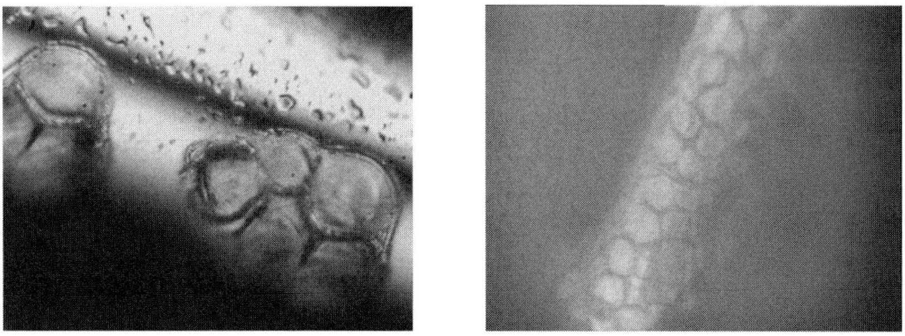

FIGURE 13 Optical micrographs of epithelial human lung cells coated on the surface of a Te–As–Se glass fiber.

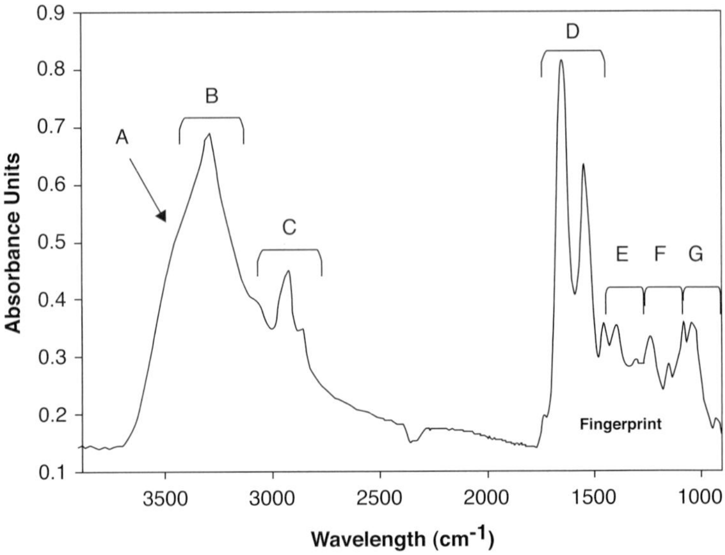

FIGURE 14 FEWS spectrum of human lung cells from a TAS fiber. The figure exhibits several spectral regions as follow: (A) O–H elongation of water, (B) amide A and B of proteins, (C) CH_2, CH_3 of lipids, (D) amide I and II of proteins, (E) fatty acids and amino acids, (F) phospholipids and nucleic acids, and (G) polysaccharides.

peaks are associated with the symmetric and asymmetric CH_2 vibrational modes of the membrane phospholipids bilayer. The surfactant Triton X-100 is a model toxic agent that disrupts cell membranes and causes rapid cell death when present at sufficiently high concentrations, but at levels below those used for total solubilization of cellular matter. Hence, the observed spectral response is

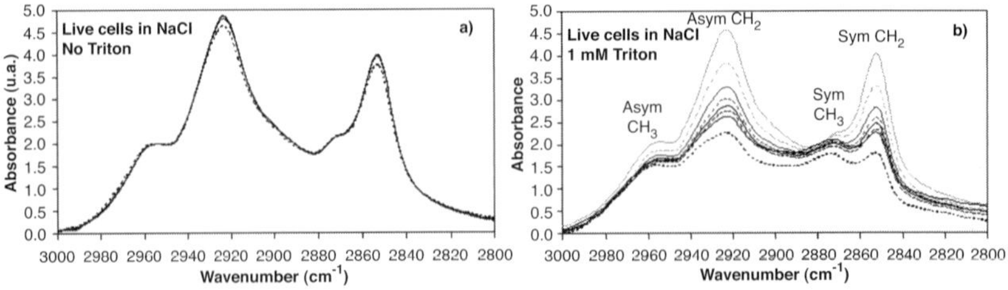

FIGURE 15 FEWS spectra of human lung cell coated on a Te–As–Se fiber collected form 5 min up to 4 h. The spectral features correspond to the CH_2 and CH_3 vibrations of the membrane phospholipids bilayer. (a) Cells in a 0.9% NaCl solution. (b) Cells exposed to Triton X-100 surfactant.

consistent with the disruption of the cell membrane expected from Triton. The FEWS technique therefore permits to monitor and identify the presence of a toxin *in situ* and in real time.

Since the cells are attached at the surface of the sensing zone, the cell membranes are in the immediate vicinity of the fiber surface and are expected to be most intensely probed by the evanescent wave. Using Figure 5, the evanescent wave penetration depth is estimated to be approximately 200 nm. The intensity of the evanescent electric field will decrease exponentially across that distance and in the case of strong absorption; the field is expected to decrease even faster. Considering that the lipid membrane is about 10 nm thick we can then consider that most of the signal is collected from the cell membrane in this wavelength range. This study therefore illustrates the capacity of FEWS for highly localized probing and emphasizes the potential of this technique for monitoring the IR signal of biointerfaces with high spatial resolution.

Monitoring the dynamic of biofilms

FEWS is also an ideal technique to monitor the dynamic of live microorganism, thanks to the high spatial resolution and good sensitivity in the aqueous solution. For example, bacteria are known to colonize surfaces through the formation of biofilms. Swarming bacteria can propagate over large areas in a matter of hours. It was shown that a FEWS sensors placed along the migration front of a swarming colony can permit to monitor their dynamic as well as differentiate vegetative and swarming phenotypes [12, 53]. Figure 16 shows the setup used for monitoring the biofilm's dynamics. The center of a Petri dish is initially spotted with a Proteus mirabilis culture (inoculums) and the dish is maintained at a constant temperature of 37°C. IR spectra are then periodically recorded for a period of 5 h as the bacterial front progresses isotropically through the Petri dish.

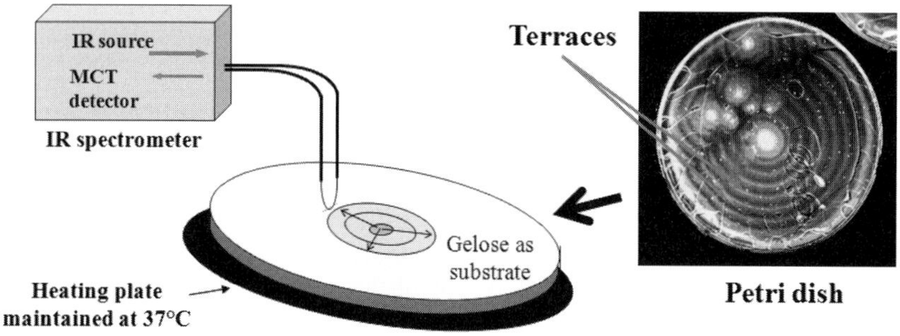

FIGURE 16 Set up for monitoring the dynamic of a bacterial biofilm. The fiber is in contact with a Petri dish colonized by Proteus mirabilis bacteria. The Petri dish exhibits terraces formed by the periodic swarming of bacteria.

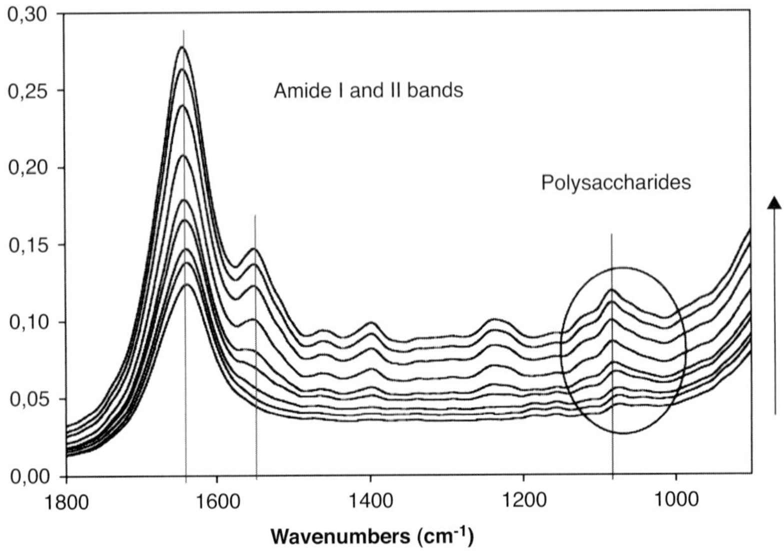

FIGURE 17 Proteus mirabilis spectra recorded in situ and in real time during swarming as they propagate through a Petri dish containing gelose.

Figure 17 shows a series of spectra acquired successively while the edge of the colony migrated across the fiber-sensing zone. The initial signal is dominated by a water peak from the gelose at 1636 cm^{-1} that overlap with the amide I of bacteria. As the bacterial front migrates and reaches the probe, the amide II and polysaccharide signal increases progressively and the water line is replaced by amide I at 1650 cm^{-1}. It is known that the edge of the colony is composed of motile swarmers cells which are localized along an annulus approximately 3 mm thick [53]. On the other end, the inner area of the biofilm disc is instead composed of dormant cells of the vegetative phenotype. Interestingly the active swarmer cells exhibit spectra distinguishable from vegetative cells at wavelengths dominated by polysaccharides. Indeed, swarming bacteria massively secrete exopolysaccharides slime used as lubricant.

Figure 18 shows the second derivative spectra of the spectral region associated with polysaccharides. The shift in spectral signature correspond to the change in bacterium phenotype as the swarmer cells migrate across the sensing zone and are later replaced by vegetative cells. While the spectral differences between the two phenotypes are not exceedingly strong, it is clear from the second derivative spectra that they can be differentiated. In addition, advanced spectral analysis techniques such as principal component analysis (PCA) can also be used to effectively identify the different bacterium phenotypes [53] as will be shown in the following. Overall, this demonstrates that the FEWS technique is sufficiently versatile to permit dynamic measurements *in situ* and

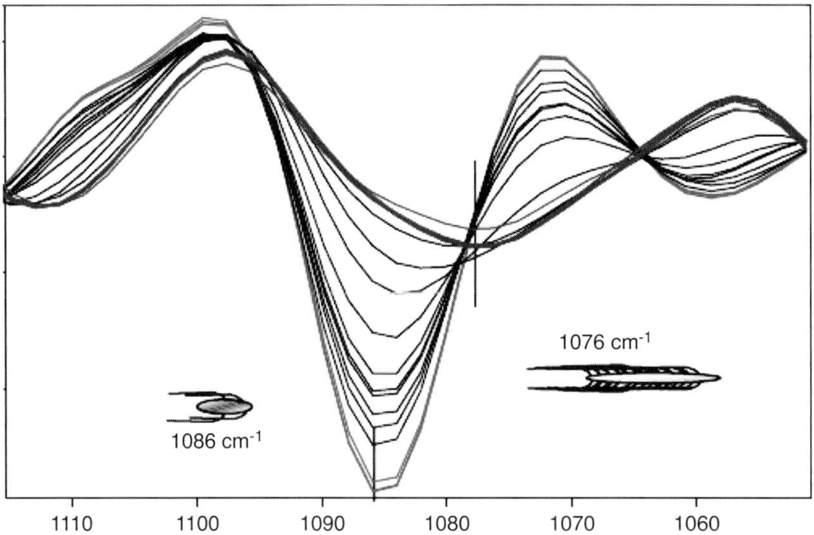

FIGURE 18 Second derivative spectra in the range of polysacharide vibrations with time. The band at 1086 cm^{-1} corresponds to vegetative cells while the band at 1076 cm^{-1} corresponds to swarming cells.

at the same time sufficiently sensitive to permit identification of different phenotypes.

PERSPECTIVES AND CONCLUSIONS

The field of IR sensing is constantly evolving and a great deal of research is ongoing to optimize sensitivity and selectivity of various sensing techniques. This can be achieved either through advances in spectra analysis techniques or through new instrumental developments. In the following, we emphasize one example of each, which are relevant to the FEWS technique and should lead to further improvement in the future.

Statistical Spectral Analysis

Infrared fingerprints of biomolecules contain much information on cells metabolism, which can help distinguish between healthy and altered tissues. In that respect, FEWS combined with adequate statistical analysis can yield sufficient spectral fingerprints to perform a rapid classification of patients with metabolic dysfunctions. This can provide relevant information for the diagnosis of specific diseases. To achieve this goal, one has to carry out some unsupervised statistical analysis method such as PCA or partial least square (PLS) regression.

For PCA, each initial spectrum Sp_i is viewed as a vector and the *n*-recorded spectra constitute a matrix X. After that, the diagonalization of the ${}^{t}X.X$ matrix provides *n* eigenvalues and *n* eigenvectors evt_i. Each eigenvector also corresponds to an IR factor spectrum. Then, one can demonstrate that each initial spectrum Sp_i can be rewritten as a linear combination of the eigenvectors:

$$Sp_i = c_{i1}\, evt_1 + c_{i2}\, evt_2 + \cdots + c_{in}\, evt_n$$

The weight of a particular factor spectrum is given by the corresponding coefficient (c_{ij}) of the combination, called score. During a PCA analysis, computer software automatically arranges the factor spectra according to these scores: the first few ones are the most significant and the higher indexed ones represent very noisy factor spectra, which do not contain any useful information.

The results of this analysis are then plotted along a PCA map where the Sp_i initial IR spectra are represented as points in two or three dimensions spaces. The axes correspond to the factor spectra and the coordinates are the corresponding scores. Generally, the first factor spectra of the combination are the most relevant to discriminate the initial spectra into nonoverlapping clusters of points. Note that generally the PCA is not implemented from the whole infrared spectra, but rather restricted to frequency ranges which present large intersample variances.

As an example, this analysis was applied to the differentiation between obese and normal mice [54]. The infrared spectral alterations related to metabolic deregulations were collected from serum of obese and normal mice using FEWS. Animals included in this study were reported to have developed obesity related to a homozygous mutation in the leptin gene, leading to hyperphagia and diabetes. In addition mice heterozygous for the mutation that did not develop obesity were used as control. After anaesthesia, blood samples were obtained by intracardiac puncture and the serum was separated by centrifugation. For infrared analysis, 10 μl of fresh serum was deposited along a few millimetres of the fibre and the spectrum then acquired.

Figure 19 displays spectra of an obese and control mouse. No evident spectral signature, such as line shifts or variation in intensities enables us to readily differentiate between both mice. PCA was then implemented in the 1100–1000 cm^{-1} range corresponding to the sugar rings vibration bands. The corresponding two first factor spectra are depicted in Figure 20(a) and the PCA map obtained by considering the two first factor spectra is presented in Figure 20 (b). The points corresponding to the control spectra were well localized, showing that the spectra are very homogeneous. This results is consistent with the fact that glycemia were similar in all these animals. On the other hand, the obese spectra were localized in a very different area. This point could be related to the various level of glycemia found in those animals. Taken collectively our results demonstrate that FEWS experiments coupled with unsupervised methods of analysis like PCA are an efficient tool to distinguish serum from obese and control mice.

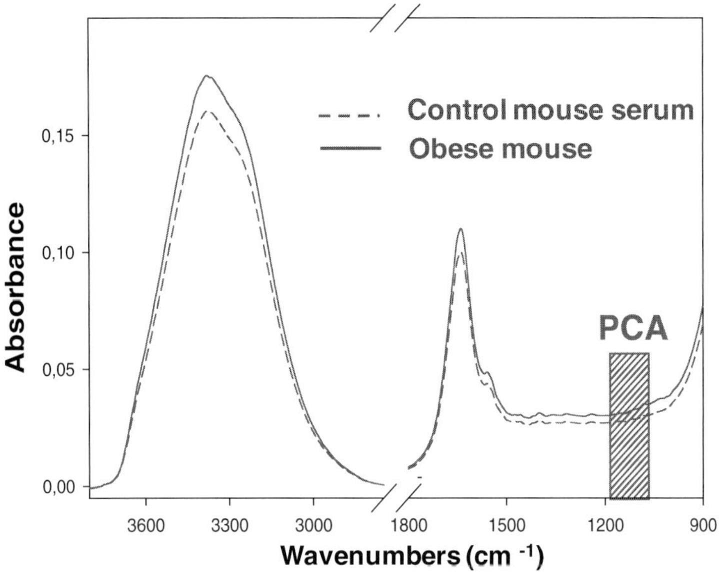

FIGURE 19 IR spectra of the serum of an obese and a normal mouse collected by FEWS. The red zone corresponds to the range where PCA was implemented.

Most recently, a PLS logistic method has been applied in a clinical trial to identify patient affected with four metabolic dysfunctions including hemochromatosis, iron depletion, cirrhosis, and dysmetabolic hepatosiderosis [55]. FEWS was used to analyze serum from a large group of patients and it was shown that

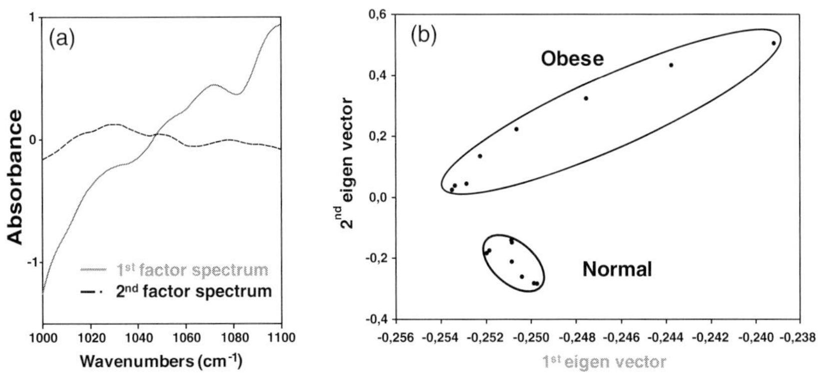

FIGURE 20 PCA analysis of mice spectra shown in Figure 19. (a) Plot of the two first factor spectrum used to generate linear combinations for each initial spectrum. (b) PCA map of 16 mice serum spectra. Eight spectra are localized in the area corresponding to ''normal'' mice and eight are localized in the area corresponding to ''obese'' mice.

PLS logistic could provide an effective mean of discriminating patients from each group. This statistical analysis method indeed appears to be very promising for medical diagnostics.

Biosensing through electrophoretic capture of charged molecules

Latest improvements in IR sensing are also driven by the development of novel instrumental approach for collection and detection of target molecules. For example it was recently shown that low bandgap telluride glasses can display fairly high electrical conductivity while retaining their transparency in the infrared [56, 57]. This unique feature has opened the way to the development of a new class of sensing probe that can simultaneously act as an electrode and as an infrared collection element. This set of properties permits to attract charged molecules at the surface of a sensing element in order to collect their infrared signature. Figure 21 shows an example of a Ge–Te–As optical element and a schematic of its use in an electrophoretic setup. A voltage is applied between an ITO counter electrode and the glass element in order to generate an electric field that induces the migration of charged molecules. The target molecules are then captured on the surface of the optical element and their infrared signature can be collected.

This type of sensing device is particularly useful in biosensing experiments because many biological molecules such as viruses, bacteria, and proteins carry a net surface charge that varies depending on solution pH. This net electrical charge is dictated by the isoelectric point, or pI, of the virus or protein. The pI is the pH at which the studied molecule carries no net charge. In a solution with pH below the pI of the molecule, the molecule carries a net positive charge. Conversely, in a solution with pH greater than the pI, the molecule carries a net negative charge.

Figure 22(a) shows an example of electrophoretic collection for a solution of bovine serum albumin (BSA). The spectra shows a rapid increase in the amide I and II peaks as well as the amino acids at 1400–1500 cm^{-1}. The migration is virtually complete within 20 min. Figure 22(b) shows that the migration rate is a function of voltage as expected from the columbic force, which drives the

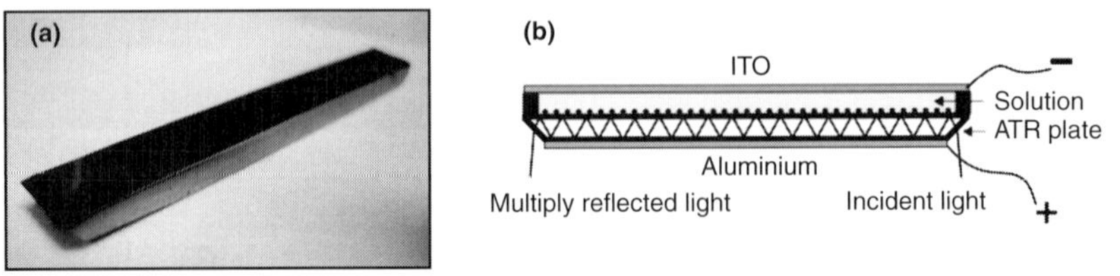

FIGURE 21 **(a) Photograph of a Ge–Te–As ATR element. (b) Schematic of a two-electrode electrophoretic sensing device.**

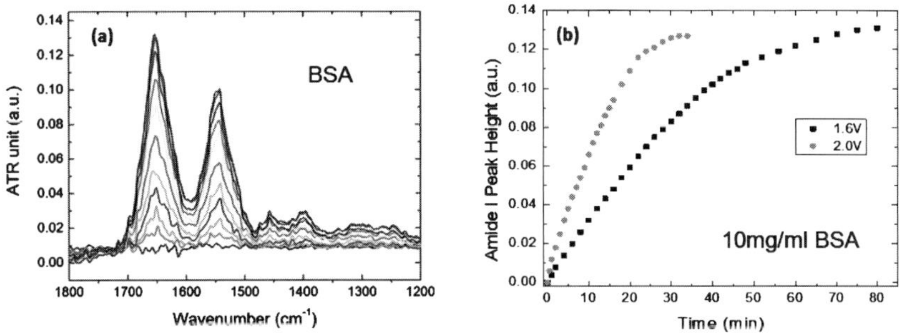

FIGURE 22 (a) Result of an electrodeposition experiment of BSA onto a Te–Ge–As glass ATR element. (b) Plot of the amide I peak height as a function of time and voltage showing a greater deposition rate at higher voltages.

particles. The greater the voltage, the higher the electric field, and the faster the molecules diffuse.

In order to ensure that the signal increase is not due to the effect of protein sedimentation under gravitational attraction, the voltage was kept off for 30 min before initiating the experiment. Figure 23 shows that no significant signal is visible until the voltage is turned on, but as soon as the voltage is applied the signal raises rapidly. The signal eventually reaches saturation once the entire population of particles has migrated to the glass surface. It can be shown that the saturated signal intensity is a linear function of the BSA concentration therefore indicating complete migration. What is more, the BSA molecule can be repealed off the surface of the optical element by simply reversing the voltage. This offers a simple way of removing the target molecules and replenishing the optical element.

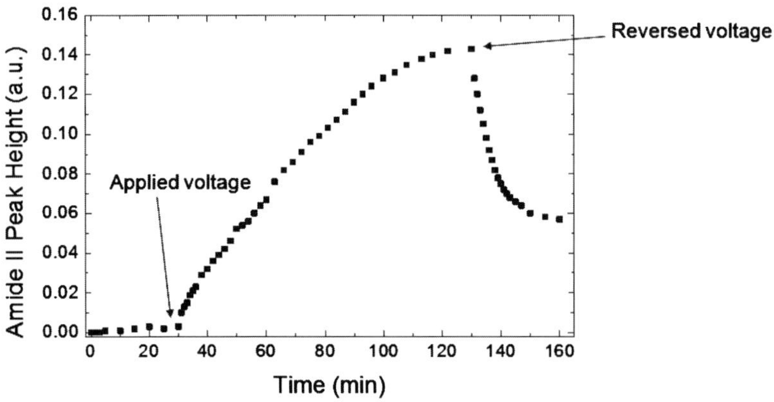

FIGURE 23 Electrodeposition of a solution of BSA with no voltage for 30 min, 1.5 V applied voltage for 90 min, and reversed voltage for 40 min.

Similar experiments have been performed with MS2 viruses, *E. coli* bacteria as well as various proteins. These experiments show the same migration profile and demonstrate the potential for electrophoretic capture of these biomolecules and their detection. Significant spectral differences can be observed between these molecules; hence, this technique permits the detection as well as the identification of the target biomolecule.

CONCLUSION

The FEWS technique is a recent but promising analytical method for the study of biointerfaces. It combines the advantages of FTIR with the versatility of fiber sensing. This provides high selectivity of FTIR as well as the ability to probe biointerfaces *in situ* and in real time. Moreover, the design of appropriate sensing probes can lead to notable increase in sensitivity and the hydrophobic nature of the glass surface promotes the signal of biomolecule versus water, making this technique a great candidate for biosensing in aqueous environments. As described in section 4, the field of FTIR is in constant development and new improvements are regularly reported. In most cases these new advances can be directly applied to FEWS technique and should further increase its sensitivity and detection limit in the future.

ACKNOWLEDGMENTS

We would like to thank our colleagues, Dr. Olivier Loreal of INSERM in Rennes, Professor Olivier Sire of Universite Bretagne Sud, Professor Mark Riley of the Agricultural and Biosystems Engineering department at University of Arizona and Professor Kelly Reynolds of the College of Public Health at University of Arizona for their critical collaborative contributions to the project described in this chapter. We also acknowledge NSF-DMR under grant #0806333 and NSF-ECCS under grant #0901069 and the CNRS International Associated Laboratory for Materials & Optics.

REFERENCES

[1] Naumann D. Infrared spectroscopy in microbiology. In: R. A. Meyers (Ed.), *Encyclopedia of Analytical Chemistry*. Chichester: Wiley; 2000 pp. 102.

[2] Naumann D. Microbiological characterizations by FT-IR spectroscopy. *Nature* 1991;351:81.

[3] Diem M, Boydston-White S, Chiriboga L. Infrared spectroscopy of cells and tissues: shinning light onto a novel subject. *Appl. Spectrosc* 1999;53:148A.

[4] Diem M, Chiriboga L, Lasch P, Pacifico A. IR spectra and IR spectral maps of individual normal and cancerous cells. *Biopolymers (Biospectroscopy)* 2002;67:349.

[5] Lucas P, Riley MR, Boussard-Pledel C, Bureau B. Advances in chalcogenide fiber evanescent-wave biochemical sensing. *Anal. Biochem.* 2006;351:1–10.

[6] Raichlin Y, Katzir A. Fiber-optic evanescent wave spectroscopy in the middle infrared. *Appl. Spectrosc* 2008;62:55A–72A.

[7] Mizaikoff B. Mid-IR fiber-optic sensors. *Anal. Chem.* 2003;75:258A–267A.
[8] Uemura T, Nishida K, Sakakida M, Ichinose K, Shimoda S, Shichiri M. Non-invasive blood glucose measurement by Fourier transform infrared spectroscopic analysis through the mucous membrane of the lip: application of a chalcogenide optical fiber system. *Frontiers Med. Biol. Eng.* 1999;9:137.
[9] Hocde S, Loreal O, Sire O, Boussard-Pledel C, Bureau B, Turlin B, Keirsse J, Leroyer P, Lucas J. Metabolic imaging of tissues by infrared fiber-optic spectroscopy: an efficient tool for medical diagnosis. *J. Biomed. Opt.* 2004;9:404–407.
[10] Afasnasyeva N, Bruch R, Katzir A. Infrared fiber optic evanescent wave spectroscopy: application in biology and medicine. *Proc. SPIE* 1999;3596:152.
[11] Lucas P, Le Coq D, Juncker C, Collier J, Boesewetter DE, Boussard-Plédel C, Bureau B, Riley MR. Evaluation of toxic agent effects on lung cells by fiber evanescent wave spectroscopy (FEWS). *Appl. Spectrosc.* 2005;59:1–9.
[12] Keirsse J, Bousard-Pledel C, Loreal O, Sire O, Bureau B, Leroyer P, Turlin B, Lucas J. IR optical fiber sensor for biomedical applications. *Vibr. Spectrosc.* 2003;32:23.
[13] Zhang X, Ma H, Lucas J. Applications of chalcogenide glass bulks and fibres. *J. Optoelectron. Adv. Mat.* 2003;5:1327.
[14] Michel K, Bureau B, Boussard-Plédel C, Jouan T, Adama LJ, Staubmann K, Baumannc T. Monitoring of pollutant in waste water by infrared spectroscopy using chalcogenide glass optical fibers. *Sens. Actuat. B* 2004;101:252–259.
[15] Charpentier F, Bureau B, Troles J, Boussard-Pledel C, Michel Le Pierres K, Smektala F, Adam J-L. Infrared monitoring of underground CO_2 storage using chalcogenide glass fibers. *Opt. Mat.* 2009;31:496–500.
[16] Houizot P, Boussard-Plédel C, Faber AJ, Cheng LK, Bureau B, Van, Nijnatten PA, Gielesen W. LM, Pereira do Carmo J, Lucas J. Infrared single mode chalcogenide glass fiber for space. *Opt. Express* 2007;15:12529–12538.
[17] Lewi T, Tsun A, Katzir A, Kaster J, Fuchs F. Silver halide single mode fibers for broadband middle infrared stellar interferometry. *Appl. Phys. Lett.* 2009;94:261105.
[18] Lucas P, Wilhelm AA, Videa M, Boussard-Plédel C, Bureau B. Chemical stability of chalcogenide infrared glass fibers. *Corrosion Science* 2008;50:2047.
[19] Wilhelm AA, Lucas P, DeRosa DL, Riley MR. Biocompatibility of Te–As–Se glass fibers for cell-based bio-optic infrared sensors. *J. Mater. Res.* 2007;22:1098.
[20] Bunimovich D, Belotserkovsky E, Katzir A. Fiberoptic evanescent wave infrared spectroscopy of gases in liquids. *Rev. Sci. Instrum.* 1995;66:2818.
[21] Gotshal Y, Simhi R, Sela B-A, Katzir A. Blood diagnostic using fiber optic evanescent wave spectroscopy and neural networks analysis. *Sens. Act. B* 1997;42:157.
[22] Vasiliev AV, Dianov EM, Plotnichenko VG, Sysoev VK, Bagrov AM, Baikalov PI, Devyatykh GG, Scripachev IV, Churbanov MF. Middle IR arsenic–sulfur and arsenic–selenium glass fibers with optical losses lower than 1 dB/m. *Electron. Lett.* 1983;19:589–590.
[23] Parant JP, Le Sergent C, Guignot D, Brehm C. Chalcogenide glass optical fibers. *Glass Technol.* 1983;24:161–163.
[24] Shibata S, Terunuma Y, Manabe T. Germanium–phosphorus–sulfur chalcogenide glass fibers. *Jpn. J. Appl. Phys.* 1980;19:L603–L605.
[25] Katzir A, Arieli R. Long wavelength infrared optical fibers. *J. Non-Cryst. Sol.* 1982;47:149–158.
[26] Compton D. AC, Hill SL, Wright NA, Druy MA, Piche J, Stevenson WA, Vidrine DW. *In situ* FT-IR analysis of a composite curing reaction using a mid-infrared transmitting optical fiber. *Appl. Spectrosc.* 1988;42:972.

[27] Heo J, Rodrigues M, Saggese SJ, Sigel GH. Remote fiber-optics chemical sensing using evanescent-wave interactions in chalcogenide glass fibers. *Appl. Opt.* 1991;30:3944.

[28] Taga K, Schwarzbach D, Stingeder G, Grassbauer M, Kellner R. Surface analysis and characterization of chemically modified chalcogenide fibers. *Anal. Chem.* 1993;65:2288.

[29] Taga K, Kellner R, Kainz U, Sleytr UB. In situ attenuated total reflectance FT-IR analysis of an enzyme-modified mid-infrared fiber surface using crystalline bacterial surface proteins. *Anal. Chem.* 1994;66:35.

[30] Katz M, Katzir A, Schnitzer I, Bornstein A. Quantitative evaluation of chalcogenide glass fiber evanescent wave spectroscopy. *Appl. Opt.* 1994;33:5888–5894.

[31] Messica A, Greenstein A, Katzir A. Theory of fiber-optic, evanescent-wave spectroscopy and sensors. *Appl. Opt.* 1996;35:2274–2284.

[32] Lecoq D, Michel K, Fonteneau G, Hocde S, Boussard-Pledel C, Lucas J. Infrared chalcogen glasses: chemical polishing and fiber remote spectroscopy. *Int. J. Inorg. Mat.* 2001;3:233–239.

[33] Eytan O, Sela B-A, Katzir A. Fiber-optic evanescent-wave spectroscopy and neural network: application to chemical blood analysis. *Appl. Optics* 2000;39:3357–3360.

[34] Spielvogel J, Lobik L, Nissenkorn I, Hibst R, Gotshal Y, Katzir A. Cancer Diagnostic using Fourier transform fiberoptics infrared evanescent wave spectroscopy. *Proc. SPIE* 1998;3262:185.

[35] Cytron SE, Kravchick S, Sela BA, Shulzinger E, Vasserman I, Raichlin Y, Katzir A. New applications of fiber optic IR spectroscopy in urologic practice. *Proc. SPIE* 2002;4614:55.

[36] Raichlin Y, Goldberg I, Brenner S, Shulzinger E, Katzir A. Infrared fiber optic evanescent wave spectroscopy for the study of diffusion in the human skin. *Proc. SPIE* 2002;4614:101.

[37] Janotta M, Katzir A, Mizaikoff B. Sol–gel-coated mid-infrared fiber-optic sensors. *Appl. Spectrosc.* 2003;57:823.

[38] Karlowatz M, Kraft M, Mizaikoff B. Simultaneous quantitative determination of benzene, toluene, and xylenes in water using mid-infrared evanescent field. *Anal. Chem.* 2004;76:2643.

[39] Mägi EC, Fu LB, Nguyen HC, Lamont M. RE, Yeom DI, Eggleton BJ. Enhanced Kerr nonlinearity in sub-wavelength diameter As_2Se_3 chalcogenide fiber tapers. *Opt. Express* 2007;15:10324–10329.

[40] Lepine E, Yang Z, Gueguen Y, Troles J, Zhang X-H, Bureau B, Boussard-Pledel C, Sangleboeuf J-C, Lucas P. Optical microfrabrication of tapers in low-loss chalcogenide fibers. *J. Opt. Soc. Am. B* 2010;27:966–971.

[41] Harrick NJ. *Internal Reflection Spectroscopy*. New York: Interscience; 1967.

[42] Bureau B, Zhang XH, Smectala F, Adam JL, Troles J, Ma HL, Boussard-Pledel C, Lucas J, Lucas P, Le Coq D, Riley MR, Simmons JH. Recent advances in chacogenide glasses. *J. Non-Cryst. Solids* 2004;345:276.

[43] Sanghera JS, Aggarwal ID, Busse LE, Pureza PC, Nguyen VQ, Kung FH, Shaw LB, Chenard F. Chalcogenide optical fibers target mid-IR applications. *Laser Focus World* 2005;41:83.

[44] Lucas J, Infrared, Glasses. *Curr. Opin. Sol. State Mat. Sci.* 1999;4:181.

[45] Wilhelm AA, Boussard-Plédel Q. CC, Lucas J, Bureau B, Lucas P. Development of far-infrared-transmitting Te based glasses suitable for carbon dioxide detection and space optics. *Adv. Mat.* 2007;19:3796.

[46] Yang Z, Lucas P. Tellurium-based far-infrared transmitting glasses. *J. Am. Ceram. Soc.* 2009;92:2920–2923.

[47] Zhang XH, Bureau B, Lucas P, Boussard-Pledel C, Lucas J. Glasses for seeing beyond visible. *Chem. Eur.* 2008;14:432–442.

[48] Shiryaev VS, Adam J-L, Zhang XH, Boussard Pledel C, Lucas J, Churbanov MF. Infrared fibers based on Te–As–Se system with low optical losses. *J. Non-Cryst. Sol.* 2004;336:113.
[49] Troles J, Shiryaev V, Churbanov M, Houizot P, Brilland L, Desevedavy F, Charpentier F, Pain T, Snopatin G, Adam JL. $GeSe_4$ glass fibres with low optical losses in the mid-IR. *Opt. Mat.* 2009;32:212–215.
[50] Lucas P, Solis MA, Le Coq D, Juncker C, Riley MR, Collier J, Boeswetter DE, Boussard-Pledel C, Bureau B. Infrared biosensors using hydrophobic chalcogenide fibers sensitized with live cells. *Sens. Actuat. B* 2006;119:355.
[51] Riley MR, Lucas P, Le Coq D, Juncker C, Boesewetter DE, Collier JL, DeRosa DM, Katterman ME, Boussard-Pledel C, Bureau B. Lung cell fiber evanescent wave spectroscopic biosensing of inhalation health hazards. *Biotech. Bioeng* 2006;95:599.
[52] Riley MR, DeRosa DL, Blaine J, Potter BG, Lucas P, Le Coq D, Juncker C, Boeswetter DE, Collier J, Boussard-Pledel C, Bureau B. Biologically inspired sensing: infrared spectroscopic analysis of cell responses to an inhalation health hazard. *Biotechnol. Prog.* 2006;22:24.
[53] Keirsse J, Lahaye E, Bouter A, Dupont V, Boussard-Pledel C, Bureau B, Adam J-L, Monbet V, Sire O. Mapping bacterial surface population physiology in real time: infrared spectroscopy of proteus mirabilis swarm colonies. *Appl. Spectrosc.* 2006;60:584–591.
[54] Keirsse J, Bureau B, Boussard-Pledel C, Leroyer P, Ropert M, Dupont V, Anne ML, Ribault C, Sire O, Loreal O, Adam JL. Chalcogenide glass fibers for in-situ infrared spectroscopy in biology and medicine. *Proc. SPIE* 2004;5459:61.
[55] Anne ML, Le Lan C, Monbet V, Boussard-Pledel C, Ropert M, Sire O, Pouchard M, Jard C, Lucas J, Adam JL, Brissot P, Bureau B, Loreal O. Fiber evanescent wave spectroscopy using the mid-infrared provides useful fingerprints for metabolic profiling in humans. *J. Biomed. Opt.* 2009;14:054033.
[56] Yang Z, Wilhelm AA, Lucas P. High-conductivity tellurium-based infrared transmitting glasses and their suitability for bio-optical detection. *J. Am. Ceram. Soc.* 2010;93:1941–1944.
[57] Wilhelm AA, Lucas P, Reynolds K, Riley MR. Integrated capture and spectroscopic detection of viruses in an aqueous environment. *SPIE Proc.* 2008;6852:68520K.

Chapter 9

AFM-IR: photothermal infrared nanospectroscopy

Application to cellular imaging

Alexandre Dazzi[1] and Clotilde Policar[2]
[1]*Laboratoire de Chimie Physique, UMR8000, Université Paris-Sud, 91405 Orsay, FRANCE*
[2]*Laboratoire des BioMolécules, UMR 7203, département chimie de l'ENS, 24 rue Lhomond, 75232 Paris Cedex 05*

INTRODUCTION

In the infrared range, the coupling between microscopy and spectroscopy, called microspectroscopy, is one of the most powerful techniques to enable chemical identification and localization of particular compounds. A number of applications in different domains of science, such as chemistry, biology, agronomy, and medicine underline the usefulness of infrared analysis. The main advantage of infrared spectroscopy is to provide direct information about molecular vibrations. Further, the intensity of light used for an IR analysis is generally weak enough to avoid sample damage [1–5] and photo bleaching that is common with organic fluorophores in the visible or UV-range.

Biointerface Characterization by Advanced IR Spectroscopy. DOI: 10.1016/B978-0-444-53558-0.00009-6

However, any microspectroscopy technique is limited by diffraction, as is the case in classical optical microscopy. In the IR-range from 4000 to 500 cm^{-1}, a maximum resolution of 2 to 20 μm is attainable. One can easily understand that this resolution is not sufficient for subcellular studies. Fortunately, lots of optical near-field techniques have been investigated and different ways to achieve high-resolution imaging can be suggested. Among those, the most known is developed by Keilmann *et al.* using apertureless near field optical microscopy with continuous infrared lasers [6, 7]. Unfortunately, the lack of CW laser tunability renders the analysis of infrared spectrum difficult and limits the potential application in biology. Moreover, the optical signal is not easy to use as the near field is convolution between dispersion effect and absorption [8].

Another issue in infrared biology studies is the similarity of infrared bands. As vibrational spectra reveal the local structure (chemical functionality), most biological samples show the similar infrared signature (OH, CH, or amide I and phosphate bands, see Fig. 20). It is thus not easy to identify subcellular structures and accurate investigations such as deconvolution method or PCA analysis are needed for unambiguous identification [9, 10]. Conventional infrared microspectroscopy does not currently allow the ability to follow or study the distribution of a specific protein or macromolecule. As a result, the best solution is often to use fluorescence microscopy after tagging of the molecule of interest.

In this chapter, we propose to expose a new approach based on a combination of atomic force microscopy (AFM) and infrared spectrsoscopy. This technique (AFM-IR) is used to enable high-resolution imaging by using a near-field photothermal technique coupled with specific infrared probes to go over the limits of optical resolution and allow specific molecular identification. We will first present the formal concept in physics to describe how the technique works. Then, we will explore the potential of such a technique by describing some applications in microbiology. To finish with, we will explain how the coupling of the AFM-IR with IR-probes, namely metal-carbonyl probes, is a powerful tool to eukaryotic cell studies.

CONCEPT AND TECHNIQUE DESCRIPTION

The principle of the AFM-IR technique is to use the tip of an AFM to detect absorption of IR radiation using the photothermal effect. When an object is illuminated by an IR light source at the wavelength corresponding to a maximum of its absorption bands then most of the energy is absorbed. This energy absorbed by the object will produce a local increase in temperature inside the object; this is the photothermal effect. At the same time, the stress induced by the temperature field inside the object will be relaxed by mechanical expansion of the object, that is, thermal expansion. This thermal expansion can be readily detected with the AFM. The AFM-IR technique uses a pulsed IR laser to produce rapid heating and thermal expansion within the sample. Each time the laser pulses, the AFM records a signal that is directly proportional to the sample absorption. The main advantage of this sort of technique is to use the sensitivity and the high spatial resolution of the AFM to overcome the spatial resolution limits of conventional optical microspectroscopy. While the focused spot of the IR laser is still limited by optical diffraction, the thermal expansion of the sample varies on much smaller length scales, and these variations can be observed by the tip of the AFM.

In this chapter, we will present the concept of AFM-IR beginning by a review of relevant issues in optical spectroscopy and thermomechanics. Then a complete model of AFM in contact mode will be proposed to explain the way that how the absorption is measured. At the end, we will demonstrate the proportionality between AFM detection and infrared absorption.

AFM-IR setup description

The AFM-IR technique was developed to go beyond the characteristics of conventional infrared microspectroscopy and reach resolutions in the nanometer range [11]. A commercial instrument based on this technology was introduced by Anasys Instruments in 2010. It is a benchtop setup with an IR laser covering the 3600–1000 cm^{-1} range.

We have chosen to use a total reflection setup to light the sample (Fig. 1). The idea is to illuminate the sample only and avoid absorption coming from the tip or the cantilever of the AFM. The infrared laser beam is aligned to be in total reflection at the prism–air interface. The angle of the prism is chosen to maintain IR light propagation into samples having an index of refraction higher than 1.2, allowing us to illuminate most of the organic samples. The tip of the AFM is in contact with the sample to study. At each laser pulse the sample's thermal expansion induces a brief force impulse on the tip, inducing ringing of the AFM cantilever's resonant frequencies. The oscillations signal of the cantilever are recorded by the 4-quadrant detector of the AFM and stored into an oscilloscope or other data acquisition system. The signal can also be simultaneously analyzed by fast Fourier transforms (FFT) to determine the amplitudes and frequencies of the cantilever vibration modes.

Finally, we can create absorption spectra by plotting the cantilever amplitude as a function of wavelength. Spatially resolved chemical images can also be

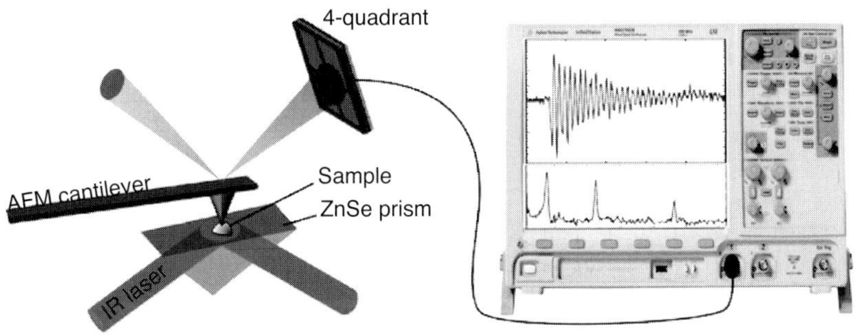

FIGURE 1 AFM-IR Set-up scheme. The IR beam is incident on a ZnSe prism in total reflection. The AFM tip is in contact with the sample and the 4-quadrant signal is recorded by an oscilloscope.

created by mapping the amplitude at one or more fixed wavelengths while scanning the AFM tip over the sample surface.

Infrared absorption and spectroscopy

By definition, spectroscopy studies interactions of a material with an incident wave, through excitation and de-excitation phenomena, as its function of the energy. In the midinfrared range (4000–500 cm^{-1}), the energy of waves usually corresponds to molecular vibrations.

In an optical point of view, materials are characterized by their optical index:

$$\tilde{n}(\lambda) = n(\lambda) + i\kappa(\lambda), \tag{1}$$

where λ is the wavelength, n is the refractive index, represents the dispersive effect, and κ is the extinction coefficient, corresponds to the absorption.

When an infrared radiation excites a molecule, the electric field of the electromagnetic wave produces a variation on the dipole moment of a chemical bond if radiation frequency matches a natural vibration frequency of the chemical bond. In this condition, the molecule will absorb the radiation resulting in oscillation of the chemical bond. The molecular vibrations also involve damping and this damping leads to the absorption of a portion of the incident radiation and this absorption is converted to heat. The degree of absorption comes from an imaginary part of the dipolar susceptibility of the molecular bond.

The principle of spectroscopy is to measure the intensity of the radiation transmitted through the sample as a function of the wavelength. In infrared domain, it is more common to use "wavenumber," (a measure of the radiation frequency) instead the wavelength. If we consider a homogeneous layer of medium with thickness z and optical index $\tilde{n}(\lambda)$, illuminated by an electromagnetic plane wave, the transmitted intensity can be given by the Beer–Lambert law (neglecting the index mismatch reflections at the interfaces):

$$I_t = I_{\text{inc}} e^{-4\pi\sigma\kappa z} \tag{2}$$

with I_{inc} being the incident intensity, σ being the wave number, and κ being the extinction coefficient, function of the wave number.

The transmittance T coefficient is defined by the ratio of the transmitted and the incident intensity:

$$T = \frac{I_t}{I_{\text{inc}}} = e^{-4\pi\sigma\kappa z}. \tag{3}$$

It is convenient to convert this to an absorbance coefficient A, which directly gives the variation of the absorption bands:

$$A = \log_{10}\frac{1}{T} = \frac{4\pi z}{\ln(10)}\sigma\kappa(\sigma). \tag{4}$$

This simple expression shows that the absorbance is in fact the representation of the extinction coefficient multiplied by the wave number.

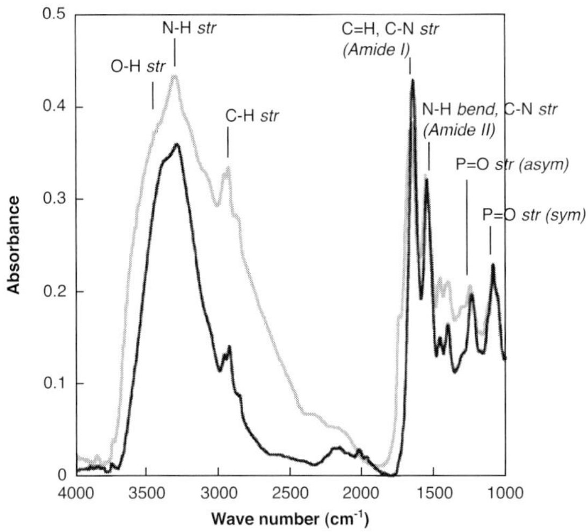

FIGURE 2 Absorbance spectrum of bacteria Rhodobacter spheroides (in gray) and eukaryotic cells (in black). All bands are identified by their chemical function and type of vibration.

Figure 2 illustrates an example of the absorbance spectrum obtained on biological samples, bacteria, and cells, with a Fourier transform spectrometer (FTIR).

One can note that both bacteria cells and eukaryotic cells display similar bands. There is no real spectral signature to identify these two different biological samples. But accurate studies purchased by the group of D. Naumann have demonstrated the ability by chemometrics analysis, the way to perfectly identify different bacteria strain based on their infrared spectra [12]. Indeed, small differences can be observed in the shape of the bands and revealed by a second-order derivation analysis.

Photothermal effect

When the sample or material absorbs the energy of the incident wave, an increase in temperature occurs, which will be driven by the Fourier law:

$$\rho C\frac{\partial T}{\partial t} - k\Delta T = \frac{Q(t)}{V}, \tag{5}$$

where ρ is the density, C is the heat capacity, k is the heat conductivity, V is the volume, $Q(t)$ is the absorbed power, and Δ is the Laplacian operator.

Let us consider that the sample illumination is produced during the time t_p. In this case, the expression of $Q(t)$ is

$$Q(t) = \Pi(t_p)*P_{\text{abs}} \tag{6}$$

with Π being the box function and * the convolution product, and P_{abs} is the power absorbed by the sample.

Using the electromagnetic laws, it is possible to express the absorbed power P_{abs} in function of the optical parameters of the sample:

$$P_{abs} = \int_{v} \pi\sigma\varepsilon_0 \operatorname{Im}(\tilde{n}^2)\left|\overset{r}{E}_{loc}\right|^2 dV \tag{7}$$

with π being the optical index, σ being the wave number, ε_0 being the vacuum permittivity, and E_{loc} being the local field in the sample.

In the case of AFM-IR setup configuration for cellular imaging, for the sake of simplicity we consider that the sample is small compared to the wavelength. The following expression gives a good estimation of the energy absorbed:

$$P_{abs} = \pi\varepsilon_0 c\sigma \operatorname{Im}\left(\frac{\tilde{n}^2 - 1}{\tilde{n}^2 + 2}\right)|E_{inc}|^2 V \tag{8}$$

with E_{inc} being the electric incident field.

In the case of weak absorption (where organic κ is about 10^{-4} to 10^{-3}), it is useful to make the following approximation:

$$\operatorname{Re}(\tilde{n}^2) \gg \operatorname{Im}(\tilde{n}^2). \tag{9}$$

This leads to a simplified expression of P_{abs}:

$$P_{abs} = \frac{6\pi c\varepsilon_0 n}{(n^2 + 2)^2}|E_{inc}|^2 V\sigma\kappa(\sigma) \propto \text{ Absorbance.} \tag{10}$$

As the absorption is weak for organic material, the relative variation of $n(\sigma)$ is really small compared to $\kappa(\sigma)$. For example, for polystyrene, the value of n at the C–H band (2950 cm^{-1}) is 1.4 and the value of κ is 5×10^{-4}. In that case, we can consider that the value of n is constant as a function of σ, showing that the absorbed power P_{abs} is finally proportional to the extinction coefficient κ multiplied by σ, that is to say the absorbance.

In the following, we consider that the period of heat relaxation inside sample takes place instantaneously during the illumination. This approximation is validated by the fact that the samples we study are of micrometer or nanometer sizes, so the time to diffuse the heat is really short generally under 1 µs, and the usual laser pulse is around 10 ns to 10 µs [13]. In these conditions, the equation of heat is really easy to solve. The temperature inside the sample is considered constant (no gradient inside), and the power absorbed is equal to the heat flow going outside the sample at the stationary state. If, for the sake simplicity, we consider the sample as a sphere of radius a, the expression of P_{abs} becomes

$$P_{abs} = -4\pi a^2 k_s \left.\frac{\partial T}{\partial r}\right|_{r=a,} \tag{11}$$

where a is the radius of the sample and k_s is heat conductivity outside the sample.

To know the temperature inside the sample (sphere) it is necessary to solve the heat equation outside the sample. As there is no heat source outside, we apply the stationary state of the Fourier law:

$$\frac{\partial^2(rT)}{\partial r^2} = 0. \tag{12}$$

The solution taking into account the boundaries conditions $T(\infty) = 0$ and $T(a) = T_{\max}$ is

$$T(r) = \frac{aT_{\max}}{r}. \tag{13}$$

Finally the temperature increase T into the sample gets instantaneously the value $T_{\max}$ as the laser illuminates and go back to 0 when it stops. Substituting (13) into (11) we find the expression of $T(t)$:

$$\begin{aligned} T_{\text{sph}}(t) &= T_{\max} = \frac{P_{\text{abs}}}{4\pi a k_s} && \text{when} \quad 0 \le t \le t_p \\ T_{\text{sph}}(t) &= 0 && \text{when} \quad t_p < t \end{aligned}. \tag{14}$$

Finally, we find that the sample heats at a constant temperature $T_{\max}$ during the illumination and the temperature is directly proportional to the absorbed power P_{abs} and so to the absorbance. This result is really interesting in terms of concept because it shows that the temperature is also a way to make the absorption measurement. The idea is to use a thermometer and record the temperature changes when illuminating the sample with the infrared light. Many experiments have been developed in this way, and more particularly Pollock *et al.* have shown the possibility of having quite good infrared spectra with the thermal detector [14–16].

Thermoelasticity

The main idea with AFM-IR is to use, not the temperature measurement, but the consequence of the heating. We have seen previously that if the sample absorbs the infrared illumination then the temperature will increase. This increase will then produce a mechanical stress inside the sample. The stress will also be combined with a strain. The temperature field is similar to the force field inside the sample, resulting deformations taking into account mechanical properties. The expression which links the displacement in function of the increase in temperature is given by Nowacki [17]:

$$(1-2\nu)\nabla^2 \vec{u} + \nabla(\nabla.\vec{u}) = 2(1+\nu)\alpha\nabla T \tag{15}$$

where ν is the Poisson coefficient, u is the displacement vector, and T is the increase in temperature.

The boundaries conditions are given for a body free of tractions on its surface bounding:

$$\begin{aligned} \sigma_{ij} n_{ij} &= 0 \\ (1+\nu)(1-2\nu)\sigma_{ij} &= (1-2\nu)E\varepsilon_{ij} + \big(\nu E\varepsilon_{kk} - (1+\nu)E\alpha T\big)\delta_{ij} \end{aligned} \tag{16}$$

where E is the Young modulus, ν is the Poisson coefficient, α is the thermal expansion coefficient, σ_{ij} is the stress tensor component, ε_{ij} is the strain tensor component, and ε_{kk} is the sum of the normal strains.

Taking into account the spherical geometry of our sample model the differential equation leads to

$$\frac{\partial u_r}{\partial r} + 2\left(\frac{u_r}{r}\right) = \frac{1+\nu}{1-\nu}\alpha T. \tag{17}$$

Considering the spherical symmetry, we assume that the expression of the displacement vector u_r at the center of the sphere is

$$u_r(r=0) = 0. \tag{18}$$

By solving equation (17) and using the limit condition (18), we obtain the expression of the thermal displacement:

$$u_r(r) = \frac{1+\nu}{1-\nu}\alpha T \frac{r}{3}. \tag{19}$$

Finally, the thermal expansion at the surface of the sphere can be written as

$$\frac{u_r(a)}{a} = \frac{1+\nu}{1-\nu}\frac{\alpha T}{3}. \tag{20}$$

The relative thermal expansion of the sample is directly linked to the increase in temperature. This shows that the relative expansion parameter is also a way to measure the absorption. The photothermal displacement expression due to a pulsed illumination is given by substituting (14) into (20)

$$\begin{aligned} u_{r=a}(t) &= \frac{1+\nu}{1-\nu}\frac{P_{\text{abs}}}{12\pi k_s} \quad \text{when} \quad 0 \le t \le t_p \\ u_{r=a}(t) &= 0 \quad \text{when} \quad t_p < t \end{aligned}. \tag{21}$$

AFM-IR formalisation

To detect the thermal expansion on the surface of the sample (sphere in our model), we use the AFM tip in contact. Under the effect of expansion, the tip will lift up and push the cantilever that begins to oscillate. We are interested here only in the flexion deformation modes of the cantilever and neglect the torsion effect of the expansion. The cantilever is model like a mechanical beam and satisfies the flexion equation of motion [18]:

$$EI\frac{\partial^4 z}{\partial x^4}+\rho A\frac{\partial^2 z}{\partial t^2}+\gamma\frac{\partial z}{\partial t}=0 \tag{22}$$

where E is the cantilever Young modulus, I is the area moment of inertia, ρ is the density, A is the cross section of the beam, and γ is the damping in the beam.

The cross section is $A = we$ and the area moment of inertia $I = we^3/12$ in the case of a rectangular beam, where w and e are, respectively, the width and thickness. The x coordinate describes the longitudinal direction and z represents the deflection of the cantilever (Fig. 3). The cantilever length is L with the clamped extremity at $x=0$ and the tip positioned at $x=L-\delta x$ (Fig. 4).

FIGURE 3 AFM cantilever scheme. The lever is embeded at $x=0$ and the length is L. The tip is positioned at $x=L-\delta x$. The contact stifness is represented by two springs, one for the vertical displacement and the other for the lateral.

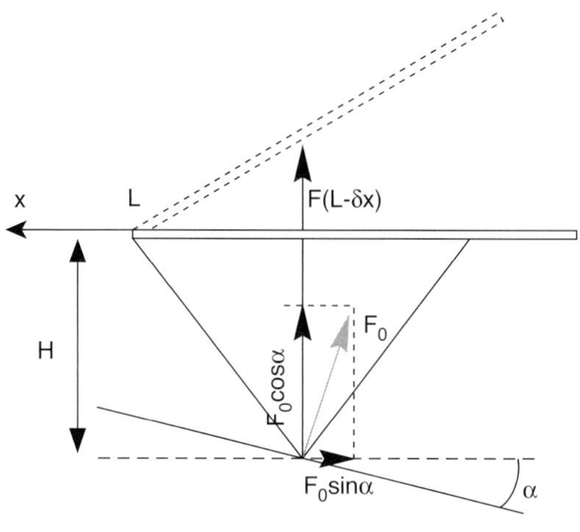

FIGURE 4 Scheme of the AFM tip in contact with the surface.

The general harmonic solution of equation (22) is:

$$z(x,t) = g(x)h(t) \tag{23}$$

with
$g(x) = A(\cos(\beta x) + \cosh(\beta x)) + B(\cos(\beta x) - \cosh(\beta x)) + C(\sin(\beta x) + \sinh(\beta x)) + D(\sin(\beta x) - \sinh(\beta x))$ and $h(t) = \exp(i\omega t)$, β being the wave vector, and ω being the complex angular frequency.

The boundary conditions of the cantilever imposed by the clamped end and the tip contact can be cast in the form

$$\begin{array}{cc} x = 0 & x = L \\ g(x) = 0 & EI\dfrac{\partial^2 g(x)}{\partial x^2} = M \\ \dfrac{\partial g(x)}{\partial x} = 0 & EI\dfrac{\partial^3 g(x)}{\partial x^3} + F_z = 0 \end{array} \tag{24}$$

where M is the bending moment and F_z is the reaction force at the lever end. The expression of F_z is found if we consider the deflection $g(L)$ at the end of the lever. The reaction force of the surface is opposite to the displacement at the end of the lever, leading to $F_z = -k_z g(L)$ where k_z is the vertical spring constant associated with the linear approximation for the small vibrations amplitude. The bending moment $M = -k_x H^2 \frac{\partial g}{\partial x}|_{x=L}$ can also be found in relationship with the moment induced by the lateral displacement of the tip and knowing the lateral spring constant k_x and the tip height H (Fig. 4). Substituting the expressions of F_z and M into equations (22), and using the modal expression of $g(x)$, we obtain the eigenvalue equation of the vibrations modes:

$$\begin{aligned} &-1 + \cos X \cosh X - VX^3(\cos X \sinh X + \sin X \cosh X) \\ &-UX(\sin X \cosh X - \cos X \sinh X) - UVX^4(1 + \cos X \cosh X) = 0 \end{aligned} \tag{25}$$

where $U = \frac{k_c L^2}{3k_x H^2}$; $V = \frac{k_c}{3k_z}$; $X = \beta L$.

This eigenvalue equation is the most general equation of vibration of the cantilever. If the tip is free ($k_x = 0$ and $k_z = 0$), we find the well-known equation of modes [19]$1 + \cos X \cosh X = 0$. In the AFMIR technique, the cantilever tip is kept in contact with the sample by the initial load force. The usual values of the cantilever spring constant k_c employed in contact mode are very small (from 0.03 to 0.2 N/m) compared to the spring constant k_z of the tip contact (around 10^5 N/m), so that the parameter V tends to 0. Moreover, we have checked that if we suppose no indentation ($V = 0$), the calculated values of frequencies by finite element method were in good agreement with those measured experimentally [20]. We will denote this specific configuration by "contact resonance." This approach is similar to that used by Yuya *et al.* on contact resonance for visco-elasticity [21], but in our case the tip can move laterally and is not allowed to produce indentation into the sample.

As a result, the general eigenvalue equation for the AFMIR configuration reads

$$-1 + \cos X \cosh X - UX(\sin X \cosh X - \cos X \sinh X) = 0. \tag{26}$$

The spatial distribution of mode n in the case of contact resonance is described by

$$g_n(x) = (\cos(\beta_n x) - \cosh(\beta_n x)) - \left(\frac{\cos(\beta_n L) - \cosh(\beta_n L)}{\sin(\beta_n L) - \sinh(\beta_n L)}\right)(\sin(\beta_n x) - \sinh(\beta_n x)), \tag{27}$$

where β_n is the wave vector of the modes solution of the eigen equation (26). One can note that the value of the mode depends on the value of the lateral stiffness. This property is interesting in a mechanical point of view, because the frequency of the mode will give us the stiffness of the sample [22]. This sort of analysis is equivalent to the phase imaging of the AFM tapping mode [23]. Combining chemical and mechanical analysis seems attractive to improve the sensitivity or the resolution of the technique.

In our case, the cantilever is excited by a source of motion which is the thermal expansion. The corresponding equation of motion can be written as

$$EI\frac{\partial^4 z}{\partial x^4} + \rho A\frac{\partial^2 z}{\partial t^2} + \gamma\frac{\partial z}{\partial t} = S(x,t) \tag{28}$$

where $S(x,t)$ is the mechanical source of motion, and where the general solution can be expanded as a sum over eigenmodes: $z(x,t) = \sum_n P_n g_n(x) h(t)$, with P_n being the amplitude coefficient of the mode n.

The source term $S(x,t)$ describes the variation of the force induced on the tip by the expansion of the object. We suppose that the tip is rigid and transmit this change of force to the cantilever. The tip contact can be characterized by the spring constant k_z, which depends on the impedance of the Young modulus of the tip and the sample, and the radius of the tip (neglecting the sample curvature compared to the tip) [24]. The force received by the tip reads

$$F(t) = k_z u_{r=a}(t) = k_z\frac{1+\nu}{1-\nu}\frac{\alpha}{3}aT(t). \tag{29}$$

The beam modes have a vanishing amplitude at $x = L$ (eqn. (27)). In this model, under the expansion of the object (sphere), the tip is not able to induce a direct displacement but creates a bending of the lever at $x = L$. Moreover, the angle between the lever and the surface is usually not null (Fig. 4). In order to account for this behavior in the source term, we split the force into two (normal and lateral) components, that are applied at a point $x = L$–δx, with $\delta x \ll L$. The normal component is $F_{\text{norm}}(t) = F(t)\cos(\alpha)$, and the lateral one is $F_{\text{lat}}(t) = F(t)\sin(\alpha)\frac{H}{\delta x}$. Therefore, the source term reads

$$S(x,t) = \delta(x - L + \delta x)F(t) = B\delta(x - L + \delta x)T(t) \tag{30}$$

with $B = \left(\cos(\alpha) + \frac{H}{\delta x}\sin(\alpha)\right)\frac{1+\nu}{1-\nu}\frac{ak_z\alpha_{sph}}{3}$.

Replacing the expression of the source term (30) into equation of motion (28) using Fourier transformation and modal orthogonality properties we finally find [21]

$$z(x,t) = \sum_n \frac{C}{\omega_n}\frac{\partial g_n(x)}{\partial x}\bigg|_{x=L} g_n(x)\left(\sin(\omega_n t)e^{-\frac{\Gamma}{2}t}\right) * T(t) \tag{31}$$

with $\omega_n = \sqrt{\frac{EI}{\rho A}}{\beta_n}^2$, $\Gamma = \frac{\gamma}{\rho A}$, $C = -(\cos(\alpha)\delta x + \sin(\alpha)H)k_z\frac{1+\nu}{1-\nu}\frac{a}{3\rho AL}\alpha_{\text{sph}}$ and * denotes the convolution product.

The signal obtained by the 4-quadrant detector, $Q(t)$, is not a complete deformation of the cantilever but only its extremity deflection. It can be expressed as the integrated cantilever slope along the diameter D of the laser diode spot on the cantilever [25]:

$$Q(t) = \sum_n \frac{C}{\omega_n}\frac{\partial g_n(x)}{\partial x}\bigg|_{x=L}\int_{L-D}^{L}\frac{\partial g_n(x)}{\partial x}dx\left(\sin(\omega_n t)e^{-\frac{\Gamma}{2}t}\right) * T(t). \tag{32}$$

As the diameter spot of the laser diode spot is small compared to the length of the cantilever, the final expression of the 4-quadrant detector is:

$$Q(t) = \sum_n \frac{DC}{\omega_n}\left(\frac{\partial g_n(x)}{\partial x}\bigg|_{x=L}\right)^2\left(\sin(\omega_n t)e^{-\frac{\Gamma}{2}t}\right) * T(t). \tag{33}$$

Most of the time, the duration t_p of the laser pulse is short (8–10 μs) considering the time response of the cantilever (15–40 μs). Taking classical cantilever parameters used to make contact imagery and polymer stiffness to model the sample, the theoretical oscillation of the lever is showing Figure 5. We clearly see several oscillations of different periods that decrease until 600 μs.

The analysis of the AFM-IR signal is easier to perform in the Fourier domain, revealing the different excited modes and their amplitude (Fig. 6). Usually the analyzed signal is the Fourier transform amplitude, that is, the modulus of the Fourier transform of $Q(t)$:

$$\begin{aligned}\tilde{Q}(\omega) &= \text{Fourier Transf}(Q(t)) = \sum_n \tilde{Q}_n(\omega)\\ &= \sum_n D|C|\left(\frac{\partial g_n(x)}{\partial x}\right)^2\bigg|x = L\frac{\left|\tilde{T}_{\text{sph}}(\omega)\right|}{\sqrt{\left({\omega_n}^2 - \omega^2\right)^2 + \Gamma^2\omega^2}}.\end{aligned} \tag{34}$$

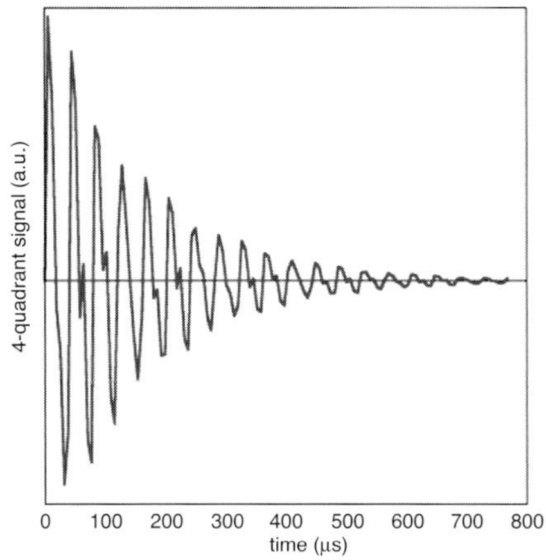

FIGURE 5 Temporal response Q(t) of the cantilever excited by a temperature increase generated by pulse laser of 1 μs.

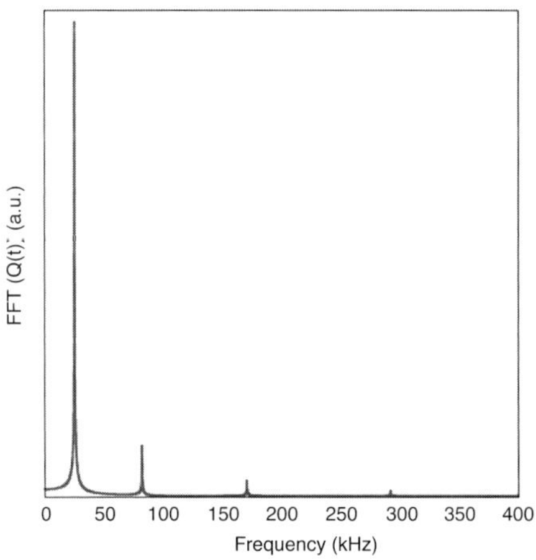

FIGURE 6 Fourier transfomation representation of the temporal signal Q(t). By this approach, the cantilever modes and their amplitudse are directly pointed.

The Fourier analysis shows that the response of the cantilever is the product of its modes response multiplying by the energy induced by the photothermal effect. In fact, following expression (34) the behavior of each modes are the same in the function of the temperature increase. Most of the time, to measure the absorption signal it is easy to study the amplitude change of the fundamental mode because it possess most of the energy. The expression of the maximum amplitude of mode n (at its resonance) is

$$\tilde{S}_n(\omega_n, \sigma) = H_m H_{\mathrm{AFM}} H_{\mathrm{opt}} H_{\mathrm{th}} \sigma \kappa(\sigma) \tag{35}$$

with $H_m = k_z \alpha a$,

$$H_{\mathrm{AFM}} = \frac{1}{\Gamma\omega_n}(\cos(\alpha)\delta x + \sin(\alpha)H)\frac{D}{\rho AL}\left(\frac{\partial g_n(x)}{\partial x}\bigg|_{x=L}\right)^2$$

$$H_{\mathrm{opt}} = \frac{n}{(n^2+2)^2}c\varepsilon_0|E_{\mathrm{inc}}|^2$$

$$H_{\mathrm{th}} = \frac{4\pi a^2}{k_s}\left(\frac{\sin\left(\frac{\omega_n t_p}{2}\right)}{\omega_n}\right).$$

Even if expression (35) seems complex, most of the parameter are constant and do not change during the study. The final expression of the amplitude of the mode demonstrates the direct proportionality between the absorption of the object (κ) and the signal recorded by the AFM tip S_n. As mentioned above, the idea to get an infrared spectrum of the sample is to record the variation of the detector signal (fundamental mode) in the function of the wave number σ. Considering the size of the tip apex in contact with the sample (a few nanometer) we clearly see the interested of such a method to make ultralocal spectroscopy. The other possibility of the AFM-IR is to perform chemical mapping by fixing the wave number at a frequency corresponding to the chemical function band and recording the amplitude in the function of the space (surface scan). The resolution of such a mapping is close to the resolution of the AFM topography (10 nm) but will also be limited by the thermal expansion and mechanical properties of the object.

APPLICATIONS IN MICROBIOLOGY: BACTERIA STUDIES

The AFM-IR technique has been used in our research center (Laboratoire de Chimie Physique, Orsay, France) for 5 years now [11]. This experiment was setup and conducted at the Centre Laser Infrarouge d'Orsay (CLIO, http://clio.lcp.u-psud.fr/clio_eng/clio_eng.htm) and now constitutes a beam line labeled AFMIR (http://www.lcp.u-psud.fr/rubrique.php3?id_rubrique=115). CLIO is conducted in a rather unusual way, which allows us to offer our systems to outside users, from other research groups. The particularity of the laser source is to be a free electron laser tunable from 3 to 150 μm. The accessibility is managed by a program comity like in all synchrotron centers. It is in this context that we have not only been able to collaborate on several projects in different areas, particularly in biology [26–29], but also in nanophotonics [30, 31]. To illustrate the main features of the AFM-IR technique by a range of experimental results, we have chosen to present here only the results obtained in biological samples.

Experimental demonstration on *E. coli*

The test sample that we have chosen to illustrate the characteristic of the AFM-IR experiment is the bacterium *E. coli*. This is because the size of these objects are well calibrated (2–6 μm long, 1 μm wide, 500 nm high), and also the absorption bands are known (idem *Rhodobacter* Fig. 2) and assigned – for example, amide I at 1650 cm^{-1}, amide II at 1550 cm^{-1}, and DNA (phosphate backbone) at 1080 cm^{-1}. To prepare the surface of the prism, a large amount of bacteria are centrifugated and washed with distilled water to eliminate their nutritive environment; a suspension of bacteria is then deposited and heated at 30°C to slowly evaporate the water. The final concentration of bacteria on the surface is typically 10–20 bacteria over 100 × 100 μm. Using this method, we obtain intact bacteria, which keep their original shape. The absorption bands in the mid-IR are still present even if the bacteria are dried. After an AFM scan on the prism surface, we can select a single bacterium and make different analysis in function of tip position and laser wavelength.

Case 1: The AFM tip is positioned on the bacterium and the laser wavelength is centered on the amide I band (Fig. 7). We observe oscillations with an amplitude of 30 nm that decreases within a time constant close to 200 μs. The periods of these oscillations are around 16 μs, effectively greater than the time scale of the laser pulse (around 8 μs). Under these conditions, the observed oscillations are evidently oscillations of the cantilever recorded by the

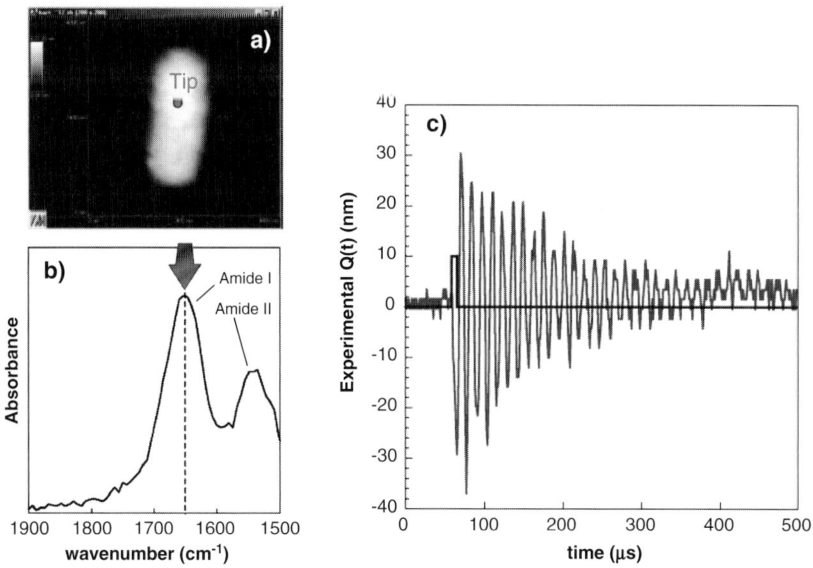

FIGURE 7 **(a) AFM topography picture of the bacterium; the position of the tip is indicated in blue. (b) FTIR spectrum, the bacterium absorption bands is drawn in green and the wave number of the CLIO laser is indicated by the red arrow. (c) Oscillations recorded by the 4-quadrant detector (in gray) in function of time superposed to the CLIO pulse laser (black).**

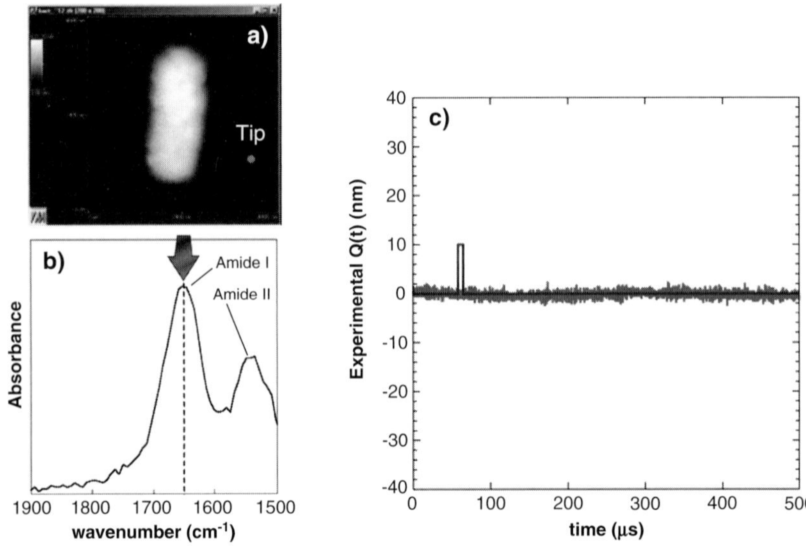

FIGURE 8 **(a) AFM topography picture of the bacterium; the position of the tip is indicated in blue. (b) FTIR spectrum, the bacterium absorption bands is drawn in green and the wave number of the CLIO laser is indicated by the red arrow. (c) Oscillations recorded by the 4-quadrant detector (in gray) in function of time superposed to the CLIO pulse laser (black).**

4-quadrant detector, as calculated in concept and technique description section. This signal is usually called deflection. The oscillations are centered at the zero value. This value corresponds to an applied static force of 10 nN (300 nm deflection) useful to make the AFM scan. The amplitude of oscillations is always smaller than the static deflection of the cantilever, allowing us to be sure that the tip stays in contact with the bacterium. We can remark here how this experimental oscillation is similar to the theoretical one (Fig. 5) confirming the approximations used in the model.

Case 2: In order to be sure that the signal is not due to direct heating of the tip or other thermal or acoustic artifacts, a control measurement (Fig. 8) was recorded outside the bacterium keeping the laser wave number at the same value. The detector signal shows only noise. This proves that the cantilever oscillations are only excited by the bacterium expansion and not by absorption and heating of the cantilever. We clearly see here the efficiency of the total reflection configuration, allowing heating sample only, which is not the case of other configuration explored by Reading *et al.* [32].

Case 3: To test whether the expansion of the bacterium is really related to the absorption, we have tuned the laser wave number out to the amide I band, at 1800 cm^{-1} (Fig. 9) and put the tip again on the bacterium. The signal shows us a small oscillation not more than 3 nm, which is in good agreement with the relatively small absorption of the bacterium.

All these experiments have shown that the oscillations recorded by the four-quadrant detector correspond to the cantilever deflection excited by the

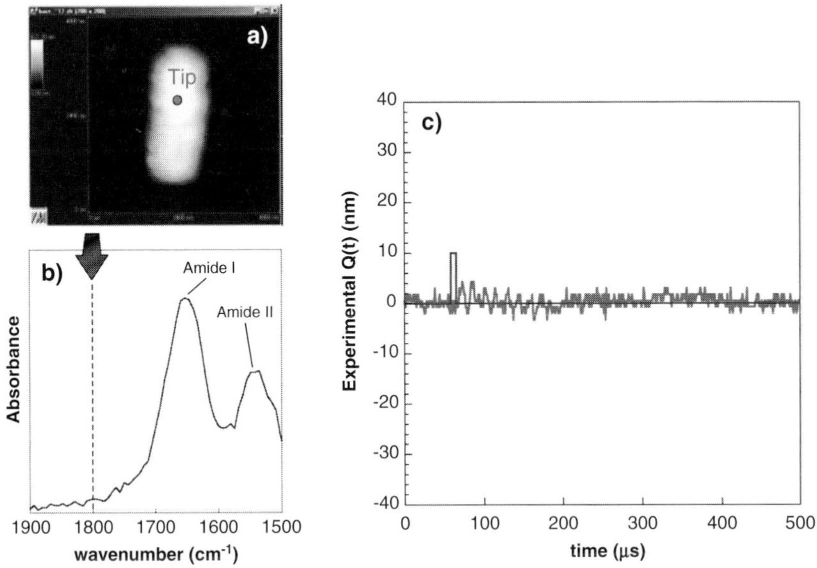

FIGURE 9 (a) AFM topography picture of the bacterium; the position of the tip is indicated in blue. (b) FTIR spectrum, the bacterium absorption bands is drawn in green and the wave number of the CLIO laser is indicated by the red arrow. (c) Oscillations recorded by the 4-quadrant detector (in gray) in function of time superposed to the CLIO pulse laser (black).

photothermal expansion of the bacterium. As demonstrated in the theoretical part (concept and technique description section), the amplitude of the cantilever oscillations is proportional to the absorption. To quantify these oscillations, the best way is to perform a Fourier analysis of this temporal signal

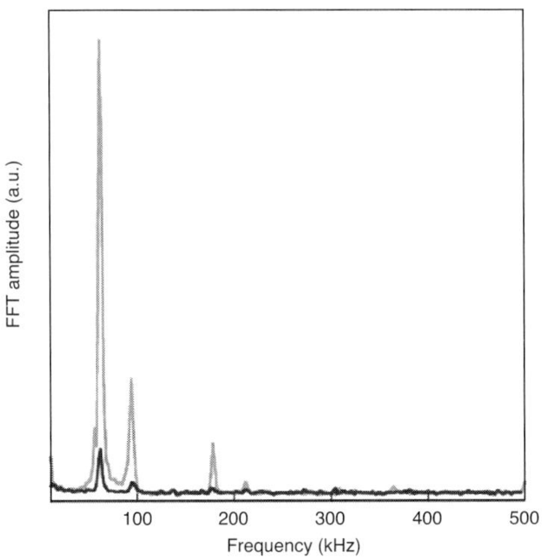

FIGURE 10 FFT representation of the temporal signal corresponding to case 1 (in red) and case 3 (blue).

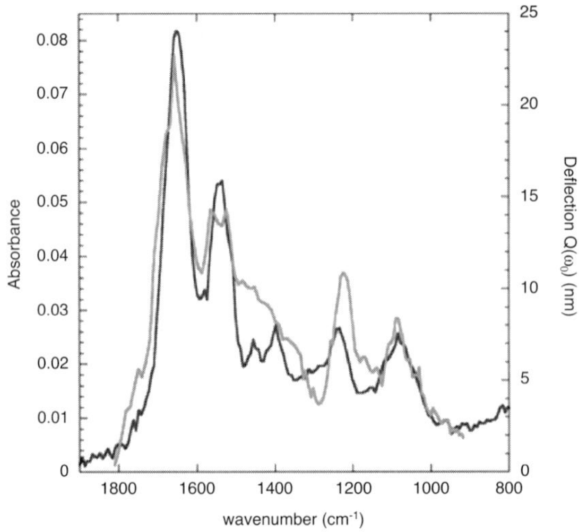

FIGURE 11 FTIR spectrum (in black) of *E. coli* culture and AFMIR spectrum (in gray) of a single *E. coli* bacterium.

The Fourier transform of the temporal signal from case 1 and 3 is depicted in Figure 10. It is easy to note that this analysis shows us immediately the different modes of the oscillations. The intensive one is the fundamental and has its resonance frequency around 62 kHz. By looking at the maximum of the first mode we can quantify the amount of absorption in the two different cases. As we have demonstrated in our theoretical section the variation of the maximum of one mode (e.g., fundamental) is related to the infrared absorption (eqn. (35)).

Then to obtain an infrared spectrum by the AFM-IR technique one just needs to record the change of the mode maximum in function of the wave number of the laser. Figure 11 shows the AFM-IR spectrum obtained with the previous bacterium compared with the FTIR spectrum obtained with a complete culture (millions of bacteria). The AFM-IR spectrum is quite similar to the FTIR spectrum, which is in good agreement with the theoretical hypothesis (concept and technique description section). The small differences probably come from the water vapor absorption, which is not eliminated during the AFM-IR measurements.

As mentioned previously chemical imaging is also possible with the AFM-IR technique. The laser wave number is fixed at a value corresponding to a specific band characterizing a chemical function (1240 cm^{-1} P=O band, 1660 cm^{-1} C–O, and C–N, etc.) and the tip is scanned over the sample. Figure 12 shows different chemical maps of *E. coli* for two different bands, amide I (at 1660 cm^{-1}) characterizing proteins and phosphate bands (at 1240 cm^{-1}) related to the DNA backbone. In the case of *E. coli,* protein and DNA are distributed homogeneously into the bacterium showing a signal directly proportional to the AFM topography signal (i.e., related to the sample thickness). The amide I

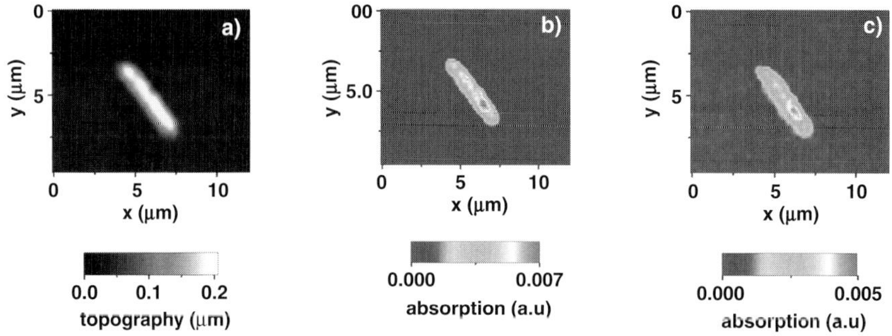

FIGURE 12 **(a) AFM topography picture of a single bacterium *E. coli*; (b) corresponding chemical mapping of the amide I band (1660 cm^{-1}); (c) corresponding chemical mapping of asymmetric phosphate band (1240 cm^{-1}).**

absorption is globally more intense than the phosphate (ratio 7/5), which is in good agreement with the spectroscopy too. When the tip is close to the border of the bacterium without touching it, there is no signal that means the expansion is about less than 10 nm. This result illustrates the sensitivity and spatial resolution of the technique. With the AFM-IR technique it seems possible to detect the object with the size of around 10 nm. The resolution limitation in the case of an isolated object on the surface is in fact the resolution of the tip itself and the sensitivity of the thermal expansion detection.

T5 bacteriophage detection inside *E. coli*

(collaboration L. Letellier, IBBMC, UMR 8619 Paris Sud, Orsay)

The bacteriophage T5 is one of the best-characterized viruses. Over the past 20 years the chemical composition and the developmental cycle of this virus have been extensively studied. The interest of using AFM-IR to detect such small objects and to follow its evolution when infecting bacteria appears to be great, since it should be possible to track it *in vivo*. Before reaching this final step, we have tested whether AFM-IR technique was able to detect the virus inside and outside of the bacterium [26].

The sample was composed of *Escherichia coli* bacteria and phage T5. Bacteriophages T5 st(0) were produced from *Escherichia coli* F and purified [33]. The final concentration of the phage stock was evaluated to 1.8×10^{13} infecting phages/ml. Bacteria *E. coli* F were grown in LB medium to the exponential growth phase (3×10^{8} cells/mL) and infected by phages with an average multiplicity of 60. The infection was stopped 20 min after the beginning by adding chloramphenicol at a final concentration of 50 mg/mL. Chloramphenicol is an antibiotic inhibiting bacterial protein synthesis, and consequently, blocking the phage replication. Infected bacteria were centrifuged and the pellet was washed three times in pure water and resuspended to a final concentration of approximately 1.5×10^8 cells/mL. A drop of the solution was deposited on the ZnSe prism and dried at room temperature. For the infected bacteria, as the infection was stopped by chloramphenicol addition after only 20 min, various stages of the virus development can be found inside the cells [34].

To detect the bacteriophage T5, we have studied the bacteria on the prism at two different wave numbers: 1660 cm^{-1} (amide I) characterizing the proteins of the capsid and 1080 cm^{-1}, which is the maximum of the DNA symmetric phosphate band. The idea is to localize mature phages (DNA inside the capsid). After studying many bacteria, we have detected a small amount of bacteria completely destroyed (broken membranes) revealing complex DNA signature and numerous clean bacteria (without specific signal). Whatever, we have found a beautiful specimen showing one isolated spot of DNA (Fig. 13).

The DNA mapping reveals the position of one hot spot of DNA surrounded by homogeneous distribution of the bacterium. This hot spot is probably a mature T5 (close to the membrane) but it seems difficult to conclude with only this first experiment. The fact that the DNA signal is not null around the phage is probably link to heat diffusion of the other part of the bacteria or also could be from other mature T5 buried deeper. The amide I mapping also reveals good indication about the presence of the phage. By looking closely to distribution of the amide I (protein) we can note that there are some localized signals with small amplitude and one is coincident to the DNA signal. This sort of contrast has never been detected with uninfected *E. coli*. All these results strongly suggest the presence of one mature phage.

This event is really rare in our sample probably for two reasons. First, the time of incubation is too short to produce enough mature T5 inside bacteria. The second comes from the AFMIR detection that is only sensitive to what is close to the bacterium membrane. Preliminary calculus has shown if the T5 is buried at distance more than its diameter (90 nm) then the detection is not really possible because the total expansion at the surface is so weak that the technique is not able to detect.

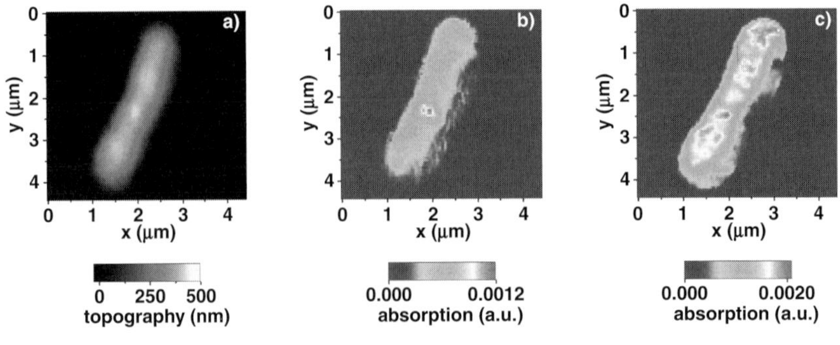

FIGURE 13 (a) AFM topography picture of a single *E. coli*; (b) corresponding mapping of the DNA (at 1080 cm^{-1}); (c) corresponding mapping of amide I (at 1660 cm^{-1}).

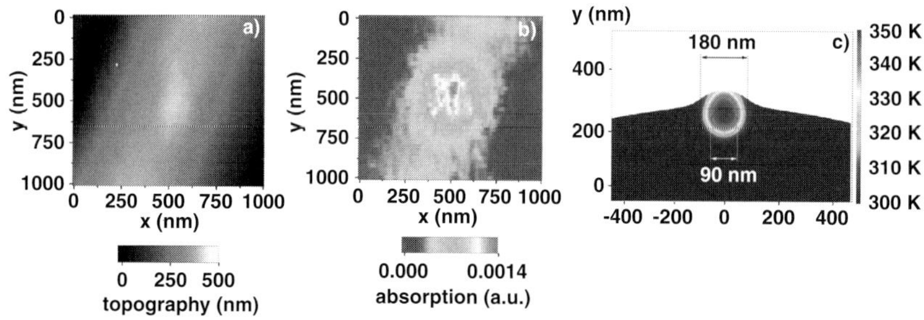

FIGURE 14 (a) AFM topography of *E. coli*; (b) corresponding chemical mapping of the DNA (at 1080 cm^{-1}); (c) numerical simulation of the phage thermal expansion inside the bacterium.

The size of the thermal expansion profile can be larger than the actual object size. Figure 14 shows a zoom scan of 1 × 1 μm of the DNA mapping. The measured width at the half maximum is around 200 nm and is greater than the real size of the phage (90 nm). This value is directly due to the elastic property of the dry bacterium and specifically the Poisson ratio that affects the transfer of stress between spatial dimensions. Figure 14(c) shows a finite element calculation (using COMSOL software) to show how the matter is displaced when the object is buried even close to the surface. The figure plots the cross section of the thermal expansion displacement with the local temperature indicated by the rainbow color table. (The deformation is magnified by more than 100 × to make it visible in the plot.) We clearly see in this model that a heated object with a size of 90 nm, appears with a surface deformation extending over roughly twice this distance, as measured at the half maximum.

To check that the hot spot is really due to DNA condensation, we obtained infrared spectra with the AFM-IR setup and compared this to the classical phosphate band of bacteria (Fig. 15). However, the phosphate band of the hot spot is very similar to the DNA signature of the bacteria and one admitted issue is that we do not have a differentiated spectral signature in the IR range to distinguish the viral DNA from the bacterial DNA.

Our next step is to make other T5 AFMIR studies jointly with transmitted electronic microscopy to know perfectly the statistic of mature phage inside the bacteria and to know more about the distribution of DNA in the case of infected bacteria.

PHB location into Rhodobacter capsulatus

(collaboration P.Sebban, LCP, UMR 8000, Univ Paris Sud, Orsay)

Rhodobacter capsulatus is a purple nonsulfur photosynthetic bacterium, which produces a polymer, the polyhydroxybutyrate (PHB), for its energy storage [35–37]. PHB belongs to the class of polyesters and has been used for several years for the production of plastics having similar mechanical and

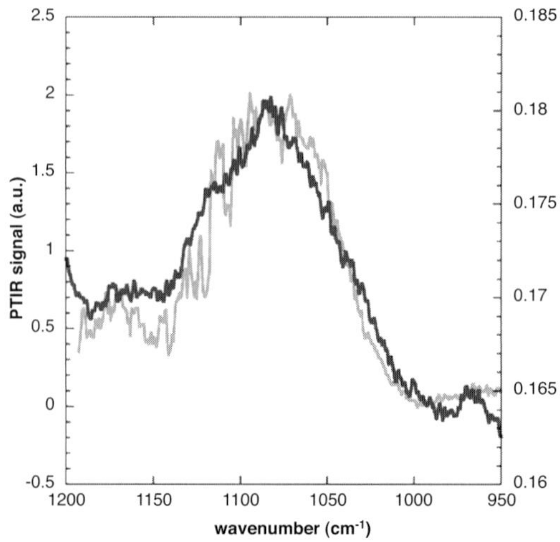

FIGURE 15 Comparison of AFM-IR spectrum of T5 phosphate band (in gray) with classical FTIR phosphate band (black).

thermoplastic properties as those of polyethylene and polypropylene but with the advantage to involve renewable resources. In bacteria, PHB is contained in granules (vesicles) with variable size and it can be used by the bacterium when its nutritive resources become limiting. The presence of PHB can be probed in the midinfrared domain by the presence of specific absorption bands, in particular around 1740 cm^{-1} (C=Os of ester) [38], easily distinguishable from other bacterial bands (amide I at 1660 cm^{-1}, amide II at 1550 cm^{-1} [38]). Many studies have already been achieved on the characterization of PHB involving different vibrational spectroscopic techniques (Raman, FTIR). However, these methods often require prior extraction of PHB [39] and do not allow an *in situ* study. Our photothermal technique allows us to study PHB directly in the cell without extraction or complex sample preparation.

Rhodobacter capsulatus wild type were grown in Erlenmeyer flasks filled to 50% of the total volume with malate yeast medium supplemented with kanamycin (20 mg/ml) and tetracycline (4 mg/ml). The cultures were grown in darkness at 30°C for 48 h on a gyratory shaker (140 rpm). The cell suspension was spun down by centrifugation at 1000 rpm for 15 min. The supernatant was removed and the pellet was diluted in distilled water. A drop of the solution was deposited on the ZnSe prism for AFMIR studies or on ZnSe glass for FTIR one and dried at room temperature for 20 min.

The top panels in Figure 16 display the topography of bacteria obtained by classic AFM. Registered variations in height give us the shape of objects; the maximum being represented in yellow. The bottom panels show the amplitude

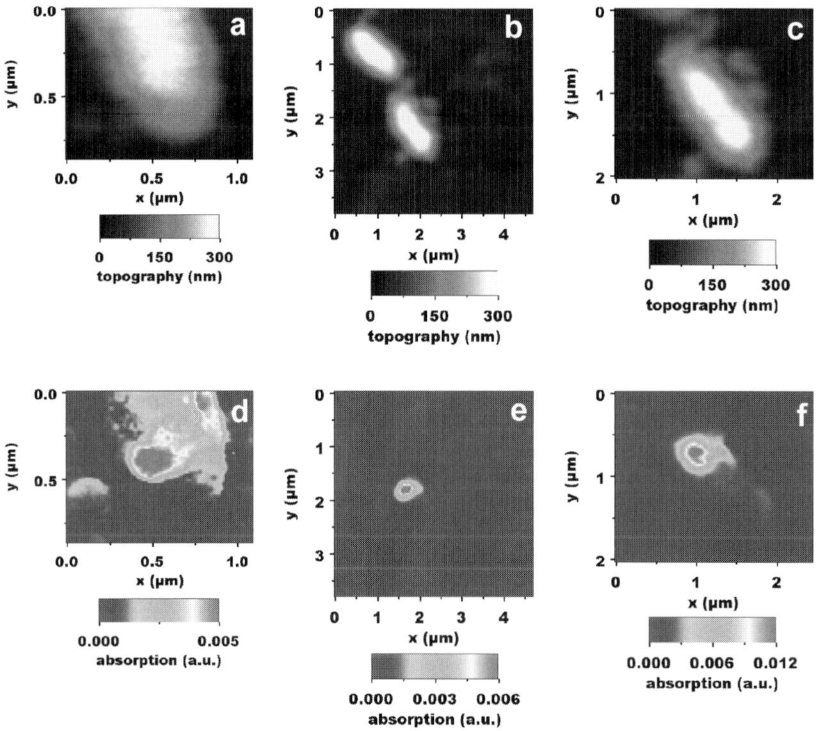

FIGURE 16 (a) AFM topography of one single Rhodobacter capsulatus. (b) AFM topography of two seperated bacteria Rhodobacter. (c) AFM zoom on the lowest bacterium localized on (b). (d) Chemical mapping of PHB (at 1740 cm^{-1}) of the corresponding topogaphy (a). (e) Chemical mapping of PHB of (b). (f) Chemical mapping of PHB of (c).

distribution of the fundamental mode of the cantilever on the surface when the laser is set at 1740 cm^{-1}, which corresponds to the chemical cartography of PHB. The scale indicates an increasing absorption when the signal shifts from purple to red. On all maps, we have localized round areas where the signal is more intense (red domains). These domains correspond to PHB granules inside bacteria (Fig. 16 (d–f)). On each map, we can estimate the size of granules by taking the width at half height (this is justified by the simulation with the finite element method described in the previous and next section). For a dozen measurements, we obtained for the round shape sizes in the range 100–400 nm, that is consistent with the MET studies made jointly [28].

Figure 16(d) reveals a big round granule of 210 nm diameter and long shape one of 50 nm (top of the image). This long shape is very likely the result of small spherical PHB vesicles lined up close to the membrane [40]. Figure 16(e) shows a bacterium without PHB vesicle. This suggests that under these growing conditions, all *Rhodobacter capsulatus* bacteria do not necessarily produce PHB.

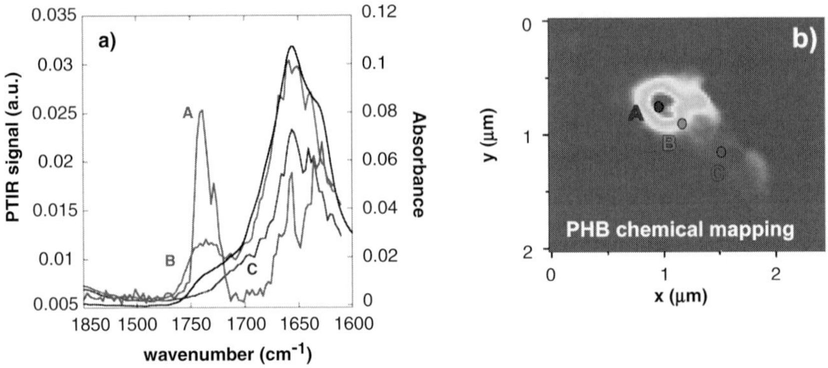

FIGURE 17 **(a) Comparison between local (A in red, B in orange, C in violet) AFMIR spectrum and FTIR spectrum (in green) of the bacteria culture. (b) bacterium chemical mapping of C=O of the ester band with corresponding position (A, B, C) of the spectra measurements.**

The PHB/proteins mass ratio is low which is in good agreement with the FTIR study (Fig. 17(a) green curve). Figure 16(f) (zoom of Fig. 16(e)) reveals that the absorbing area is in fact composed by two adjacent vesicles of different sizes. The estimated size and separation distance will be studied in the simulation coming next.

One must be careful interpreting AFM-IR chemical images alone because the contrast can not only come from IR absorption but also thermomechanical properties of the sample. To check that the observed contrast results from chemical differences, we have studied the spectroscopy response of the one vesicle and compared with the FTIR spectrum of the bacteria culture (Fig. 17).

The spectrum on a single bacterium (in red Fig. 17(a)) was measured by positioning the tip of the AFM directly on the maximum of the signal using the chemical mapping of PHB (as pointed by A Fig. 17(b)). We observe an intense band of PHB (centered at 1740 cm^{-1}) whereas the amide I band at 1660 cm^{-1} appears weaker and noisy. As the tip is perfectly positioned on the vesicle this should lead to no amide I signal. But there are many proteins surrounding the vesicle and a part of the bacterium that respond to the amide I absorption. This external heating (out from the vesicle) probably creates an impulse to the vesicle of PHB resulting in a nonzero AFM-IR signal. However, concerning the C=O band of ester, we obtained a clear signal demonstrating the PHB nature of the hot spot mapping from Figure 16. The second spectrum was recorded by positioning the tip at point B (Fig. 17) at the border of the absorption vesicle. The spectrum shows a better signal for the amide I that is similar to the FTIR spectrum of the bacteria culture (Fig. 17 in green). The intensity of the PHB in that case has decreased compared to the previous position, which is consistent with the PHB chemical mapping. When the tip is positioned at C (Fig. 17), out of the vesicle, the AFM-IR spectrum does not show the C=O band (in violet Fig. 17).

In order to determine the vesicles diameter and to estimate the resolution of our technique, we decided to study in more detail a zoom of Figure 17(b), where two PHB granules of different size and close to each other seem to be observed. We made a cross-section of this absorption image, previously converted in a gray scale, shown with a green dashed line in Figure 18(a) and obtained the corresponding thermal expansion curve (green curve on Fig. 18(e)).

We used finite element analysis (using the commercial software COMSOL) to make all our absorption simulations. The bacterium is modeled by a cylinder in which two spheres of different sizes are included in order to represent the two PHB vesicles (Fig. 18(b)). We simulated the PHB absorption by the heating these two spheres. The thermal expansion is calculated during the laser pulse considering that the heat diffusion after the pulse is negligible (eqn. (35)). The 3D calculation (Fig. 18(c)) shows the bacterium expansion resulting from the spheres heating. For an easy comparison with the experimental mapping, we represent the calculated expansion projected on 2D (Fig. 18(d)). We can note that this simulated picture is really similar to the experimental chemical mapping making us confident about the way of modeling. From this calculating picture, we extracted as previously the thermal expansion curve and compared it to the experimental one (Fig. 18(e)). We have adjusted manually the vesicles sizes and distance parameter to obtain the best fit between the two curves. Finally, the best fit is obtained for vesicles of 280 and 100 nm in diameter respectively and distant of 110 nm with an uncertainty of 10 nm on each value. The vesicles are also 10 nm under the surface of the bacterium. This example

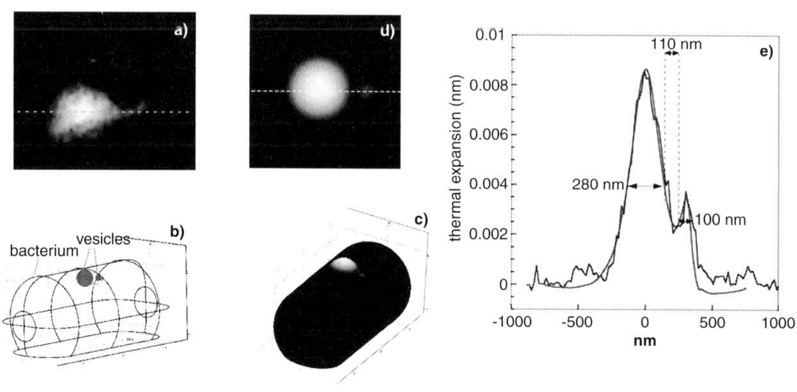

FIGURE 18 **(a) chemical mapping zoom of PHB vesicle same as Figure 17(b) converted in a gray scale, the green dashed line indicates the position of the cross section represented in (e). (b) The model structure of the bacterium with the two vesicles. (c) 3D photothermal simulation of the two vesicles expanding. (d) 2D projection mapping corresponding to the 3D simulation, blue dashed line indicates the position of the cross section represented in (e). (e) Comparison with experimental vesicles thermal expansion (Fig. 18(a)) and the corresponding COMSOL numerical simulation (Fig. 18(d)).**

demonstrates that the AFM-IR technique possesses a very good resolution, probing buried object of 100 nm with a high sensitivity. We clearly see that the 100 nm resolution is not the limit of this technique. But it seems difficult to make samples test with controlled size objects to reach the limit.

MAPPING EUKARYOTES USING AFM-IR

Imaging eukaryotes require the mapping of the subcellular structures and their content. For that purpose, the AFM-IR technique is very attractive for several reasons. IR-excitations concern vibrational levels and, as no electronic excited state is involved, no photo-bleaching is induced on the contrary to what is observed with organic fluorophores excitation in the visible or UV-range. Vibrational spectroscopies, such as IR, but also Raman or CARS (coherent antistokes Raman scattering), are already used for topological analyses of tissues [41–43] and in a few cases for cellular mapping, but always with resolution above several μm, using FTIR [44] or Raman microspectroscopy [45–49]. However, in classical optical microscopy, submicrometric resolutions are not attainable in the IR-range as the Abbe limit imposes a resolution higher than $\lambda/2$ (i.e., 2.5 μm at 2000 cm^{-1}) [50]. Due to cellular dimensions (a few μm), this limit renders IR-optical spectroscopy not well suited for intracellular mapping. But submicrometric resolutions are possible using near-field techniques such as AFMIR [51–53].

Detection of exogenous molecules and the determination of their localization in subcellular structures are of great importance in biology both for fundamental studies and for the development of novel drugs. It requires localization of both the subcellular structures and of the exogenous molecule of interest. In this part of the chapter, we will first exemplify the possible use of AFMIR for label-free imaging of a subcellular structure, namely the nucleus. In the second place, we will describe the application of AFM-IR to the mapping of an exogenous molecule.

Localization of the endogenous structure by AFMIR

IR spectroscopy is a powerful technique for simultaneous analysis of lipids, proteins, carbohydrates, and phosphorylated molecules. Chemical function presents vibrational modes that can be identified in an infrared spectrum. For example, as shown in Figure 20, phosphate from the DNA shows symmetric and antisymmetric stretching bands of the P=O bond are, respectively, absorbing around 1080 cm^{-1} and 1240 cm^{-1}, amide from proteins show two characteristics signals with amide I at 1650 cm^{-1} and amide II at 1550 cm^{-1}, and C–H bonds from lipids around 3000 cm^{-1}. As exemplified previously (note for page about *E. coli*), DNA can be located by irradiation in the phosphate symmetric band.

Subcellular structures do not present the univocal single FTIR signature, and differences in the FTIR-spectra of isolated organelles (mitochondria, nucleus)

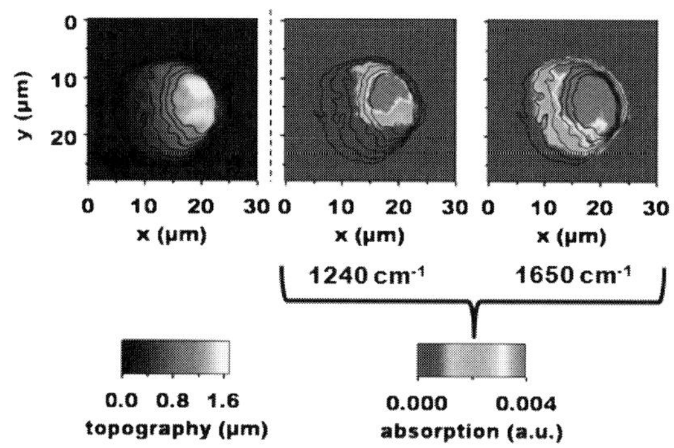

FIGURE 19 **MDA-MB231 cells. AFM-topography and chemical mapping by AFM-IR under irradiation at 1240 cm^{-1} (phosphate) and 1650 cm^{-1} (amide) [55].**

and cytosol are subtle [54]. However, it has been shown possible to identify nucleus, cytosol, and organelles by a fine comparison of the relative intensities of the IR-bands in different regions of the IR-spectrum (lipids, amide I and II, phosphate) and to map them using SR-FTIR [41]. One interest here is the access to label-free subcellular imaging. Based on these previous studies in classical optical IR-spectroscopy, we have recently shown AFM-IR enables the label-free imaging of the nucleus in human cells [55].

Example of localization of an endogenous structure, the nucleus

MDA-MB-231, hormone-independent breast cancer cells obtained from the human tumour cell bank were used for the experiments. They were grown under usual conditions, and then they were suspended in NaCl (0.9% solution) and ca. 10000 cells were deposited onto the ZnSe prism [55]. They were allowed to settle for a period of 15 min before the excess saline solution was removed by careful blotting with paper towel and then allowed to air-dry. AFM showed cells dispersed between NaCl crystals. The first step was to locate cells and record their topography using the AFM (see Fig. 19). The nucleus was located using irradiation at 1240 cm^{-1} (antisymmetric PO_2^- vibration) and at 1650 cm^{-1} (amide I), as a high phosphate and amide content is characteristic for the nucleus (phosphate for DNA and amide I for histones) [41]. A region with a high AFMIR-signal at 1240 and 1650 cm^{-1} can be seen in Figure 19.

Example of the localization of exogenous compounds

The next challenge of this technique is the intracellular detection and mapping of exogenous diluted compounds. Their signature needs to be revealed out of the strong signals of concentrated endogenous components. Figure 20 shows the IR-window for which cells are transparent and where exogenous compounds can easily be detected.

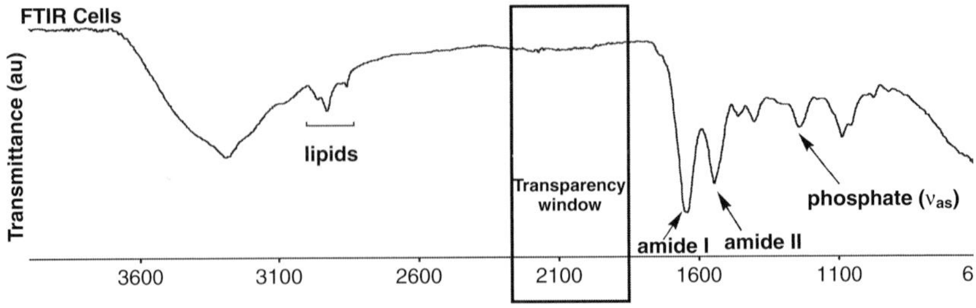

FIGURE 20 FTIR of human cells (MCF7) showing the transparency window where exogenous compounds can be easily detected.

To test the potentiality of the AFMIR microspectroscopy to map exogenous compounds inside cells, a metal–carbonyl derivative has been selected. As a matter of fact, metal–carbonyl compounds have been developed in the literature as IR-probes for their many advantages (foremost, stability in biological environments, and intense absorption in the 1800–2200 cm^{-1} region where biological samples are transparent) (RENVOI Chapitre M. Salmain). Moreover, organometallic compounds are known to be lipophilic [56, 57], thus ensuring a good intracellular pénétration [58].

(collaboration A. Vessières, ENSCP)

The selected reorganometallic conjugate **1** (see Fig. 21) with a cyclopentadienyl rhenium tris-carbonyl ($CpRe(CO)_3$ or cyrhetren) in a hydroxytamoxifen-like structure is robust in biological media, and has been shown to inducing a proliferative effect on hormone dependent breast cancer cells (MCF7) by interaction with estrogen receptors [58]. The organometallic moiety $Re(CO)_3$ presents a C_{3v} symmetry (ref Salmain, Chapitre and [59]) and the elongation vibration, symmetric (A_1) and antisymmetric (E) bands, are IR-active. Intense IR-signal are recorded in the desired area, respectively at 2020 and 1925 cm^{-1}

As a proof of principle, the cellular uptake of **1** and its distribution have studied in nonhormonal dependent breast cancer cells (MDA-MB-231). The cells were incubated for 1 h with **1** at concentrations in the micromolar range (10 μM) typically used for biological studies [58]. They were then washed, suspended in a NaCl solution (0.9 % solution), and ca. 10000 were deposited on the ZnSe prism. Cells were first located at the surface of the ZnSe prism using the AFM (see topography and black contours showing the AFM topology in Figs. 21–23) and were then mapped using the AFMIR setup at several wavelengths. The mappings were compared with those of control cells processed in a similar way (see above, localization of the nucleus). Figure 21 displays a series of chemical mappings both for a control cell (C) and cell treated with **1** (T). The AFMIR-signal is shown in color (increasing from blue to red).

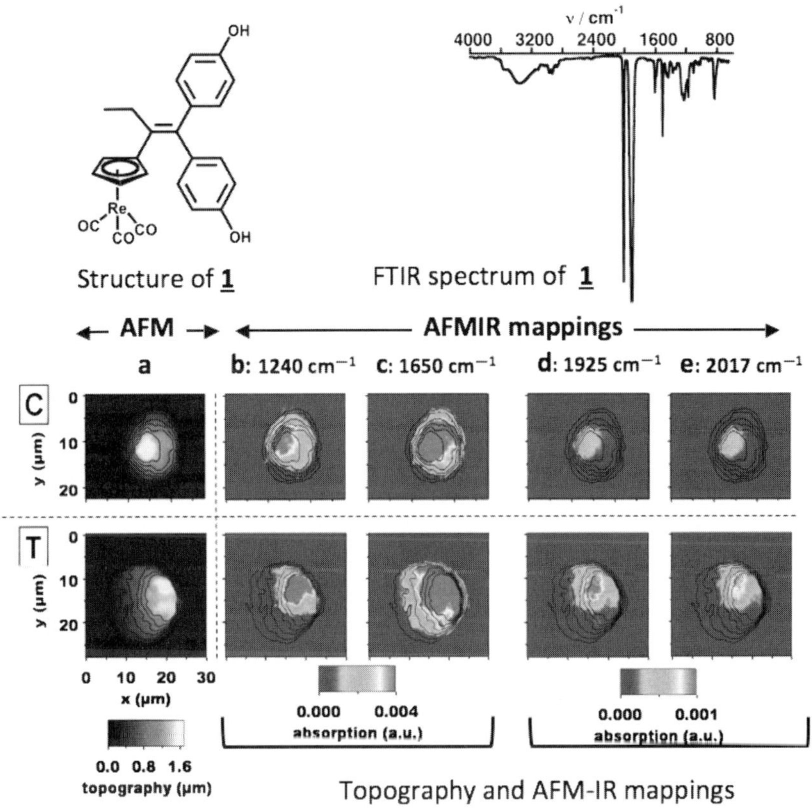

FIGURE 21 **Structure of 1, FTIR spectrum and AFMIR mappings of MDA-MB-231 cells. C: control cells; T: treated cells for 1h at 37°C with 10 μM of 1. AFM topography and AFMIR-mappings at different wavelengths, with the AFM contours superimposed as black lines [55].**

We will first consider the images recorded upon irradiation at 1925 and 2017 cm^{-1} (IR-bands of **1**). The control cells showed a weak AFMIR-signal at these three frequencies (Fig. 21), which was assigned to residual water. In the case of the treated cells, clear AFMIR-signals were recorded at 1925 and 2017 cm^{-1}. The mappings recorded at the two bands from **1** correspond to each other, further confirming that they are due to **1**. Interestingly, the cell showed an uneven distribution of the signals at 1925 and 2017 cm^{-1}, revealing an uneven distribution of **1**, with a spot showing an intense AFMIR-signal (hot spot). Figure 22 presents a cell that has been mapped at 1925 cm^{-1} at several magnification scales: these two images show that the same pattern is observed.

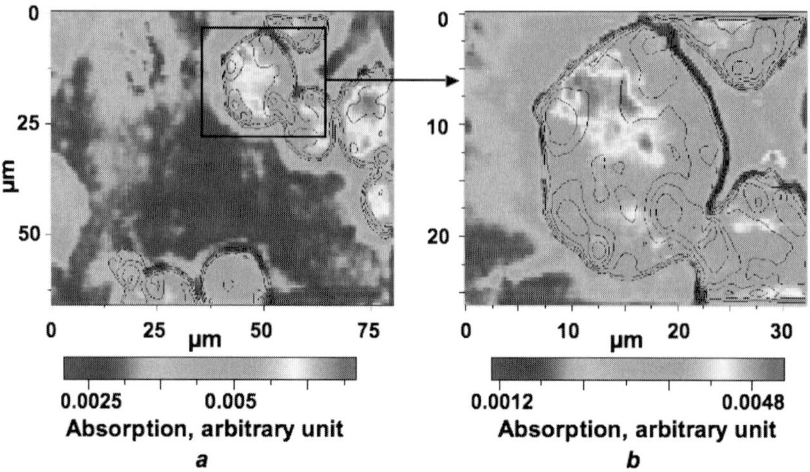

FIGURE 22 Localization by IR of metal carbonyl moiety of 1 in MDA-MB-231 cancer cells after a 1 h incubation at 37°C with 10 μM of 1. AFMIR mappings at 1925 cm^{-1}. In black: AFM contours showing individual cells. (a) First view (b) New image after zooming at the squared cell.

This uneven distribution within the cells is of interest. The hot spot at 1925 and 2017 cm^{-1} for the treated cells T is co-localized with the hot spot at 1240 and 1650 cm^{-1}, strongly suggesting a nuclear localization of **1**. This is consistent with previous observation suggesting that the presence of the organometallic moiety in the conjugate may favour nuclear accumulation [60–62].

Microspectroscopy using AFM-IR

The AFM-IR setup gives an access to local IR spectra through spectromicroscopy experiments: it is possible to record a spectrum at a well-defined location, that is, the location of the AFM tip on a sample surface. This is of utmost importance to validate the chemical images and to verify that the AFM-IR signal corresponds to the molecule of interest, and not due to background sources. This is exemplified in Figure 23. Spectra were recorded at two locations. Spots A and B are respectively inside the hot spot at 1925 cm^{-1} and outside. At spot A, a spectrum clearly showing the band at 1925 cm^{-1} was recorded (see arrow), whereas at spot B no such band was observed. This observation unambiguously indicates that the AFMIR signal at spot A is due to **1**. This is also shown in Figure 24 on a broader energy scale (2200–1200 cm^{-1}).

These experiments suggest an accumulation in the nucleus. It should be noted that the presence of the organo-metallic moiety in the conjugate may favor nuclear accumulation, as evidenced for some other organometallic bioconjugates [61, 62].

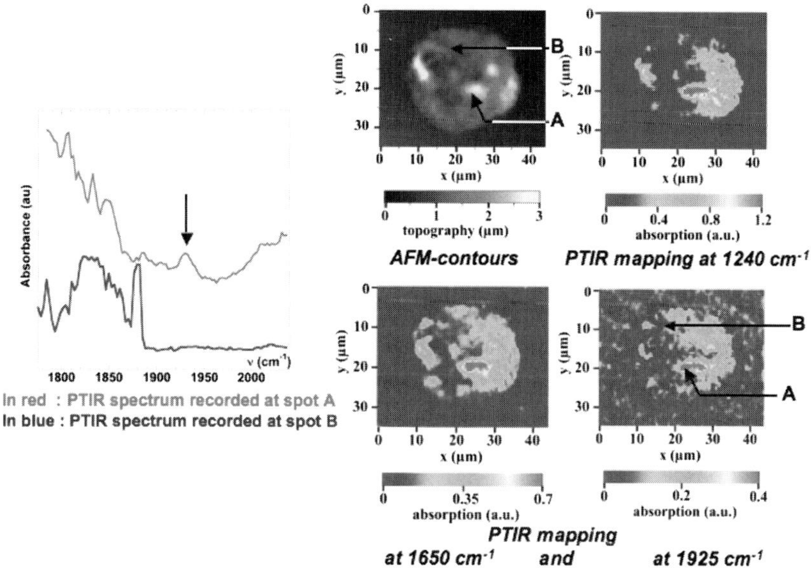

FIGURE 23 **Localization of IR-metal carbonyl moiety of 1 / MDA-MB-231 incubated 1 h at 37 °C with 10 μM P89. On the right: AFM contour and AFMIR mappings at 1240 cm^{-1} (DNA), 1650 cm^{-1} (amide I), 1925 cm^{-1} (probe). On the left: microspectroscopy in the 2050–1880 cm^{-1} region at position A or B.**

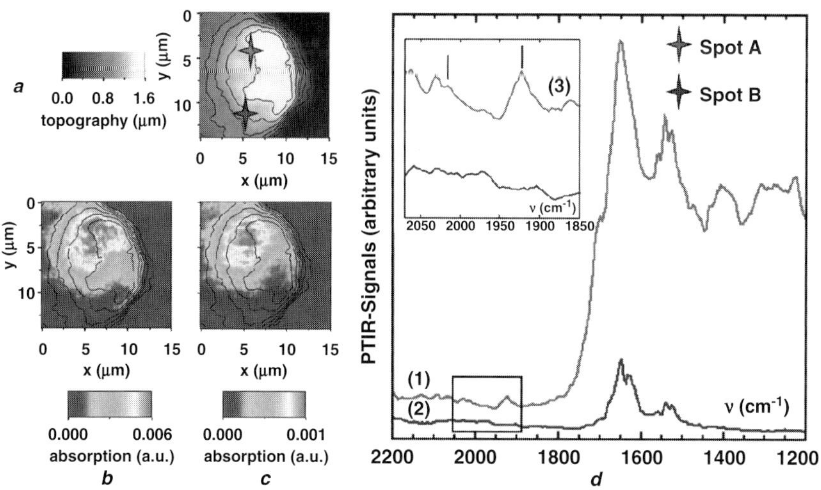

FIGURE 24 **Zoom at the nucleus of a cell treated with 1 (T) and spectromicroscopy. (a) AFM topography. (b,c) PTIR(/AFMIR)-mappings with the AFM contours superimposed as black lines. (b) 1240 cm^{-1}; (c) 1925 cm^{-1}; (d) AFMIR-spectromicroscopy with an average incident power of 10 mW. (1) spectrum at spot A (red); (2) spectrum at spot B (blue) (3) zoom at the 1850–2050 cm^{-1} region; the two ticks indicate the position of the characteristic peaks of 1[55].**

CONCLUSION

These successful examples, imaging procaryotes and eukaryotes, endogenous component, and exogenous compound, illustrate the wide possibilities that AFM-IR offers to answer biological questions with regard to topology. Nanoscale chemical mappings allow for the intracellular localization of both substructures and exogenous compounds. AFMIR was used in this example to localize the nucleus without the use of any nucleus-tracker: label-free mapping is possible in the IR-range. This is an important breakthrough in IR-based microspectrosocpy. It should be noted however that the different organelles, besides the nucleus, are difficult to distinguish based on their IR-signature [41, 54]. This limitation is due to the close similarity of IR-spectra of most biomolecules. But the success of the mapping of compound **1** opens the possibility of using metal-carbonyl probes, already used for labels [63–65], for immunoassays [59, 66, 67], or as local pH probes [66], to be designed as trackers to tag organelles [68, 69] and map their distribution inside cells. More generally, the development of metal-carbonyl IR-probes may be of interest for use in tagging other molecules for AFM-IR imaging, as they are small stable units.

These results are of utmost interest as AFM-IR is a noninvasive technique that can be applied directly to single cells. Thanks to this technique, nanoresolution is now attainable for imaging using IR radiation. This renders possible IR imaging at the subcellular scale, a breakthrough in IR-mapping. Spectromicroscopy represents a powerful tool to determine ultralocal composition *in cellulo*.

ACKNOWLEDGMENTS

For all collaborators: Sylvain Clède, François Lambert, Anne Vessières, Gérard Jaouen, Marie-Aude Plamont, Jenny-Birgitta Waern, Jean-Michel Ortega, Rui Prazeres, François Glotin, Céline Mayet, Ariane Deniset, Rémi Carminati, Marta de Frutos, Pascale Boulanger, Pierre Sebban, and Valérie Derrien.

For careful reading and useful advises Craig Prater and Sylvain Clède.

REFERENCES

[1] Dumas P, Carr GL, Williams P. *Analysis* 2000;1:68.

[2] Guilhaumou N, Dumas P, Carr GL, Williams GP. *Appl. Spectrosc.* 1998;52:1029.

[3] Jamin, N., Dumas, P., Montcuit, J., Fridmann, W. H., Carr G. L., Williams, G.P. *SPIE Accelerator-Based Infrared Sources and Applications* (G. P. Williams; P. Dumas, Eds.). 1997, pp. 133.

[4] Bhargava R, Levin I. *Spectrochemical Analysis Using Infrared Multichannel Detectors.* Oxford: Blackwell Publishing; 2005.

[5] Levenson E, Lerch P, Martin MC. *Infrared Phys. Techno.* 2006;49:45–52.

[6] Knoll B, Keilmann F. *Nature* 1999;399:134–137.

[7] Brehm M, Taubner T, Hillenbrand R, Keilmann F. *Nanoletters* 2006;6:1308–1310.

[8] Dazzi A, Goumri-Said S, Salomon L. *Opt. Commun.* 2004;235:351–360.
[9] Wold H. *Nonlinear Estimation by Itérative Least Square Procédures.* London: Wiley; 1996 411–444.
[10] Lasch P. *Biomédical Vibrational Spectroscopy.* New York: Wiley-Interscience; 2008.
[11] Dazzi A, Prazeres R, Glotin F, Ortega JM. *Opt. Lett.* 2005;30:2388.
[12] Naumann D. Infrared spectroscopy in microbiology. R. A. Meyers In: *Encyclopedia of Analytical Chemistry*. Chichester: Wiley; 2000 pp. 102–131.
[13] Ortega JM, Glotin F, Prazeres R. *Infrared Phys. Technol.* 2006;49:133–138.
[14] Hammiche A, Pollock HM, Reading M, Claybourn M, Turner P, Jewkes K. *Appl. Spectrosc.* 1999;53:810–815.
[15] Bozec L, Hammiche A, Tobin MJ, Chalmers JM, Everall NJ, Pollock HM. *Meas. Sci. Technol.* 2002;13:1217.
[16] Hammiche A, Walsh MJ, Pollock HM, Martin-Hirsch PL, Martin FL, Biochem J. *Biophys. Methods* 2007;70:675.
[17] Nowacki W, *Thermoelasticity. International Series of Monographs on Aeronautics and Astronautics* Oxford: Pergamon Press; 1962.
[18] Boussinesq J. *Application des potentiels à l'étude de l'équilibre et du mouvement des solides élastiques. Notes complémentaires.* Paris: Gauthier-Villars; 1885.
[19] Sarid D. *Scanning Force Microscopy.* Oxford: Oxford University Press; 1991.
[20] Dazzi A, Prazeres R, Glotin F, Ortega JM. *Ultramicroscopy* 2007;107:1194.
[21] Yuya PA, Hurley DC, Turner JA. *J. Appl. Phys.* 2008;104:074916.
[22] Dazzi A, Glotin F, Carminati R. *J. Appl. Phys.* 2010;107:124519.
[23] Schmitz I, Schreiner M, Friedbacher G, Grasserbauer M. *Appl. Surf. Sci.* 1997;115:190–198.
[24] Landau, L.; Lifchitz, E. (Eds). Théorie de l'élasticité, 1967.
[25] Putman C. AJ, De Grooth BG, Van Hulst NF, Greve J. *J. Appl. Phys.* 1992;72:6.
[26] Dazzi A, Prazeres R, Glotin F, Ortega JM, Alsawaftah M. *M. De Frutos Ultramicroscopy* 2008;108:635–641.
[27] Mayet C, Dazzi A, Prazeres R, Allot F, Glotin F, Ortega JM. *Opt. Lett.* 2008;33:1611–1613.
[28] Mayet C, Dazzi A, Prazeres R, Ortega JM, Jaillard D. *Analyst* 2010;135:2540–2545.
[29] Lasch. P., Kneipp. J. (Eds.), *Biomedical Vibrational Spectroscopy* New York: Wiley; 2008.
[30] Houel J, Sauvage S, Boucaud P, Dazzi A, Prazeres R, Glotin F, Ortega JM, Miard A. *Phys. Rev. Lett.* 2007;99:217404.
[31] Houel J, Homeyer E, Sauvage S, Boucaud P, Dazzi A, Prazeres R, Ortega M. *J. Opt. Exp.* 2009;17:10887–10894.
[32] Hill GA, Rice JH, Meech SR, Craig D. QM, Kuo P, Vodopyanov K, Reading M. *Opt. Lett.* 2009;34:15.
[33] Bonhivers M, Ghazi A, Boulanger P, Letellier L. *EMBO J.* 1996;15:1850–1856.
[34] Zweig M, Rosenkranz HS, Morgan C. *J. Virol.* 1972;9:526–543.
[35] Kranz R, Gabbert K, Locke T, Madigan M. *Appl. Environ. Microbiol.* 1997;63:3003–3009.
[36] Zhao Q, Cheng G. *J. Mater. Sci.* 2004;39:3829–3831.
[37] Wong HH, Lee SY. *Appl. Microbiol. Biotechnol.* 1998;50:30–33.
[38] Naumann D. *FT-IR and FT-NIR Raman Spectroscopy in Biomedical Research. The 11th Int. Conf. Fourier Transform Spectroscopy* 1998;96–109.
[39] Hahn SK, Ryu HW, Chang YK. *Korean J. Chem. Eng.* 1998;15:51–55.
[40] Jendrossek D, Selchow O, Hoppert M. *Appl. Environ. Microbiol.* 2007;73:586–593.
[41] Gazi E, Dwyer J, Lockyer NP, Miyan J, Gardner P, Hart C, Brown M, Clarke NW. *Biopolymers* 2005;77:18–30.

[42] Dumas P, Sockalingum GD, Sulé-Suso J. *Trends Biotechnol.* 2006;25:40–44.

[43] Fabian H, Thi N. AT, Eiden M, Lasch P, Schmitt J, Naumann D. *Biophys. Biochem. Acta* 2006;1758:874–882.

[44] Kong KV, Chew W, Lim L. HK, Fan WY, Leong WK. *Bioconjugate Chem.* 2007;18:1370–1374.

[45] Meister K, Niesel J, Schatzschneider U, Metzler-Nolte N, Schmidt DA, Havenith M. *Angew. Chem. Int. Ed.* 2010;49:3310–3312.

[46] Hildebrandt P. *Angew. Chem. Int. Ed.* 2010;49:4540–4541.

[47] Matthaüs C, Kale A, Chernenko T, Torchilin V, Diem M. *Mol. Pharm.* 2008;5:287–293.

[48] Matthaüs C, Bird B, Miljkovic M, Chernenko T, Romeo M, Diem M. *Biophys. Tools Biologists: Vol.2 in Vivo Techniques* 2008;89:275.

[49] Puppels GJ, Demul F. FM, Otto C, Greve J, Robertnicoud M, Arndtjovin DJ, Jovin TM. *Nature* 1990;347:301–303.

[50] Lasch P, Naumann D. *Biophys. Biochem. Acta 2006.*, 814–829.

[51] Keilman M, Taubner T, Hillenbrand R, Keilmann F. *NanoLetters* 2006;6:1307–1310.

[52] Dazzi, A. In: *Thermal Nanosystems and Nanomaterials* (S. Volz, Ed.). Berlin: Springer, 2009, 469–503.

[53] Dazzi A, Glotin F, Carminati R. *J. Appl. Phys.* 2010;107:124519.

[54] Lasch P, Boese M, Pacifico A, Diem M. *Vibr. Spectrosc.* 2002;28:147–157.

[55] Policar C, Waern JB, Plamont MA, Clède S, Mayet C, Prazeres R, Ortega JM, Vessières A, Dazzi A. *Angew. Chem. Int. Ed.* 2011; 860–864.

[56] Top S, Hafa HE, Vessières A, Quivy J, Vaissermann J, Hughes DW, McGlinchey MJ, Mornon JP, Thoreau E, Jaouen G. *J. Am. Chem. Soc.* 1995;117:8372–8380.

[57] Ott I, Schmidt K, Kircher B, Schumacher P, Wiglenda T, Gust R. *J. Med. Chem.* 2005;48:622–629.

[58] Hillard EA, Vessières A, Top S, Pigeon P, Kowalski K, Huché M, Jaouen G. *J. Organomet. Chem.* 2007;692:1315–1326.

[59] Salmain, M. Vessières, A. *Bioorganometallics–Biomolecules, Labeling, Medicine, Vol. 1* (G. Jaouen, Ed.). Weinheim: Wiley-VCH, 2006, 263–302.

[60] Waern JB, Harris HH, Lai B, Cai Z, Harding MM, Dillon CT. *J. Biol. Inorg. Chem.* 2005;10:443–452.

[61] Noor F, Wuestholz A, Kinscherf R, Metzler-Nolte N. *Angew. Chem. Int. Ed.* 2005;44:2429–2432.

[62] Neundorf I, Hoyer J, Splith K, Rennert R, Peindy N'Dongo HW, Schatzschneider U. *Chem. Commun.* 2008;5604–5606.

[63] Vessières A, Top S, Ismail AA, Butler IS, Louer M, Jaouen G. *Biochemistry* 1988;27:6659–6665.

[64] Metzler-Nolte N. *Angew. Chem. Int. Ed. Engl.* 2001;40:1040–1043.

[65] Peindy N'Dongo HW, Neundorf I, Merz K, Schatzschneider U. *J. Inorg. Biochem.* 2008;102:2114–2119.

[66] G. R. Stephenson, in *Bioorganometallics—Biomolecules, labeling, medecine* (Ed.: G. Jaouen), Wiley-VCH, Weinheim, 2006, pp. 215-262.

[67] Salmain M, Vessières A, Varenne A, Brossier P, Jaouen G. *J. Organometal. Chem.* 1999;589:92–97.

[68] Vitor RF, Correia I, Videira M, Marques F, Paulo A, Pessoa JC, Viola G, Martins GG, Santos I. *Chem. Biochem.* 2008;9:131–142.

[69] Amoroso AJ, Arthur RJ, Coogan MP, Court JB, Fernández-Moreira V, Hayes AJ, Lloyd D, Millet C, Pope S.J. A. *New J. Chem.* 2008;32:1097–1102.

Chapter 10

Sum-frequency generation spectroscopy of biointerfaces

C. Humbert and B. Busson

Laboratoire de Chimie Physique, Université Paris-Sud; CNRS, Orsay, France

INTRODUCTION

Infrared absorption spectroscopy directly measures the absorption of the investigated materials, and the previous chapters have illustrated how difficult it is to extract information from an interface surrounded by overwhelming amounts of

Biointerface Characterization by Advanced IR Spectroscopy. DOI: 10.1016/B978-0-444-53558-0.00010-2

bulk material. Discriminating between surface/interface and bulk requires additional sensitivity. The usual way to overcome this problem is to perform a differential spectroscopic study in order to subtract the disturbing signals from the bulk and keep only the key information from the surface. Infrared-visible sum-frequency generation (SFG) vibrational spectroscopy offers another solution as it directly extracts the vibrational information from the surface or the interface with no, or only minor, perturbation from the surrounding bulk. As a consequence, SFG analysis does not require the use of a blank sample or material, but absolute information can be readily obtained. This becomes particularly interesting in the case of biointerfaces in an environment of water and salts, where the amounts of matter involved may become very small so that the signals extracted from differential procedures drop below the noise level. Other common applications of SFG cover *in situ* interfacial electrochemistry, surface heterogeneous catalysis and self-assembly processes of monolayers.

SFG is a nonlinear optical probe that aims at describing *in situ* the surface properties of materials or molecules at an interface between two media, including specific adsorption processes, chemical evolution, and molecular organization as a function of time or externally tuned parameters. As the other techniques based on infrared, it relies on the excitation of vibrational transitions. SFG spectroscopy is generally used to afford complementary or specific structural information when IR absorption or Raman spectroscopies become inadequate or not sensitive enough in a demanding environment like those encountered at the interface between two interacting media.

The spectroscopic tool borrows its name from the nonlinear optical process that it uses. As the name suggests, the SFG process requires two input light beams to create a third one, which energy (i.e., frequency or color) is the sum of the two incoming energies. In other words, two photons are destroyed in the process to produce a new one with the same total amount of energy. Ordinary optics, called linear by reference to the formal relations between the quantities which define it, deals with the propagation and interference of light waves, but does not allow changing the colors of photons. SFG therefore belongs to the world beyond classical optics, which is referred to as nonlinear optics. It was performed for the first time in 1987 by Shen group (Berkeley) on a Langmuir film of pentadecanoic acid deposited on water under various compression, from the liquid expanded (LE) phase to the liquid condensed (LC) phase in order to evidence the structural reordering of the molecules at the air/liquid interface [1]. As all nonlinear optical processes, SFG has a low yield and accordingly requires a high density of photons, both in time and space, in other words short-pulsed lasers. Its history is therefore correlated to the invention of the laser in the last century [2]. In fact, the first nonlinear optical effect was ultraviolet (347 nm wavelength) second harmonic generation (SHG) by the passage of the intense red beam of the ruby laser (694 nm) through a quartz crystal [3]. From that time, it has been quickly demonstrated that the laser could provide the power necessary to reach the magnitude of the Coulombian electric field reigning in the atoms between the nucleus and the electrons, leading to an enhancement of anharmonic (hence nonlinear) effects inside the matter. Subsequently, theoretical and experimental developments were established by N. Bloembergen who described the interaction of light at the boundary of nonlinear media. In this manner, he became the most famous contributor to the development of laser spectroscopy in the nonlinear optics field as he shared the 1981 Nobel Prize in Physics with A. L. Schawlow, the father of the optical maser theory (now called laser in common language), for their contribution to the development of laser spectroscopy [4]. Fifteen years prior to the first experimental evidence of SFG spectroscopy, the submonolayer sensitivity of nonlinear optics to the interface between two media was demonstrated in 1973 by the enhancement of a SHG signal of Ge(1 0 0) and Ge(1 1 1) surfaces covered by Na atoms [5]. During this period, SHG spectroscopy developed and applied to the study of monolayers adsorbed on surfaces in order to probe molecular adsorption or desorption because, contrary to SFG, it does not provide the vibrational fingerprint of an interface but enlightens rather the modifications of its surface electronic properties. The SHG signal ($\omega_{SHG} = 2\omega$) from an interface is detected through the visible photons generated in reflection or transmission by the incident laser (usually in the green at 532 nm for a Nd:YAG pump at 1064 nm) as illustrated in Figure 1. The following interfaces have been investigated with SHG spectroscopy: metal and semiconductor surfaces, liquid and electrochemical interfaces, and biological systems to which one must add surface microscopy. In particular, a current research field in SHG relates to the imaging of biological systems upon, or even without, excitation of chromophores with a submicrometer resolution [6].

Thereafter, SFG spectroscopy was achieved thanks to the development of IR tunable picosecond-pulsed laser sources with sufficient peak-power to excite nonlinear effects. SFG spectroscopy can be regarded as the generalization of the particular case of SHG by the simultaneous use of two pump laser sources differing in energy to probe an interface ($\omega_{SFG} = \omega_1 + \omega_2$). Infrared-visible SFG becomes a vibrational spectroscopy as the incident IR frequency is scanned, whereas SHG is confined to remain an electronic spectroscopy [7]. Historically,

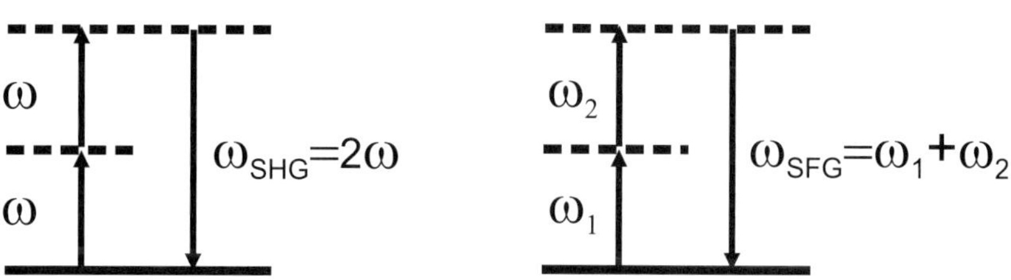

FIGURE 1 Energy-level diagrams for the second harmonic generation (SHG) and sum-frequency generation (SFG) nonlinear second-order processes.

ω_1 refers to a tunable IR frequency ($\omega_{IR} = 2500$–$4000\ cm^{-1}$) and ω_2 to a fixed visible frequency (ω_{Vis}), in the earliest times the green 532 nm doubled YAG. In these conditions, SFG spectroscopy allows us to probe the chemical signature of OH, NH, and CH bonds. These features constitute the basic technical requirements encountered in the SFG setups present in the major part of the laboratories performing SFG spectroscopy worldwide. Contrary to IR spectroscopy based on thermal black body sources, SFG requires the use of intense pulsed IR tunable laser sources, which rapidly increases the cost of the setup. Nevertheless, where IR spectroscopy requires the use of IR sensitive cooled detectors, SFG spectroscopy only requires photomultipliers which easily detect the SFG photons. With the continuous development of tunable IR laser sources, SFG spectroscopy extends its IR spectral range from the near-IR to the mid-IR ($\omega_{IR} = 1000$–$4000\ cm^{-1}$) in more and more laboratories, giving access to the chemical signature of CO, CN, Si–H, Pt–H, NO_2, amide, amine and ring bonds. SFG spectroscopy at $\omega_{IR} \leq 1000\ cm^{-1}$ is restricted yet to a few research groups working with complex and expensive top IR tabletop laser sources or with the intense IR beam delivered by a free electron laser (FEL) in a large-scale facility. The highest IR wavelength for which SFG spectroscopy was performed to date is 24 μm ($\sim 420\ cm^{-1}$) by the probing of the C-S stretching mode of thiophenol adsorbed on Ag(1 1 1) in 1999 [8]. Theoretically, with FEL radiation, it should be possible to reach $\sim 100\ cm^{-1}$ but the experimental constraints related to the filtering of the SFG photons from the large diffusion of the incident visible beam limit, in practice, SFG spectroscopy down to $250\ cm^{-1}$ for the moment.

With the gradual lifting of the limitations inherent to tunable IR laser sources, SFG spectroscopy has progressively extended in the last 20 years to the study of the chemical and electronic fingerprints of any interface from the molecular microscopic scale (SAMS, Langmuir films, electrochemistry, and catalysis) to the biological scale (biomolecular recognition, DNA strands, chromophores, and biosensors) and at present to the nanoscale (functionalized gold/silver/platinum nanoparticles and quantum dots). In fact, another supplementary specific advantage of SFG with respect to IR spectroscopy is the routine use of a tunable visible laser source [9], in the optical range from 420 (violet) to 700 nm (red), in order to also probe the electronic structure of various systems and its coupling to the vibrational signature: Soret band of porphyrins, surface plasmon of metal nanoparticles, excited electronic states of chromophores, metal interband transitions, electron-hole excitations in semiconductors, and surface states in catalysis and electrochemistry.

Eventually, in the last decades, SFG spectroscopy has experienced a spectacular evolution, becoming a more versatile technique thanks to the development of femtosecond (fs) laser sources. In fact, SFG spectroscopy, confined hitherto mostly in frequency-domain studies, becomes at present also efficient as a probe of temporal evolution of vibrations, which leads to the implementation of multiple scanning techniques aimed at probing selected surface properties. This fs broadband SFG tool allows studying either the structure and kinetics of interfacial molecules (as does a picoseconds setup) or their dynamics as reported in a recent comprehensive review [10]. In the latter case, for instance in a pump–probe experiment, the thermal equilibrium is disturbed by an additional ultrashort laser pulse (infrared or visible) and the chemical dynamics (chemical transformation), substrate–adsorbate coupling dynamics (energy transfer) and vibrational dynamics (vibrational energy and phase relaxation, vibrational inter- and intra-adsorbate coupling) can therefore be probed by SFG at the fs scale.

THEORETICAL BACKGROUND, RELEVANT PROPERTIES, AND EXPERIMENTAL SETUPS FOR SFG SPECTROSCOPY

A complete and comprehensive description of the numerous phenomena spanning the wide scientific domain of nonlinear optics can be found in the books of Boyd and Shen [11, 12]. We detail in this section the principles of SFG vibrational spectroscopy at an interface as far as necessary to understand how to probe a biointerface with this tool and which information one may extract.

SFG basics

The photonic description of the SFG phenomenon is rather straightforward. Considering a molecule lit by two monochromatic laser beams, one in the infrared and one in the visible, several processes involving vibrational transitions may occur. An infrared absorption (Fig. 2(a)) is depicted as the annihilation of an infrared photon and the excitation of a vibration mode of the molecule. Another way to excite a vibration mode is through a Raman Stokes process. Less favored is the desexcitation of a vibration mode through an anti-Stokes Raman process (Fig. 2(b)). The SFG process (Fig. 2(c)) combines an infrared absorption

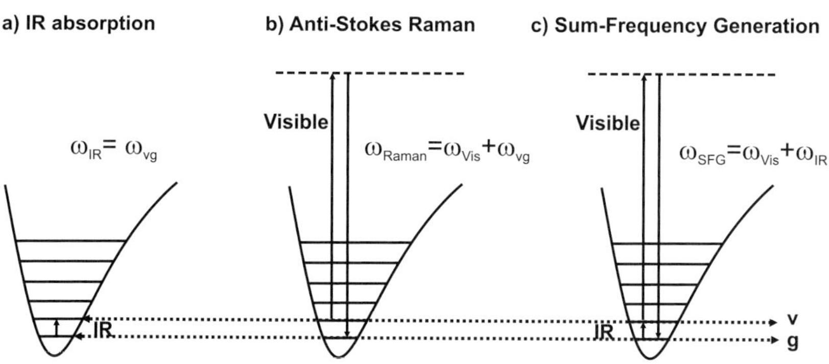

FIGURE 2 Molecular energy schemes of the infrared absorption (a), anti-Stokes Raman (b), and sum-frequency generation (c) resonant processes. v and g are two vibrational levels of the molecule.

to an anti-Stokes Raman process, with an overall outcome of two photons destroyed and one higher energy photon created. The resonance of the IR and Raman transitions with a vibration mode of the molecule enhances the SFG response as compared to the nonresonant process (which always exists but is hardly measurable by usual SFG spectrometers). As a consequence, when the IR energy is tuned to that of a vibration mode, the measured SFG signal goes from zero to a measurable value. The simplest SFG spectra correspond to the SFG intensity plotted as a function of the tunable IR wavenumbers and is made of peaks corresponding to the energies of vibration modes. The quantum calculation of the SFG response (see below) shows that the simple picture described here accounts rather well for the actual process. As a consequence, a vibration mode must be both IR and Raman active to become SFG-active and the SFG cross-section is proportional to the IR and Raman amplitudes of the vibration mode involved. In addition, IR and Raman selection rules apply, which has consequences on the symmetries allowed for the molecule. It is easily checked [13] that a molecule with inversion symmetry has no vibration mode, both IR and Raman active. As a consequence, IR-resonant SFG is forbidden for such molecules. This important symmetry rule applies in fact to any SFG process and is intrinsic to all second-order nonlinear processes. In order to understand its origin, we shall first present the macroscopic description of SFG.

In the previous paragraph, the IR-resonant SFG response of a molecule was briefly described. In a general manner, it must be emphasized that all processes must be mediated by matter. In other words, the transformation of two photons (ω_{vis}, ω_{IR}) into one (ω_{SFG}) fully relies on the interaction of light with matter and may not happen without it. In the electromagnetic theory of optics, the convenient frame to describe the interaction of light with matter is through the polarization. This quantity describes the response of matter to the excitation of light, represented by an oscillating electric field (frequency ω). In return, the polarization acts as a source of electromagnetic radiation. In linear optics, the polarization response is proportional to the excitation (i.e., linear) with coefficient $\chi^{(1)}$ called susceptibility. The light emitted by the polarization has the same frequency as the excitation; it therefore interacts with the incoming light. Interaction processes include absorption and give birth, among others, to refraction. For an isotropic medium, $\boldsymbol{P}_{\text{linear}} = \varepsilon_0\chi^{(1)}\boldsymbol{E}$, with $\chi^{(1)} = 1 - n^2$, where n is the complex refractive index, in which real and imaginary parts quantify refraction and absorption, respectively. For a generic medium, $\chi^{(1)}$ is a 3×3 matrix, that is, a two-dimensional tensor.

Higher order terms in the polarization response give birth to nonlinear (NL) optics. The lowest order of nonlinear interaction is quadratic in the electric fields (second-order nonlinear optics): $\boldsymbol{P} = \boldsymbol{P}_{\text{linear}} + \boldsymbol{P}^{(2)}_{\text{nonlinear}} = \varepsilon_0\chi^{(1)} : \boldsymbol{E}_{\text{IR/vis}} + \varepsilon_0\chi^{(2)} : \boldsymbol{E}_{\text{IR/vis}} : \boldsymbol{E}_{\text{IR/vis}}$, where subscript IR/vis means that all combinations of IR and visible beams must be used. The contraction sign (:) is more easily understood when expressing the above quantities in Cartesian coordinates in a given referential: $P_i^{(2)} = \varepsilon_0\chi^{(2)}_{ijk}E_jE_k$ for $i, j, k = x,$

y, or z. Restricting to the SFG process, P_i may oscillate at the sum-frequency $\omega_{SFG} = \omega_{vis} + \omega_{IR}$ and generate a new beam (i.e., new photons) at this frequency. Quantity $\chi^{(2)}$, a third-rank tensor with 27 components, is called the second-order nonlinear susceptibility. Similarly to the linear susceptibility, it contains the response of matter to the electromagnetic excitation, as a function of the three frequencies involved. For the SFG user, it quantifies the SFG process and its resonances; it is therefore the key quantity to monitor. We may note at this stage that the generated photon has a frequency ω_{SFG}, which was not present in the incoming beams. Thus, contrary to linear optics, the collection of the light emitted at this frequency gives an absolute measurement of the response of matter to the excitation by beams (ω_{vis}, ω_{IR}), namely the SFG intensity $I_{SFG}(\omega_{vis}, \omega_{IR})$. This explains why, as soon as SFG becomes measurable, information can be obtained for very low amounts of matter, optimally far less than a molecular monolayer. We mention here the existence of a twin beam at the difference frequency (DFG), which gives birth to a similar spectroscopy. DFG is often neglected experimentally because, having less energy than the incident visible, its detection competes with fluorescence. In addition, for most fs experimental setups (see below), it is simply not accessible. The information brought by DFG is often redundant with SFG, but in some cases the comparison of both techniques helps extracting additional data [14].

Before describing the macroscopic SFG intensity, we focus on an important point: the relation between the symmetries of a sample and its $\chi^{(2)}$. Suppose that the sample possesses inversion symmetry of its electronic properties. Then $\chi^{(2)}$ tensor, which is characteristic of the sample only, remains unchanged under the inversion transformation. On the contrary, $\boldsymbol{E}$ and $\boldsymbol{P}$ fields change sign under such a transformation. According to equation $P_i^{(2)} = \varepsilon_0 \chi_{ijk}^{(2)} E_j E_k$ above, component $\boldsymbol{P}_i$ and its opposite both equal $\varepsilon_0 \chi_{ijk}^{(2)} E_j E_k$, leading to the consequence that $\chi^{(2)}$ vanishes. The SFG process is forbidden in a medium with inversion symmetry. This property is thus not limited to the resonant case with vibration modes, but a very general property of second-order nonlinear optics. As a consequence, considering two media with inversion symmetry (either gas, liquid, or solid), the only source of SFG radiation is the interface between them, where the inversion symmetry is broken. In such systems, SFG spectroscopy represents an intrinsically interface-specific tool providing information through an absolute measurement. At an interface, the number of nonzero components of $\chi^{(2)}$ depends on the additional in-plane symmetries.

In the following, we treat the most-encountered case of an isotropic interface. In-plane anisotropy is seldom needed apart from very specific cases [15], which are irrelevant to the biointerface issue. The typical experimental setup appears in Figure 3. It may be interesting to tune the incident angles in order to gain more information on the geometry of the interface [16], but most setups use fixed angles around 60°. Incident beams are linearly polarized, that is, their electric fields are either parallel to the interface (s-polarization) or parallel to the

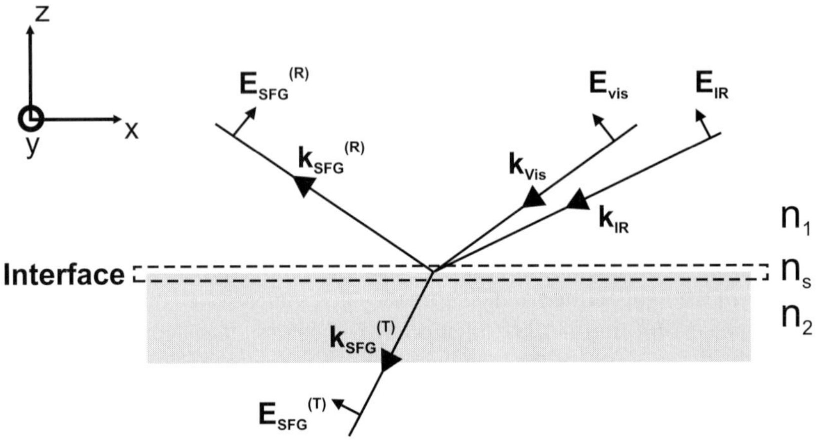

FIGURE 3 Classical SFG spectroscopic configuration of an interface within the three layers model for p-polarized electric fields. R and T refer to the reflected and transmitted SFG beams, respectively. The incidence angles of the visible and infrared beams are tilted from 55° and 65° with respect to the surface normal (z-direction).

plan of incidence (p-polarized). Combinations of these states of polarization (e.g., 50% p- and 50% s-polarized) may also prove useful but shall not be described here [17]. SFG is emitted either in a reflection or transmission geometry at the point of the interface where the two incoming beams overlap spatially and temporally. Reflection geometry is mandatory when the lower medium is opaque to visible light (e.g., metal), and mostly preferred at the surface of liquids.

Vibrational SFG consists in recording the SFG intensity as a function of the IR wavenumber. The practical way to perform such measurements shall be described later. The reflected intensity of the SFG signal reads therefore for an azimuthally isotropic medium [18]:

$$I_{\mathrm{SFG}} = \frac{8\pi^2\omega_{\mathrm{SFG}}^2}{c^3 n_1(\omega_{\mathrm{SFG}}) n_1(\omega_{\mathrm{Vis}}) n_1(\omega_{\mathrm{IR}}) \cos^2\theta_{\mathrm{SFG}}} \left|\chi_{\mathrm{eff}}^{(2)}\right|^2 I_{\mathrm{Vis}} I_{\mathrm{IR}}$$

where I_{Vis} and I_{IR} are the intensities of the incident visible and IR laser beams, respectively; θ_{SFG} is the reflection angle of the SFG beam with respect to the surface normal; $n_1(\omega)$ is the refractive index of the upper medium at frequency ω; $\chi_{\mathrm{eff}}^{(2)}$ is the effective nonlinear second-order susceptibility of the probed interface. The latter consists of a combination of the nonvanishing components of $\chi^{(2)}$, weighted by optical factors (Fresnel factors) describing the different reflectivities (resp. transmittivities) of the beams as a function of polarization. Their contribution has to be taken into account in the case of metallic or highly reflecting surfaces, or in the need of an accurate quantitative description of the SFG intensity. In most of the SFG publications on liquid surfaces and biological systems, however, they are neglected or considered as constant with the involved energies.

At this stage, it is convenient to keep in mind that there are roughly two levels of analysis of vibrational SFG spectra: the first one, essentially macroscopic, is rather straightforward, and the other goes deeper into the microscopic organization of the molecules at the interface and often requires additional steps: complementary experiments (IR absorption and Raman scattering), hypotheses on the molecular organization and order, molecular modeling and calculations (*ab initio*, DFT), careful analysis of optical factors (Fresnel factors, local field effects, and dipole–dipole couplings).

At the first level of analysis, the practical modeling of the SFG signal only depends on the square of the nonlinear second-order susceptibility of the interface depicted in Figure 3:

$$I_{\mathrm{SFG}}(\omega_{\mathrm{IR}}) \propto \left|\chi_{\mathrm{eff}}^{(2)}\right|^2$$

on which we shall focus now. The complex quantity $\chi_{\mathrm{eff}}^{(2)}$ separates generally into a term resonant (R) with the IR wavenumber and a second one called nonresonant (NR), as it remains constant within the probed infrared range. The term nonresonant is rather misleading and the full expression nonresonant with the IR in the probed range should be preferred. As a matter of fact, this term is seldom not resonant, because it would be hardly measurable. It originates from the substrate or sometimes from the molecules themselves and becomes measurable because resonant processes in the visible range enhance it: intraband and interband transitions in metals [14], bandgap transitions in semiconductors, and HOMO–LUMO transitions in molecular systems. SFG users concentrate mostly on the resonant term, as it carries the vibrational information on the molecules at the interface:

$$\chi_{\mathrm{eff}}^{(2)} = \chi_{\mathrm{NR}}^{(2)} + \chi_{R}^{(2)} = \left|\chi_{\mathrm{NR}}^{(2)}\right| e^{i\phi_{\mathrm{NR}}} + \sum_n \frac{a_n e^{i\varphi_n}}{\omega_{\mathrm{IR}} - \omega_n + i\Gamma_n}$$

The resonances with vibration modes appear as Lorentzian contributions described by four parameters: position (resonance energy) ω_n, amplitude (oscillator strength) a_n, phase φ_n, and line width (damping constant) Γ_n. In the harmonic model of vibration modes, phases φ_n only take values 0° or 180°, in other words a_n is either positive or negative. However, experience shows that it is sometimes difficult to adjust data within such a constraint. An originality of SFG as compared to RAIRS appears here: the intensity being proportional to the square modulus of $\chi_{\mathrm{eff}}^{(2)}$, the resonant and nonresonant terms do not simply add but rather interfere with each other. Depending on the relative phases and amplitudes, the interference pattern may produce peaks, dips, or a derivative shape [14] in the spectrum. This represents sometimes a drawback. A classical example shows that if ϕ_{NR} and φ_n equal 90° and 0° respectively, then the interference produces a strict flat line for $a_n = 2\Gamma_n\left|\chi_{\mathrm{NR}}^{(2)}\right|$ and the vibrational information disappears from the spectrum. DFG provides the solution in such a case. In most cases however, this interference is an advantage as it allows determining the sign of the oscillator strength a_n and the phase of the NR term, which are usually not experimentally measured. In addition, the interference pattern is a local proof that both contributions (R and NR) originate from the same interface and accordingly that the molecules are adsorbed indeed on the substrate. A fit of the experimental spectrum is usually sufficient to obtain the values of the parameters, which gives the following elements: position ω_n allows assigning the vibration mode; amplitude a_n is directly related to the product of its IR and Raman activities, whereas the sign contained in the phase φ_n relates to the orientation of the molecule. Finally, width Γ_n gives an image of the order inside the molecular layer.

The second level of analysis aims at linking the macroscopic measurements of the first level to the microscopic and molecular SFG response. Sum-frequency generation of an individual molecule is depicted in the same way as before: the interaction of the two laser beams with the molecule generate a second-order dipole moment $p_I^{(2)} = \beta_{IJK}E_JE_K$, where β is the hyperpolarizability, a three dimension tensor equivalent to the second-order susceptibility, evaluated in a molecular frame $(\hat{X}, \hat{Y}, \hat{Z})$. The local electric fields felt by the molecules differ from the macroscopic fields as a consequence of screening and dipole–dipole coupling, but such effects are often omitted. Strictly speaking, only the hyperpolarizability (i.e., the individual molecule) relates to the photonic processes described at the beginning of this section. It is possible to show [19], using standard approximations in molecular physics (Born-Oppenheimer, harmonic vibrations, small vibration amplitudes, and SFG energy far from molecular electronic transitions), that

$$\beta_{IJK} = \sum_n \frac{1}{2\,\hbar\,\omega_n} \frac{\left(\frac{\partial \alpha_{IJ}}{\partial Q_n}\right)_{Q_n=0} \left(\frac{\partial \mu_K}{\partial Q_n}\right)_{Q_n=0}}{\omega_{\mathrm{IR}} - \omega_n + i\Gamma_n}.$$

In other words, for a given vibration mode n, the IJK hyperpolarizability component is proportional to the IR transition moment along K and the Raman anti-Stokes polarizability tensor component IJ, proving the intuitive rule deduced from the photon diagrams above. This relation is often transferred in the literature to the macroscopic level and applied as is to the second-order susceptibility; although not rigorously grounded, this method justifies the first-level approximation described above. All the dipole moments emit light at the SFG frequency, which sum up coherently to produce the macroscopic signal. The macroscopic surface polarization is defined as an averaged dipole moment per surface unit. However, the averaging encompasses the various configurations adopted by the molecules at a given time. The most disturbing effect of the averaging lies in the geometric factor, which sums up the distribution of molecular orientation and has to be taken into account. As a consequence, the macroscopic and microscopic levels relate by

$$\chi_{ijk}^{(2)} = N_s \sum_{IJK} \langle T_{IJK}^{lmn} \rangle \beta_{IJK}^{(2)},$$

where N_s is the surface density of molecules. Transformation T relates the coordinates in the molecular frame to that in the laboratory frame; it usually makes use of the three Euler angles. The bracket denotes averaging as a consequence of orientation distribution of molecules, that is, averaging over Euler angles. At this step, hypotheses must be introduced on the molecular distribution of orientations. Shallow hypotheses make the analysis rather cumbersome [20], and the usual hypothesis is: a uniaxial molecule with full isotropy in the plane perpendicular to its principal axis. In this way, only the tilt distribution (θ) of molecules remains unknown, for which either a single discrete value or distribution (Gaussian, rectangle) may be considered. Determination of the orientation of a simple molecule at an interface by comparing SFG intensities of several vibration modes under various polarization combinations is therefore a rather well-established procedure. For complex biological systems however, a deeper data analysis is required and several usual approximations drop. We emphasize here, as a fully investigated example, the case of the amide I mode of the α-helix for a protein/peptide in interaction with a lipid bilayer. The full hyperpolarizabilities of the amide I vibrations must be taken into account, and a completely unknown tilt angle θ distribution must be considered. Z. Chen group performed a theoretical analysis of this complex problem and set out a rather generic method to extract the orientation parameters from the SFG polarization analysis. The orientation averaging on the tilt angle θ implies that the distribution function for θ is only indirectly known by its moments (averaged values of successive powers of its cosine). Combining several spectroscopic methods gives access to different moments (orders 2 for FTIR, 2 and 4 for Raman and CARS, 1 and 3 for SFG). In particular, SFG provides the key odd-order moments. The orientation function is then reconstructed from its moments through the maximum entropy function method, which provides the most probable distribution function for a given set of moments [21].

Having related the molecular hyperpolarizability to IR and Raman activities, and the susceptibility to the hyperpolarizability, the last step is to sum the susceptibility components into the effective susceptibility. For an isotropic achiral surface in the $(\hat{x}, \hat{y})$ plane as depicted in Figure 3, only four independent components of $\chi^{(2)}$ remain. In details, $xxz = yyz$, $xzx = yzy$, $zxx = zyy$, and zzz. In Figure 3, s-polarized light has an electric field along $\hat{y}$ while p-polarization projects onto $\hat{x}$ and $\hat{z}$. By choosing correctly the polarization combinations of the three beams, it is possible to separate the four components. For example ssp, sps, pss, and ppp are enough for that respect (the classical notation for polarizations, respectively SFG, visible and infrared, will be used throughout this chapter).

Despite several levels of approximations and hypotheses, such a detailed analysis allows us to correlate the IR transition dipole orientation, which may be known from experiment, molecular modeling, or DFT calculations, to the polarization-dependent SFG intensities. The simplest use of such a work is to guess the orientation of molecules inside a molecular monolayer. This has been applied by several teams on various kinds of layers for more than a decade [18].

Experimental SFG setups

A spectrometer dedicated to SFG spectroscopy is made of several stages: laser sources, SFG photons production, and detection. It is possible to define three families of setups, which differ mostly in their first stage, more precisely in their temporal structure. As explained in Introduction section, nonlinear optics requires short-pulsed and high-power lasers. The first family (broadband SFG) works in the fs time domain, and the two others in the picosecond range. SFG has also been demonstrated with nanosecond lasers [22], but the relatively long nanosecond pulses deposit a lot of energy on the surface in order to reach the SFG production threshold, which usually leads to the alteration or destruction of the sample. In any case, biomolecules cannot stand nanosecond SFG spectroscopy.

Laser sources

All families rely on a laser chain composed of an oscillator delivering a fundamental radiation with specific characteristics (wavelength, repetition rate, pulse width, and peak power), which pumps subsequent devices in order to produce the two beams necessary for the production of SFG photons. The delicate step is the production of infrared radiation in the fingerprint region of molecules (2.5–10 μm $\leftrightarrow$ 4000–1000 cm^{-1}), which always relies on an optical parametric process. Such a frequency conversion takes advantage of the properties of a nonlinear crystal to split the fundamental beam in two beams of lower energy (called idler and signal). Depending on the fundamental frequency, the output of the optical parametric device is either used directly or further frequency converted. The tuning parameter is the phase matching between the three beams, which depend on the relative orientations of the three wave vectors with respect to the optical axes of the nonlinear crystal. In other words, wavelength tuning is obtained by changing the angles of propagation of the light waves inside the crystal, or much better by rotating the crystal around one of its axes.

A typical fs broadband SFG setup of the first family relies on a Ti:sapphire oscillator, which produces a fundamental beam around 800 nm with a repetition rate in the kHz range (or lower) and a pulse duration of several tenths of fs (10–200 fs). Due to the relation between time and frequency, such short laser pulses carry a broad frequency spectrum. Part of the fundamental beam is used after amplification and spectral narrowing (i.e., temporal stretching) as the visible source for SFG. This explains why DFG spectroscopy is unreachable in this case, as DFG energy goes far out of the visible range and is not as easily detected as SFG. The remaining amplified fundamental beam is used to pump an optical parametric amplifier (OPA) or generator (OPG) to produce idler and signal in the near-infrared range. A last stage mixes both beams by DFG in a second nonlinear crystal to generate the IR light. The IR beam still possesses the broadband characteristics of the source and usually covers around 100 cm^{-1} in one laser pulse. Mixed with the narrowband visible beam, this pulse generates a broadband SFG pulse, which carries all the necessary spectral information. It is necessary to narrow the visible pulse in order to avoid a convolution of the two spectral distributions, which would result in a loss of resolution. In other words, the infrared spectral range is ensured by the infrared beam, whereas the visible beam carries the spectral resolution. The spectrum contained in the SFG pulse must be dispersed by a grating monochromator, which converts spectral range into angular dispersion. After proper filtering of the fundamental beam, this allows the recording on a CCD camera, in which each pixel corresponds directly to a given infrared spectral range, whose width (i.e., spectral resolution) depends on the characteristics of the grating, the spectral width of the visible beam, and the range of infrared source. As the spectrum is produced in one pulse, such broadband setups may allow for a short acquisition time of SFG spectra (in favorable case as low as 0.2 s [23]) and consequently monitor fast dynamics. Finally, the fs structure matches a high peak power with a low mean energy, which makes it possible to work on rather fragile systems with minimal damage (e.g., the surface of a liquid and, obviously, biological systems) due to laser-induced infrared heating, even if damage due to the visible beam is still possible [10]. The interest of using a 800 nm visible pump resides in the ability to minimize the nonlinear contribution from most of solid substrates when a weak and specific molecular response of the interface has to be selected: only constructive interferences, that is, peak-shaped vibration modes, are observed on a zero baseline in SFG spectra.

The second family works in the picoseconds range. The source is usually a mode-locked Nd:YAG (or equivalent) delivering high-energy single-pulses of a few tenths of picoseconds at 1064 nm. The repetition rate is lower than that with fs lasers. Two main differences arise from the time scale and energy range of the fundamental beam. First, the latter may not conveniently be used as the visible beam, which is preferentially obtained after frequency doubling to reach 532 nm. As a consequence, SFG is produced in the blue–green region of the spectrum, whereas DFG reaches the green–yellow region. Both are measurable within a common spectrometer and photomultipliers. The second difference relates to the picoseconds time scale, which in turn implies a high resolution on the spectral dispersion (i.e., all beams are monochromatic). Therefore, broadband methods are not usable. The fundamental beam pumps an OPA/OPG/DFG system as before, whose output is a monochromatic IR beam. Therefore, SFG radiation is produced for one IR wavenumber at a time, and SFG spectra result of a point-by-point scan over the desired wavenumber range. The IR wavenumber is tuned in the OPA/OPG/DFG system for each point in a calibrated way. Detection in that case does not need dispersion onto a CCD detector, but rather high-yield photomultipliers after filtering of the residual visible incoming beam. The great advantage of picoseconds technology is therefore the high sensitivity of detection, the narrow spectral resolution, and large tunability of the IR source. On the other hand, spectra require more time than in the broadband case, and consequently a high stability of the lasers during a scan. This kind of setup is well adapted to probe liquid, aqueous, and more generally soft matter interfaces because an efficient thermal relaxation is possible between their enlightening by the successive giant single pulses, which have a sufficient temporal separation.

The third family also belongs to the picosecond world. The main difference lies in the repetition rate of the Nd:YAG 1064 nm source in the MHz range, with shorter pulse length (between 5 and 15 ps). This very high-repetition rate implies that each pulse carries less energy than in the previous case and needs amplification. In order to optimize amplification, only a few pulses are sent to the amplifier. The temporal structure therefore bears two levels: individual micropulses at high-repetition rate (up to 100 MHz) grouped in bunches (macropulses) containing from several tenths to several thousands of micropulses (depending on the MHz rate and the length of the macropulses). Finally, macropulses repetition rate is of a few tenths of hertz. The visible line is again a frequency doubler. The specific temporal structure makes it possible to pump a synchronous optical parametric oscillator (OPO). Contrary to the OPA/OPG case, the OPO delivers a new laser beam after its pumping by the fundamental beam. The nonlinear crystal is installed in a laser cavity (i.e., two mirrors) in which the signal beam oscillates. The synchronicity imposes that a signal micropulse produced in the nonlinear crystal by one fundamental micropulse oscillates exactly once in the OPO cavity before the next pump micropulse arrives. The oscillating signal stimulates the splitting of the pump into signal + idler, leading to a progressive growth of the signal and idler output until the laser threshold is reached. The idler is extracted from the cavity and may be used as the IR beam for SFG. There is no need of a DFG stage as the IR idler is directly produced at the required wavenumber. As above, tunability is obtained by rotating the nonlinear crystal and changing the phase-matching conditions. An OPO relies on a cumulative optical parametric process; therefore, it may not be pumped by single-pulse picoseconds sources. The detection part is very similar to the one dedicated to the single-pulse picosecond family. Although this is not mandatory, this high-repetition rate family of laser sources has the same temporal structure as IR-free electron lasers [24]. Therefore, it may synchronize with such a source, which replaces the IR OPO, and offers two assets: the first one is the high power (around a hundred times more peak power) and the second one the wide tunability: a typical FEL may cover in principle an extended range, only limited by the electron energy provided by its electron accelerator and the designs of its undulator and laser cavity which produce the laser light. As an example, the CLIO FEL [25] is able to produce light from 6 to 150 μm, from which the range

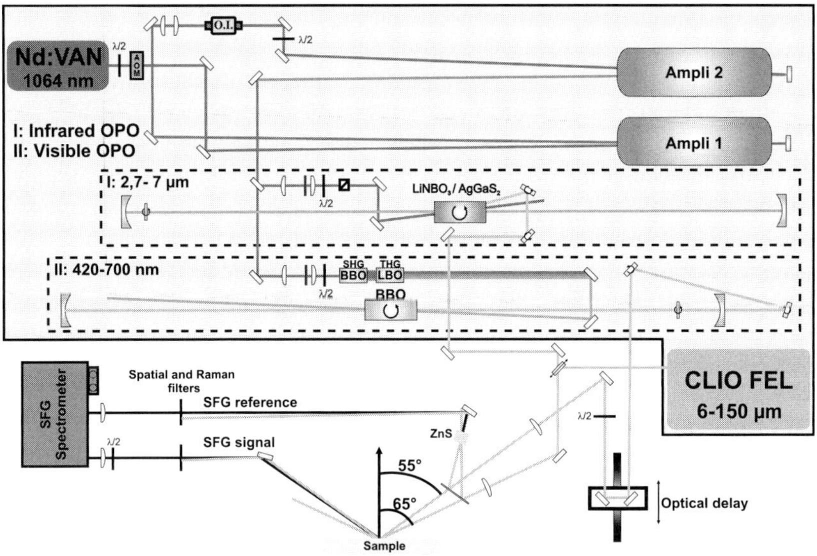

FIGURE 4 Sketch of the picosecond SFG setup tunable both in the infrared and visible spectral ranges from 2.7 to 150 μm and 420 to 700 nm, respectively, coupled to the CLIO free electron laser.

spanning from 6 to around 40 μm is available for SFG analysis as mentioned in the Introduction section. A complete SFG setup of this family is described in the Figure 4 above.

A summary of the advantages and constraints for each family of SFG setups presented in this section is given in Table 1. The damage threshold

TABLE 1 Main characteristics of the three kinds of experimental setups for SFG spectroscopy of biointerfaces

	Broadband (fs)	Single pulse (ps)	Pulse train (ps)
World diffusion	High	Medium	Medium
IR range	2.5–20 μm	2.5–20 μm	2.5–150 μm
Usual visible wavelength	800 nm	532 nm	532 nm
Tunable visible	Possible	Possible	Yes
Biological damage threshold	High	Medium	Low
Kinetics	Yes	Yes	Yes
Time-resolved dynamics	Yes	No	No
Pump-probe SFG	Yes (fs)	Not favored	Yes (ps)
DFG	No	Yes	Yes
Liquid interfaces	Yes	Yes	Not favored

remains a key characteristic to check for applications to biomaterials. The main drawback in using SFG spectroscopy lies indeed in the principal characteristics of nonlinear optics: the evidence of higher order optical effects in matter requires the use of powerful laser sources. Such lasers may have serious consequences on the damages to the samples when probing a thin molecular monolayer at an interface. As for biointerfaces, liquid–aqueous solutions or cell membranes are fragile materials and photon absorption, for instance by water molecules in the bulk, may disturb their structural integrity as a consequence of a high-repetition rate of energy deposition. In summary, the SFG process is a permanent competition between the damage threshold of the sample and the detection threshold of the nonlinear effects.

SFG photons production and detection

We detail hereafter a picosecond SFG setup having mandatory characteristics to perform most of the user-specific applications (Fig. 4) from the nanoscale to the biological scale. A diode-pumped Nd:VAN laser source (1064 nm) delivers continuously 7.5 ps micropulses at a 62.5 MHz repetition rate with an average power of 5 W. An acousto optic modulator (AOM) diffracts a small part of this intense beam to generate a pump beam with a similar temporal structure for the micropulses but constituted of 1.25 μs macropulses at a 25 Hz repetition rate in order to be synchronized with the IR-FEL of the CLIO (Centre Laser Infrarouge d'Orsay) European Facility [25]. The weak pump beam is therefore amplified by double passage through two Nd:YAG amplifiers in order to have sufficient energy for the simultaneous generation of tunable IR and visible laser beams. As depicted in the scheme, while one part is used as a pump in an IR OPO by passage of the fundamental beam through a $LiNBO_3$ or $AgGaS_2$ nonlinear crystal, the other part is frequency-doubled (532 nm) and frequency-tripled (355 nm) through BBO and LBO nonlinear crystals, respectively. The latter UV beam acts as a pump of the visible OPO built around a BBO nonlinear crystal. The pump beams (1064 nm; 355 nm) allow generating the IR 2.5–8 μm (~3 cm^{-1} spectral resolution) and visible 420–700 nm (~6 cm^{-1} spectral resolution) domains, respectively. The CLIO FEL generates IR beam from 6 to 150 μm (~2 cm^{-1} spectral resolution) and can be used instead of the IR beam of the OPO by just moving a mirror.

The production of the SFG photons requires the spatial and temporal coherent mixing of the IR and visible beams at the same point of the probed interface. Their incidence angles are 65° and 55° with respect to the surface normal. An optical delay line is added on the visible path in order to ensure temporal matching between the picoseconds micropulses of the IR and visible beams. As we deal with a coherent process, the experimental energy and direction of emission of the SFG beam are easy to calculate from the energy and momentum conservation rules. The conservation rules write, neglecting dispersion:

$$\omega_{SFG} = \omega_{Vis} + \omega_{IR}$$

$$k_{SFG}^{||} = k_{vis}^{||} + k_{IR}^{||} \Leftrightarrow \omega_{SFG}\sin\theta_{SFG} = \omega_{Vis}\sin\theta_{Vis} + \omega_{IR}\sin\theta_{IR}.$$

A part of the IR and visible beams is sent through a reference ZnS crystal to generate in parallel bulk nonresonant SFG in order to take account of potential laser fluctuations or absorption during the point-by-point IR scan of ~20 min for ~300 cm^{-1} spectral range. The role of this reference is therefore quite different from the bare reference used in IR spectroscopy for differential absorption. As the weak SFG beam is geometrically close to the intense visible beam, especially in the mid- and far-IR region, it has to be spatially and spectrally filtered through a double-grating monochromator before detection by a photomultiplier. The SFG spectrum intensity of the sample is usually recorded and displayed as a function of the IR frequency.

EXAMPLES OF APPLICATIONS TO BIOLOGICAL INTERFACES

In its early years, SFG spectroscopy was adopted by chemical physicists to help improve their understanding of planar interfaces in scientific fields like heterogeneous catalysis or interface electrochemistry. Besides such fundamental studies involving mostly simple molecules, interest has grown in time for more complex systems (thin films, self-assembled monolayers), while experiments on molecules adsorbed at the surface of water appeared. Finally, recent years have seen an increase in direct addressing to questions about biological objects and their interactions at the molecular level. In such issues, the assets of SFG summarized in the previous part are often put forward in order to get information complementary to other experimental techniques. We therefore present below a few examples which emphasize the specificities and complementarities of IR-visible SFG spectroscopy as compared to conventional IR techniques. The novel aspects of dynamic studies at the subpicosecond time scale introduced by broadband SFG will be overviewed in Prospects section.

In biology, the cell is the fundamental component of life. The cell membrane acts as a barrier: on one side, it protects the nucleus and the organelles from external intrusion while on the other side it permits the matter and energy exchanges required to sustain cell-life and reproduction. These transfers are monitored through membrane proteins (channel proteins) located inside the lipid bilayers constituting the frontier between two aqueous environments. The hydrophilic head groups face the inside and outside of the cell while the hydrophobic interactions of the acyl chains of each lipid layer maintain the membrane cohesion. SFG spectroscopy is well suited to challenge and complete the description of specific mechanisms related to membranes with respect to their overwhelming surroundings (water molecules and proteins/peptides in solution) because of its intrinsic interfacial sensitivity, as underlined in the previous sections. It has naturally extended over the years to several other specific issues related to biomolecules. The investigated mechanisms include at present the dynamical structural properties of lipid layers as a function of pressure and temperature, the molecular recognition processes, the interfacial protein/peptide interactions, the analysis of protein secondary structures (α-helix and β-sheet), and the molecular chirality. Recently, SFG has additionally started investigating biological interactions in complex materials such as DNA (hybridization and biosensors).

Lipid mono- and bilayers

Lipid monolayers: sensitivity boosted

Lipid monolayers, as elemental building blocks of cell membranes, have been extensively investigated in various environments. Practically, biological model membranes are built from Langmuir films deposited on water or transferred onto a solid substrate (by Langmuir–Blodgett or Langmuir–Schaeffer methods). The substrate includes silicon, glass, mica, quartz, and calcium fluoride. We focus here on the air–water interface, for which the low reflectivity and high absorption of the liquid phase limit the sensitivity of conventional techniques. PM-IRRAS has however proved able to tackle such complex issues (see Chapter 2). In addition to absolute sensitivity, SFG has added to previous characterizations its capability to measure order and disorder inside the monolayer, allowing quantifying the disturbance induced in a Langmuir monolayer under pressure by various parameters: layer composition, ionic content of the subphase, coadsorbants, and molecules in interaction.

The lipid layers constituting the building blocks of membranes are made of polar heads interacting with water molecules and bound to tail moieties containing alkyl groups. As a consequence, the 2500–4000 cm^{-1} IR spectral window of the OH and CH stretching vibration modes provides information on the main constituents [26]. Some efforts have been made to extend the range to the mid-infrared by probing OD and CD stretches around in the 2000–2500 cm^{-1} region [27, 28], and phosphate groups around 1100 cm^{-1}[29]. More recently, SFG was also applied in the spectral range of NO_2 stretching (1275–1375 cm^{-1}) and CH bending (1000–1100 cm^{-1}) vibration modes [30] of antigenic polar heads. Schematically, considering a lipid monolayer at the air/water interface, the CH region probes on one hand the structural evolution of the hydrophobic tails, first in static mode (for example as a function of the surface pressure induced by the moving barriers of the Langmuir trough), and then as a consequence of an interaction with a molecule (salt, acid or more complex, up to proteins, as shown below). On the other hand, information is gained in the OH stretching range from the structural evolution of the water molecules in contact with the hydrophilic side of the monolayer. Specifically, the first spectroscopic study of the top layer of water-free surface has become an archetype of the power of SFG spectroscopy [31], evidencing the direct measurement of the dangling (free) OH bonds. A natural evolution leads to the study of the variation of the signal from water (and thus of its structure) when the free surface is covered by a monolayer, among which amphiphiles like phospholipids play a specific and strategic role.

As for pure DPPC lipids in a Langmuir layer under compression, Bonn [32] has evidenced an unknown phase transition, sharp and reversible, at very low compression, which could not be seen by usual tools (Langmuir isotherms and fluorescence imaging). Following the Langmuir isotherm by SFG in the CH_2 and CH_3 spectral range from the LE (molecularly disordered) phase to the LC (highly ordered and oriented) phase, they enlightened this new transition at low compression where the surface density and chemical content of the monolayer remains undisturbed, but the hydrophobic alkane chains uncurl upon compression, creating a first stage of order in the LE phase. Revealing such a behavior requires a method, which directly quantifies order/disorder properties.

The behavior of water below a compressed Langmuir layer of lipids also appears more complicated than expected. For simple surfactants, electrostatics governs the orientation of water molecules, which remain in the subphase. For more complicated molecules like phospholipids, a debate has arisen on this orientation. The determination of their absolute orientation requires extracting the sign of the SFG resonant contributions from the spectra, which is not straightforward. Using different kinds of SFG apparatus and data analysis, two teams first disagreed on their conclusions. For [29], water molecules strictly obey the electrostatic rules and orient their dipoles as a consequence of the charge of the lipid head groups. On the contrary, a second group first stated that the orientation of the top layers of water was not simply related to the electric charge of the head group, orienting their dipoles in the same direction whatever the charge of the head group [28], which appears contrary to electrostatic laws. The debate ended when it was shown that their conclusions were attributed to an incorrect data analysis. Beyond this controversy, SFG proves its selectivity to a low amount of oriented water molecules, which in fact carry the most precious information at the interface. However, this case also illustrates the difficulty to correctly interpret SFG data, which, for tricky situations, requires care and a tinge of specific know-how.

Of course, SFG may also follow more conventional phase transitions in lipid monolayers. Ordering upon compression, condensation or fluidization effect of additives in the subphase (SCN^-[33] and Ca^{2+}[34]), or of coadsorbants inside the layer (SDS [27], palmitic acid [29], cholesterol [35]) has been addressed by several teams. In particular, Allen [36] have analyzed a model layer for lung surfactant, composed of a mixture of up to three lipids (DPPC, POPG, and palmitic acid PA) and a model peptide. They have shown the condensation effect (i.e., ordering) of PA on the DPPC layer, the opposite fluidization effect (i.e., disordering) of POPG in the DPPC layer, and displayed the balance between both in the ternary mixture as a function of molar ratios.

Lipid bilayers: symmetry properties

Lipid bilayers provide one of the most striking applications of symmetry properties of second-order nonlinear optics to investigate structural properties. As explained in Theoretical Background section, SFG is interface-sensitive because the inversion symmetry is broken at the interface along the normal. However, if the molecules at the surface adopt a configuration which creates additional symmetry along the surface normal, their specific SFG signal vanishes. A bilayer of a single amphiphilic molecule is an archetype of such a geometry and early studies have illustrated the fact that it is indeed transparent to SFG as a consequence of its up-and-down structure [37]. On the contrary, asymmetric bilayers may be studied by SFG spectroscopy, in an IR spectral range adapted to the chemical groups breaking the symmetry. It is therefore possible to choose by construction whether to see the lipid bilayer or not, as a function of the property one wants to investigate. Bilayers in the following examples are supported by a substrate. We emphasize two research thematics as examples: the first one relates to the flip–flop mechanism in a homogeneous bilayer, and the second one tackles the gel to liquid crystal phase transition in a hybrid bilayer.

One of the important structural parameters to monitor in the behavior of lipid molecules in cell membranes is their movements, that is, lateral translocation or interleaflet displacement, the latter (also called flip–flop) being involved in cell growth and intercellular signaling [38]. To address this issue, one of the two leaflets constituting the membrane is perdeuterated to break the inversion symmetry of the lipid bilayer. The SFG signal of the CH_3 moieties is therefore measurable. When complete flip–flop is achieved, the symmetry has been regained as the bilayer comprises two layers of 1:1 mixture of perhydrogenated and perdeuterated molecules, and the CH_3 signal vanishes. Therefore, by following in time the decrease of methyl group SFG intensity, one obtains direct information on the flip–flop kinetics, as illustrated in Figure 5, and by extension its thermodynamics.

Several parameters can be tuned to measure their effects on the flip–flop process: temperature, layer pressure during LB transfer [38], as well as the influence of membrane proteins. In the latter case, it has been shown that inclusion of small proteins inside the layer hardly changes its symmetry properties, implying that the method still applies in this situation. In fact, both hydrophilic and hydrophobic proteins may facilitate the flip–flop process, with a greater efficiency for melittin (hydrophilic). Therefore, scenarios for the action of the proteins inside the lipid bilayer can be proposed [39].

In this study, SFG was applied to follow the gain of symmetry inside an asymmetric bilayer. Accordingly, it is possible to monitor the symmetry appearing and disappearing inside a symmetric bilayer. The same group has followed in this way the temperature-induced transition from the gel phase to the liquid crystal phase in a planar supported lipid bilayer. The bilayer is totally symmetric

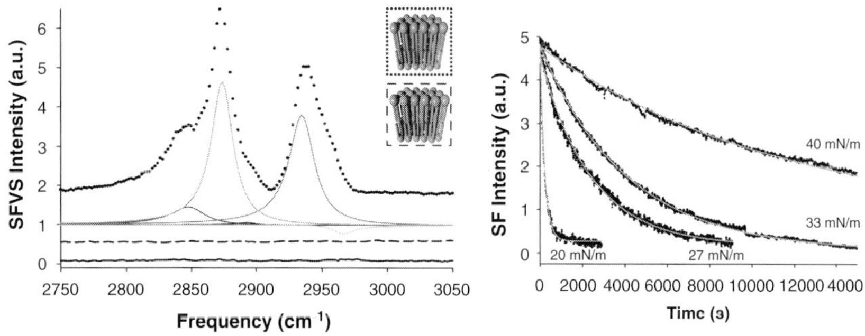

FIGURE 5 (Left) SFG spectrum of an asymmetric DSPC/DSPC-d70 bilayer at room temperature (black points). Individual vibrational contributions, as determined by data fitting, are shown in gray. The observed peaks correspond to CH_2 and CH_3 vibrations. The sample was held at 46.8 °C until the SFG signal reached a minimum following the flip–flop process, and then returned to room temperature (dashed line spectrum). For comparison, a symmetric bilayer consisting of a 1:1 ratio of DSPC and DSPC-d70 was prepared and a spectrum obtained at room temperature (solid black line). The spectra are offset for clarity. (Right) Representative decays of the SFG intensity for the CH_3 symmetric vibration at 2875 cm^{-1}. The above decays were collected for DSPC/DSPC-d70 bilayers at 46.8 °C. Data are shown in black. Gray lines represent the fit to the data for deposition pressures of 40, 33, 27, and 20 mN/m. From [38].

before and after the transition, but during the transition the symmetry is broken by the asymmetric distribution of domains of the two phases inside each layer. By following again the intensity evolution of the stretching modes of the fatty acid methyl groups as a function of the temperature, the transition temperature is easily determined for several lipid chains [40], with values comparable to DSC measurements. A strong correlation is noted with the transition temperature for a supported monolayer in water. Similar research work was performed on hybrid bilayers membranes (HBM) as an alternative to lipid bilayers. The asymmetric HBM is made of one single lipid layer deposited on top of a hydrophobic self-assembled monolayer (SAM) adsorbed on a metal substrate (gold). In the gel phase, the disordered alkyl chains exhibit many gauche defects, leading to a measurable CH_2 signal; on the contrary, a perfectly straight alkyl chain is said to show no methylene SFG signal for local symmetry reasons, although this point is still debated [41]. The phase transition temperature T_m is therefore measured by the steep variation of the ratio of CH_2 over CH_3 stretch intensities. The temperatures appear higher by an average of 10 °C to those evaluated in vesicles for several lipids with phosphocholine heads [42]. To account for this result, the crystallinity of the SAM is tuned, and it is shown that the transition temperature decreases toward values analogous to those in vesicles when the SAM underlayer becomes disordered. The high T_m is therefore correlated to the interaction of the top lipid layer with the underlying SAM [43]. Finally, the same method allows monitoring the phase transition in binary mixtures of phospholipids and cholesterol as a function of cholesterol content. This leads to the evaluation of the size of ordered and disordered coexisting domains during the transition, which are proved to remain small for planar layers as compared to vesicles, as confirmed by AFM measurements [44]. Nevertheless, HBMs are not suitable to perform studies including processes related to the inner leaflet (flip–flop, insertion of interfacial proteins). Fundamental mechanisms modifying the biomimetic lipid bilayers are induced by membrane proteins or peptides. Their mutual interaction as investigated by SFG is presented below.

Peptides and proteins

The sensitivity of SFG combined to the transparency of lipid symmetric bilayers to the technique makes it an interesting tool for the study of membrane proteins *in situ*. However, proteins are much more complex than lipid layers, and in particular exhibit no overall specific symmetry properties. Their study by vibrational spectroscopy is therefore less straightforward. As above, the energy region comprising the CH stretches of proteins and OH stretches of water are focused on, whereas some groups tackle the more delicate, but more informative, domain of amide vibration modes [45]. We illustrate here the typical analysis of protein conformation in various environments, at liquid interfaces, solid interfaces and inside lipid bilayers. Prior to studying these complex interactions within membranes, SFG works are performed on peptides/proteins alone in contact with a liquid/aqueous or liquid/solid interface. In this case, the structural effect of protein adsorption is investigated either on pH-dependent water surface (salt ions and electrochemical potential) or on biocompatible materials (biopolymers).

Liquid–air and liquid–liquid interfaces: protein conformation

We choose to illustrate three types of environment and molecular interactions involving amino acids, peptides, and proteins. Again, the high sensitivity of SFG makes it possible to tackle difficult issues in various geometries and to access orientation parameters of molecules. The Richmond group has a long-time speciality in analyzing the behavior of various molecules at the water/CCl_4 interface. Such a liquid–liquid interface, although not relevant for biological systems, is a model for water-hydrophobic liquid systems. Specifically, L-forms of phenylalanine, lysine, leucine, isoleucine, methyonine, tryptophan, threonine, and tyrosine have been under study [46]. Although the complexity of molecules and functional groups in the CH and OH stretching region makes the analysis difficult, the results show that threonine and tyrosine do not adsorb at the interface, whereas tryptophan does not produce enough SFG signal. On the contrary, polarization effects bring some information about the orientation of the symmetry axes of the molecules (or the chemical moieties under study) with respect to the surface normal. At the air/water interface, a more detailed study of a leucine monolayer was performed in [47]. A more complete analysis allowed to calculate the tilt and rotation angles of the alkyl part of the amino acid (pointing out of the water subphase) as a function of concentration (hence surface density) and pH. The dependence in both parameters is rather low, whereas the structure of interfacial water under the leucine layer is strongly modified by the pH as a consequence of electrostatic interactions. Finally, it is possible to study by SFG the interaction of a small molecule with a monolayer of protein at the air/water interface. The case is illustrated in [48] with urea interacting with the well-known bovine serum albumine (BSA). Urea is known to denature proteins but the interaction between them is partly unclear. SFG sensitivity to low concentrations and orientation allows us to study the way urea orientates when interacting with a protein, as a function of pH (hence protein surface charge). The analysis of CH stretches from the hydrophobic BSA amino acids, OH stretches from interfacial water, and NH stretches of urea prove that urea and water tend to remain oriented in the same direction at the surface of the BSA molecules, with a random orientation at the isoelectric point of the protein.

Liquid–solid interfaces: protein interaction with surfaces

Immobilization of proteins at a solid interface may appear less compatible with biology. However, for fundamental studies, it may be a way to work on a sample of better-known structure and reproducibility. In addition, controlled adsorption on a dedicated surface may boost the sensitivity of the detection technique. Finally, some processes are inherently linked to the presence of a solid, like fibrinogen adsorption or interactions linked to biomolecular recognition.

We stress the huge work performed by the Chen group on the analysis of protein adsorption through SFG spectroscopy. This group has pioneered the use of SFG as a probe of the key amide I band [45], whose position and intensity is strongly related to the secondary structure of a protein (e.g., α-helix, β-sheet, and random coil). To reach that purpose, an original SFG setup was developed, near total reflection through a CaF_2 prism and taking advantage of the high optical reflectivity favored by optimizing the Fresnel factors. Amide I band is routinely used in IR absorption spectroscopy for secondary structure analysis in the bulk. Coupled to ATR-IR absorption spectroscopy, SFG allows performing the same kind of studies on a monolayer at an interface [49] because, as before, the randomly dispersed proteins in solution are not SFG-active while highly oriented interfacial proteins are easily detected. As for α-helices, the relation between orientation distribution and SFG spectrum requires a careful calculation of the hyperpolarizability of the helix. With proper analysis and the help of coupled ATR-IR measurements, valuable information is extracted from polarization combinations about the structural changes of peptides and proteins upon adsorption [50, 51]. In addition, chiral-specific SFG signals from β-sheet structures have been evidenced in [52]. An analogous analysis of the hyperpolarizability of the anti-parallel β-sheet is under way, with promising applications to the interaction of peptides and proteins with lipid bilayers and surfaces by SFG spectroscopy of the Amide I modes [53].

Fibrinogen is a key blood protein involved in coagulation through its transformation into fibrin. It has been extensively used as a model protein for adsorption studies by SFG on solid surfaces like polymers or SAMS, for fundamental purposes [54, 55] as well as biocompatibility issues (in the case of synthetic implants in contact with blood) [56]. Fibrinogen is also the model protein used to rationalize the extraction of α-helix orientation of a protein adsorbed onto a surface from spectroscopic data in the amide I infrared range, using the fact that the molecule in its linear conformation has a rather symmetric shape, which disappears upon molecular bending. On polystyrene, fibrinogen adsorbs along a bent configuration, with a broad distribution of tilt angles [50]. The conversion to fibrin in the presence of thrombin enzyme results in a strong conformational change [55].

In a general way, the behavior of amino acids, peptides, and proteins upon adsorption on a solid substrate strongly depends on numerous factors: pH, ionic strength, orientation, charge distribution, chemical nature of the adsorbing surface, and protein size. SFG may help investigating the effects of several of them, bearing in mind the applications to molecular recognition or biosensing. To reach that purpose, it is proposed to use model short synthetic peptides with designed properties. The Somorjai group has investigated the adsorption of such model peptides with classical α-helical residue pattern (regular alternation of hydrophilic and hydrophobic residues) on hydrophilic silica and hydrophobic polystyrene. With a 14-residue lysine-leucine peptide (LK_{14}), the adsorption processes appear very different. CH stretch modes show up on hydrophobic substrate, independently of the ionic strength of the buffer. On the contrary, only a NH mode attributed to amide A can be measured on polystyrene. This mode depends indeed on the ionic strength, which is to be related to the secondary structure of the peptide: at high ionic strength, LK_{14} adopts an α-helical structure and the mode is present. On the contrary, when the peptide takes the form of a random coil at low ionic strength, no amide A signal is measurable (although quartz microbalance data show that proteins adsorb in all cases onto the surface). This evidences a different adsorption geometry in each case, with stabilizing interactions between the substrate and the peptides involving hydrophobic leucine on polystyrene and hydrophilic lysine on silica, respectively [57, 58]. The behavior is confirmed by a similar study on analogous synthetic 14-residue peptides showing the same repetition scheme with various natures of hydrophobic (leucine, alanine, and phenylalanine) and hydrophilic (lysine and arginine) residues. On hydrophobic surfaces, the vibration modes of the hydrophobic residues are all detectable. On silica, the only measurable signal comes from the LK_{14} peptide, which is also the only one stable in solution as an α-helix. This proves in a different way the behavior evidenced above: when peptide adsorption is driven by hydrophobic interactions, the SFG response does not depend on the secondary structure, whereas adsorption driven by hydrophilic interaction is associated in SFG with a secondary structure signature, thus to a well-ordered geometry [59].

Interfacial proteins in lipid bilayers: in situ *behavior revealed*

Proteins have been shown above to play a facilitating or inhibiting role in the flip–flop process in lipid layers. In the present part, we focus on the interplay between the peptides/proteins and the model membranes, the way they interact and modify their mutual structures and ordering. In order to have a comprehensive view of the numerous biological interactions related to peptides/proteins within membranes studied by SFG, the reader can refer to reviews [60, 61].

Interactions between lipid layers and proteins/peptides are in a first step carried out in the CH and OH spectral ranges. In this region, vibration modes of both lipids and peptides coexist, which allows us to monitor in parallel their evolution during interaction. For example, in a partially deuterated Langmuir monolayer of lipids with deuterated alkyl tails, reference [62] follows the effect of the insertion of the gramicidin A model peptide when orientation processes take place upon layer compression. At low peptide concentration, peptide and lipids orientate together, whereas the peptide remains unchanged at high concentrations. For intermediate concentrations, it is interesting to note that the peptide does not orientate as a whole, probably starting by its tryptophane rich C-terminal part. As for bilayers, studies take place on planar supported lipid bilayers or hybrid bilayers in contact with water. For example, the highly reversible perturbation of a hybrid bilayer (deformation and loss of order) by a signal peptideless protein is evidenced in [63]. The literature proves that it is possible to monitor in real time the interaction of peptides with supported lipid bilayers and determine their orientation in the layer through their distinctive secondary structures: α-helix [64], β-sheet [61], and even 3_{10}-helix [65].

In order to illustrate the possibilities opened by SFG on complex lipid/peptide interactions, we develop here the melittin case, which is one of the most studied membrane proteins for its anti-microbial properties, on supported DPPG bilayer on CaF_2. The issue of the process of melittin-induced pore formation inside the lipid bilayers has been much debated and investigated [66]. A summary of this rather complete study can be found in [60, 61, 67]. It uses the state-of-the-art SFG data collection and analysis: three spectral windows contribute to the data (CH/OH, CD for perdeuterated lipids, amide I stretching ranges). Both symmetric (perdeuterated dDPPG/dDPPG) and asymmetric (perdeuterated/perhydrogenated dDPPG/DPPG) lipid bilayers are investigated in order to access the symmetry properties while differentiating the structures of the distal (in contact with the peptide solution) and proximal (in contact with the substrate) lipid leaflets. A careful analysis of amide I vibrations leads to the orientation of α-helices inside the lipid layers. The time evolution of CD and CH stretches monitors in time the melittin-induced order and disorder, symmetry breaking, and recovering inside the distal and proximal leaflets. We summarize below the information on the interaction collected from the SFG experiments (Fig. 6),

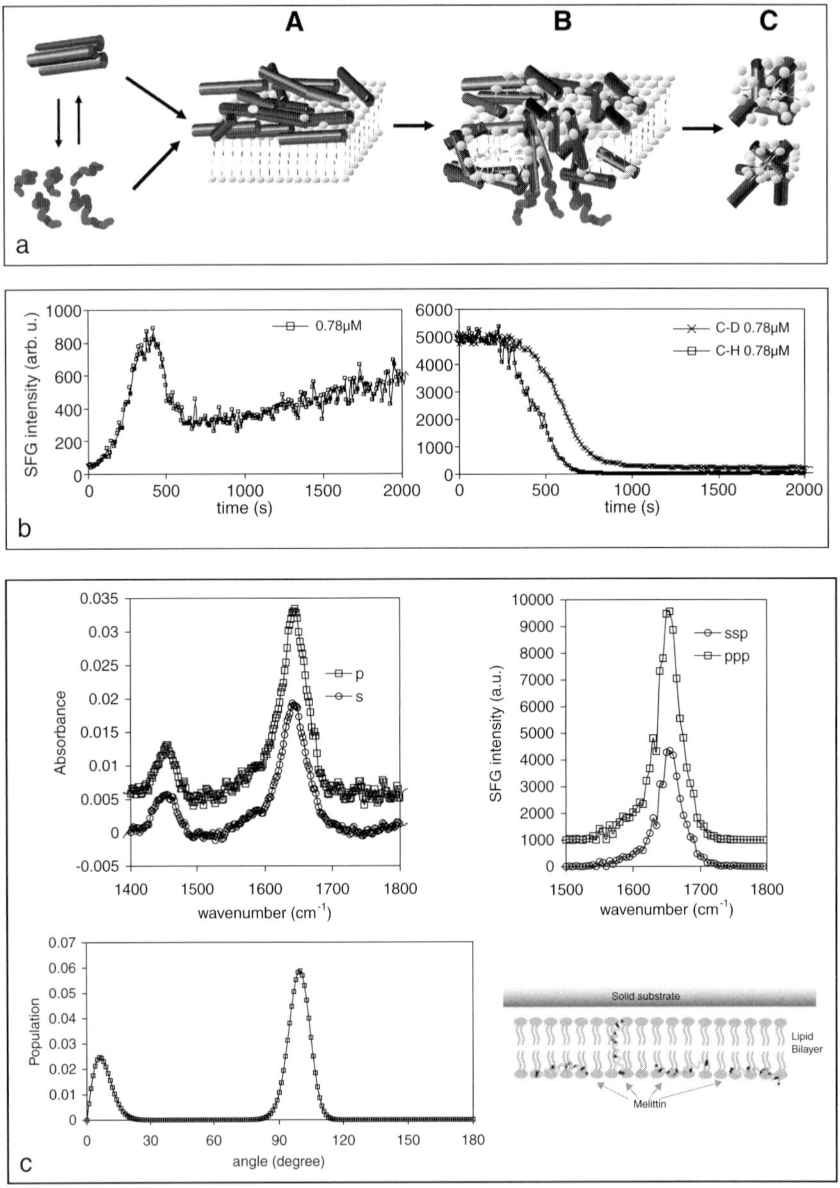

FIGURE 6 (a) Cartoon illustrating model suggested for membrane permeation. The peptides reach the membrane either as monomers or oligomers, and then bind to the surface of the membrane with their hydrophobic surfaces facing the membrane and their hydrophilic surfaces facing the solvent (step A). When a threshold concentration of peptide monomers is reached, the membrane is permeated and transient pores can be formed (step B), a process that can lead also to membrane disintegration (step C). From [68]. (b) SFG intensity of the CD and CH symmetric stretches of the system DPPG/DPPG bilayer in interaction with a melittin solution at intermediate concentration. Left: symmetric perdeuterated DPPG bilayer, CD

along with the way they have been extracted from the analysis of various kinds of SFG data with the help of scenarios from the literature.

Scenario	Evidence by SFG
• At low concentration in melittin (below 0.5 μM), the peptide switches from its solution structure (random coil) to an ordered conformation (involving two bent α-helices) while adsorbing onto the distal leaflet parallel to the surface (carpet model). This creates disorder inside the layer as the number of adsorbed peptides increases, and screens or neutralizes the surface charge of the leaflet. The proximal leaflet remaining mostly undamaged, the cell survival does not seem endangered (Fig. 6(a), step A).	• The amide I signals proves the adsorption of the peptide in an α-helix conformation (Fig. 6(c)). The amide A signal also appears while the water OH stretches disappear, indicating a vanishing of surface charge leading to the disorientation of the previously dipole-oriented water. Disorder in the perhydrogenated distal leaflet is evidenced in asymmetric bilayers by its CH drop of intensity, while the change in CD stretches of the proximal leaflet are limited. In symmetric layers, the CD stretch signal slowly increases in time [67].
• Increasing the concentration of melittin in solution (between 0.5 and 3 μM) allows us to reach the critical surface concentration above which the peptides dig into the distal leaflet while reorienting perpendicularly to the surface. Perturbation of the distal leaflet increases. The next step is a disturbance of the proximal leaflet by the peptides dug into the layer and formation of melittin pores inside the bilayer. At this stage, the flip–flop process starts: this implies that the pores	• A careful analysis of the amide I signals (Fig. 6(c)), coupled to ATR-FTIR data, evidences a dual orientation for the α-helices: three quarters remain parallel to the surface while one quarter adopts a perpendicular orientation. At this stage, the proximal leaflet disturbance is evidenced in asymmetric layers by a drop in the CD intensity (Fig. 6 (b)). In symmetric perdeuterated layers, the CD stretch first increases when the distal layer is perturbed,

(continued)

stretch, melittin injected at t = 0 s. The intensity rises from zero (full symmetry) to a maximum (symmetry breaking through perturbation of the distal leaflet, step A) and then drops (disorder induced in the proximal leaflet, step B). Right: asymmetric perdeuterated (proximal leaflet)/ perhydrogenated (distal leaflet) DPPG bilayer, melittin injected at t = 200 s, both signals normalized to the same starting value. Intensity drops first from the distal layer (step A), and then from the proximal layer (step B) as a consequence of leaflet disorder. Both intensities vanish as a consequence of flip–flop induced symmetry (step B). From [67]. (c) Top: ATR-IR (left) and SFG (right) spectra in the amide I region of the melittin/DPPG symmetric bilayer complex. Polarizations are indicated on the curves. Bottom: the maximum entropy function analysis on the intensity ratios from the spectra lead to the dual peaked distribution function (left). This function matches the carpet scenario for the peptide/lipid interaction (right, step B). From [64].

(continued)

Scenario	Evidence by SFG
facilitate the rapid lipid transfer from one leaflet to the other, favoring a toroidal pore model (porous to lipids) as opposed to the hermetic barrel stave model. The time constants of these processes depend on the peptide concentration in solution. This scenario is illustrated in Figure 6(a), step B.	and then decreases when the perturbation reaches the proximal layer. The formation of pores is evidenced by the start of the flip–flop process, which leads to a symmetrization of asymmetric bilayers and a fast decrease of CD and CH stretch signals. Flip–flop is also proved by the comparison of absolute intensities of CD stretches after melittin interaction in the symmetric and asymmetric cases: without it, intensities should always be stronger for the latter. On the contrary, flip–flop reduces its deuterated lipid content in the proximal layer by 50%, which should lead to a fourfold CD stretch intensity decrease. The experimental factor (around 3) evidences a strong but incomplete flip–flop.
• At higher concentrations (above 3 μM), the previous processes take place much faster as the critical surface concentration is reached sooner, leading to a strong perturbation and eventually dislocation of the bilayer (step C).	• The time constants characteristic of the vanishing of the CD and CH stretches drop with concentration, from a few tenths of minutes to a few minutes for concentrations above 3 μM [67].

Biosensing

In the previous paragraphs, we have seen that SFG spectroscopy could address fundamental biological processes *in vitro*. As it is able to probe selectively the chemical fingerprint of the thin molecular barrier between two biological media, it should naturally be used as a noninvasive optical probe in biosensing-related applications: characterization of the biosensors efficiency to track pollutants, protein-ligand recognition, antigen-antibody recognition in medical applications, and DNA strands hybridization. The application of SFG is even more straightforward in such applications, as most of them require, and are already performed on, flat solid surfaces with designed surface properties, either in a gas or liquid environment. However, the number of studies on such subjects remains surprisingly scarce, even if one may expect a growth in the forthcoming years. We present below two aspects on biosensing: specific biomolecular recognition and hybridization of DNA strands attached to surfaces.

Specific biomolecular recognition

In the case of biosensors, the recognition process is often performed at liquid/solid interfaces because the ligand–receptor biomolecules are naturally present in an aqueous phase. While surface plasmon resonance or fluorescence techniques detect the recognition process, only IR-based spectroscopy (FT-IRRAS, PM-IRRAS, SFG) provide chemical information. The unique advantage of SFG over other techniques is that it does not require fluorescent labels, controlled surface roughness, or IR markers to be efficient: it allows remaining as close as possible to realistic biological conditions.

A first significant biomolecular recognition process put in evidence by SFG coupled to FT-IRRAS was carried out on the ligand-protein archetype model of biocytin/avidin [69] as described below (Fig. 7). The specificity is proved by comparing the changes in vibrational resonances due to several interacting protein and checking that only avidin provokes a specific SFG response. The signature of the recognition appears in two ways: a reordering of biocytin molecules as a consequence of interaction with avidin and the appearance of specific signals from the protein/biocytin complex (therefore partially ordered at the interface). This proof of interaction is in fact very close in its principle to that observed above in the melittin/dDPPG-DPPG case [67].

In more details, a biocytin ($C_{16}H_{28}N_4O_4S$) self-assembled monolayer is adsorbed onto the base of a CaF_2 prism closing a Teflon cell. The latter is then filled with a HEPES buffer solution containing test proteins: avidin, avidin presaturated with biotin (PSA), or bovine serum albumin (BSA). After 16 h of contact with the biocytin film, the cell is emptied from the solution and the prism base is rinsed with the HEPES buffer and dried before the SFG measurements. SFG experiments could have also been performed *in situ* (i.e., in contact with the solution), but it would have required some experimental modifications [45]. In these conditions, significant specific spectral changes in the SFG signal of the interface in the 2800–3400 cm^{-1} spectral range are only observed in the Biocytin + Avidin case (Fig. 7(B)), whereas the changes are not protein-specific in the other situations (even if they prove that proteins perturb the monolayer). Interaction with avidin causes ordering of the sloppy biocytin monolayer, evidenced by a drop in the intensities of CH_2 vibrations modes of the biocytin backbone due to symmetry cancellations in an all-trans chain, as explained previously. The most striking feature is the appearance of a sharp CH band (3050 cm^{-1}) and a broad NH band (3150 cm^{-1}), the former being related to the

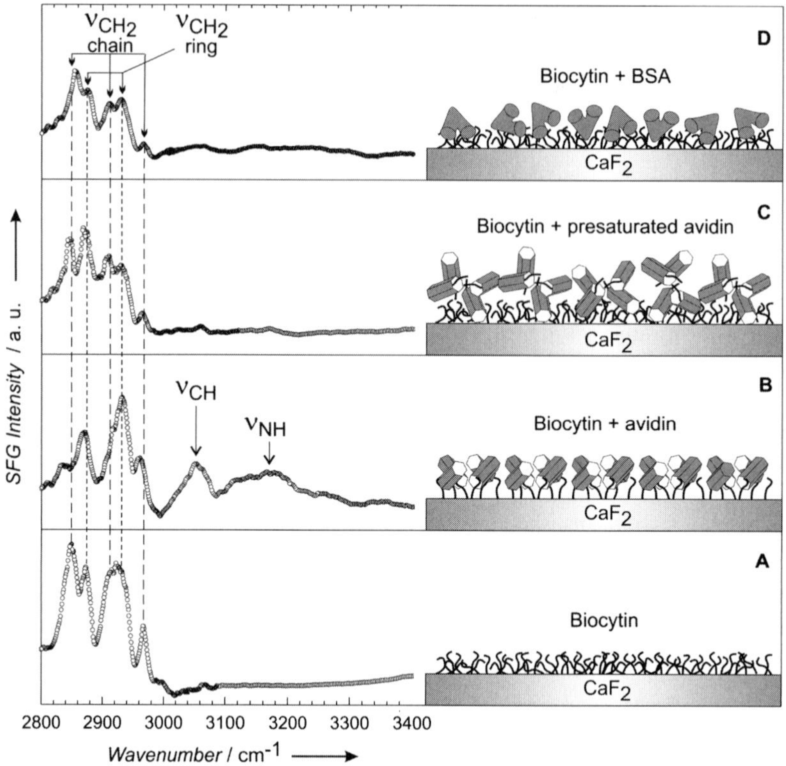

FIGURE 7 SFG spectra of a biocytin monolayer on CaF_2. (A) bare monolayer; (B) after interaction with avidin; (C) after interaction with avidin presaturated with biotin; (D) after interaction with BSA. From [69].

ring CH groups of the biocytin molecules and/or to the aromatic CH entities of the avidin amino acid residues Tyr33, and the latter to the amine/amide vibration modes of all amino acids implied in the docking process. It is an indication of a reorganization of all the molecular groups in the binding pocket, especially with strong H-bonding environment, which favors coherence in orientation and thus SFG activity. Complementary FT-IRRAS measurements used as reference in the amide I and amide II spectral range allow refining the interpretation. They also enlighten the presence of the nonspecific amide A band at 3330 cm^{-1} whatever the protein in solution. They reveal the presence of a protein at the surface (i.e., proteins stick to the biocytin monolayer) while SFG detects the specific biomolecular recognition process.

In the case of common biosensors based on a gold substrate, the excitation of s-d interband electronic transitions by the visible or SFG beams may mask the molecular response because it generates a strong nonlinear response from the metal, as underlined in Theoretical Background section. It becomes embarrassing indeed if destructive interferences dominate and hamper the molecular vibrational nonlinear response. More generally, this problem exists for any interface built on a substrate having a strong optical activity in the visible spectral range. Solutions are at present developed to skip these potential problems. The use of nanostructured interfaces in biosensing affords an original solution. With spherical gold nanoparticles, SFG acts as a powerful local probe at the molecular scale [70] as a consequence of the excitation of their surface plasmon resonance (SPR) by the incident visible beam. It favors indeed the molecular response of the interface to the prejudice of the s–d interband contribution. Moreover, when combined to SFG-ATR geometry, the unique association of the SPR excitation of gold nanoparticles with visible and/or SFG evanescent fields at the probed interface enhances the molecular response with respect to the nanoparticle one by at least two or three orders of magnitude. This has recently been demonstrated in the biomolecular recognition process of the biotin/avidin on a closed-packed array of gold nanoparticles in SFG-ATR geometry, constituting the first step toward a new and extremely sensitive platform for biosensing [71]. The main advantage of such nanostructured devices and SFG-ATR experiments resides in the use of label-free biosensors: no IR or fluorescent marker is required to have a great sensitivity to the biomolecular recognition process, even with a low amount of matter on a biochip.

DNA conformation and pairing at surfaces

From the previous experiment, it is rational to consider that SFG spectroscopy must apply to characterize *in situ* biosensors based on olignucleotides or DNA strands. Several available commercial biosensors consist for example of single-stranded DNA (ssDNA) covalently attached to glass or silica surfaces. The key process in medical, material, photonic, and electronic applications and devices related to DNA is the hybridization, that is, the fundamental molecular recognition process in which base paired double-stranded DNA (dsDNA) is formed. Hybridization efficiency and selectivity at interfaces depends on molecular orientation, packing density, and electrostatic interactions. Knowledge still misses on the detailed molecular mechanisms of hybridization. SFG could give a boost to this field, as other label-free methods suffer from low sensitivity to the few amounts of matter used. Few teams have successfully tackled this issue in the past 5 years; most of the published results are summarized in [72].

Two different kinds of DNA strands have been used in the literature. In the first scheme, films of deposited oligonucleotides mixing the four types of bases are studied in air. Even if such objects are close to real DNA strands, their lack of repeated patterns and the absence of solution buffer represent a severe drawback. Internal disorder in the DNA strands and low specificity to vibrational signals from the bases prevent from measuring a signal either from the backbone or from the bases. As a consequence, only the signal from the linker, intercalated between the DNA strand and the surface, can be seen.

However, this already provides information on the structure of the DNA layer (inferred from the order in the linker layer) after immobilization on the surface. The single DNA strands appear rather disordered in most cases [73], but their structural organization is better when the buffer contains monovalent rather than divalent cations [74].

Another approach relies on designed model oligonucleotides with uniform base content. Again, one may compare this situation to the use of model peptides with periodic amino acid sequences as detailed above [58, 59]. The right choice for the bases makes it easier to study pairing processes while keeping individual signals from each strand. Examples in the literature focus on thymine (T) and adenine (A) oligonucleotides [75]: thymine possesses a distinct CH_3 vibrational signature, absent in the chemical structure of its fellow adenine and of the backbone, but easily accessible to conventional SFG setups. The most illustrating study [76] shows that it is possible to extract several kinds of information on the hybridization process of a surface bound DNA strand. First, on T_{15} ssDNA surfaces, no specific CH_3 signals can be measured. After hybridization forming T_{15}/A_{15} dsDNA, the CH_3 signals come out, even though the amount of CH_3 species remains constant. This is again a proof of ordering in the surface strands upon hybridization, hinting at the formation of the DNA double helix. In a second step, the dsDNA is analyzed with a different polarization combination, which allows extracting only the chiral contribution in an isotropic surface. Two contributions appear (Fig. 8), attributed to the thymine bases (asymmetric CH_3 stretch) and the backbone (deoxyribose CH stretch). The thymine contribution changes sign when switching from $3'\text{-}T_{15}/5'\text{-}A_{15}$ to $3'\text{-}T_3A_{12}/5'\text{-}A_3T_{12}$, which means that the thymine moieties reverse their direction upside down and is therefore a direct proof both of the existence and of the antiparallel character of the double helix. The thymine moiety being chemically achiral, this chiral-specific signal originates from the helical structure of its local environment. On the contrary, the deoxyribose signals do not change sign when reversing the direction of the helix, because they are equally part of both strands. Such an analysis thus allows us to prove the reordering of the strands upon hybridization, while directly evidencing the senses of the DNA single helices involved. This example perfectly summarizes the interest of SFG as a label-free investigation technique in biosensing for biological applications, giving access not only to binary responses over hybridization or recognition, but additionally to ordering, orientation, and chiral-specific information on the processes at the molecular scale.

As a conclusion to this section, we would like to stress the high versatility of SFG spectroscopy applied to biological objects. However, whatever the molecules under study, its originality (in addition to chemical assignation through vibration modes) relies on the strengths summarized in Theoretical Background section: sensitivity to low amounts of matter; recording of absolute spectra; specificity of surfaces and interfaces (including buried, liquid/gas, and liquid/

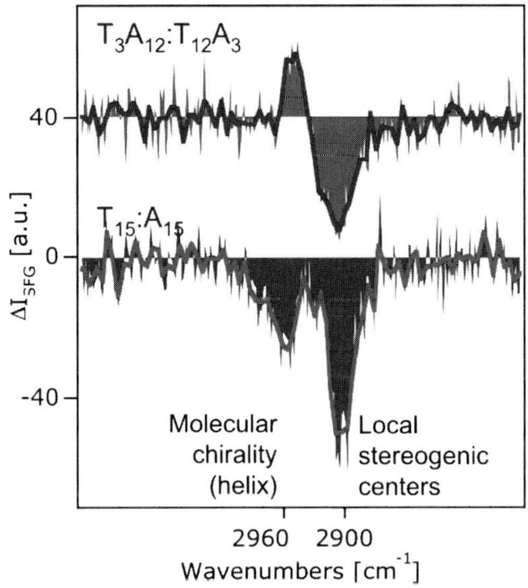

FIGURE 8 **Chiral-specific SFG spectra of 15-mer double-stranded DNA bound to a glass substrate. The base sequence is indicated above each curve. Thymine signal (2960 cm^{-1}) change sign when the corresponding helix is reversed, whereas backbone signals (2900 cm^{-1}) don't. From [76].**

liquid interfaces); high yield to molecular and supramolecular symmetry, order and orientation; extraction of chiral-specific information; and label-free sensing method. We enlighten in the next part how these characteristics have been recently boosted by bringing the technique to a higher level through doubly resonant spectroscopy, SFG microscopy, and time-resolved spectroscopy.

PROSPECTS

Dynamics and kinetics: from fast to ultrafast time resolution

SFG has been used rather early to measure the evolution and dynamics of chemical systems over various time scales (from picoseconds to minutes) [77]. Nevertheless, at that time, the subpicosecond level of chemistry was still unreachable, whereas acquisition times made it difficult to record successive spectra in a short time. Additionally, the signal-to-noise ratio was not optimal because of laser intensity fluctuations during wavelength scans. The emergence of fs broadband SFG setups has opened an era where both problems have been overcome. An extensive report on technical properties and applications of fs-broadband SFG spectroscopy is given in [10, 26]. In fact, the major part of the works on biointerfaces cited in the present chapter is based on fs-broadband SFG.

Firstly, acquisition times are shorter, as the whole spectrum is recorded for each laser pulse; longer acquisition times improve the signal-to-noise ratio and the sensitivity to weak resonances. As an illustration, impressive rates of five spectra per second have been published [23]. Such rates allow kinetic studies of the order of seconds by following over time the evolution of spectral features, caused by a perturbation at $t = 0$ s or by continuous tuning of an external parameter (pressure, temperature, and electrochemical potential). Such studies remain scarce on biological samples at the second time range; as an example in the minute time scale, the hydrolysis of a planar supported-bilayer catalyzed by an enzyme is described by [78].

Secondly, the fs duration of the IR pulses ($\sim$100 fs) allows performing ultrafast spectroscopy at this scale. Several experimental procedures may be reviewed in [10], for which pulse lengths, and accordingly frequency domains, are smartly tuned as a function of the information sought. We focus on the time-resolved SFG and pump-probe SFG spectroscopies, from which two vibrational relaxation times may be measured. In a time-resolved experiment, the usual SFG configuration is used, but the visible pulse is delayed with respect to the infrared one. The IR-resonant SFG process is only possible when the visible excitation applies to a system excited by a vibrational quantum provided by the infrared photon. Such a vibrational excitation has a finite lifetime, also known as the dephasing time T_2. For a given vibration mode, the maximal SFG signal is therefore measured for a vanishing delay, whereas an exponential decay quantified by T_2 follows for positive delays. Another characteristic lifetime may be measured for the same vibrational mode by an IR pump-SFG probe experiment. In that configuration, an additional intense IR pump beam is sent to the interface prior to the simultaneous probe IR and visible pulses. As a result of the pump, IR absorption occurs and the ground state is partially bleached while one vibration state is populated. When the resonant SFG process takes place, its intensity has decreased as compared to the unpumped situation as a consequence of the depopulation of the ground state. Delaying the SFG process allows the system to recover the ground state population from the vibration state and the SFG signal increases with the delay. The signal recovery is a function of the population differences between ground and vibration excited states: the time constant measured in such a scheme is therefore the population relaxation time T_1. Finally, measuring both lifetimes allows evaluating the pure dephasing time T_2* through the relation:

$$\frac{1}{T_2} = \frac{1}{2T_1} + \frac{1}{T_2^*}.$$

Applying these new techniques to biological interfaces is still rare, but we mention two distinct studies, one on water bound to a lipid monolayer at the surface of water and the other on the vibrational dynamics in the lipid monolayer itself. As previously mentioned, SFG is able to extract vibrational properties of the top monolayer of water at the air/water, solid/water, and lipid/water interface. Apart from the signal from dangling OH bonds, which disappears when the surface is covered by a monolayer, the water SFG signal is made of two broad maxima from OH stretches involved in hydrogen bonding network. Specific dynamics of such water molecules, performed at four different wavelengths of excitation inside the maxima (between 3100 and 3500 cm^{-1}) reveal that their vibrational relaxation schemes differ between the air/water and lipid/water interfaces. For the former, vibrational relaxation occurs through ultrafast energy transfer (resonant vibrational relaxation and thermalization) from the top layer to the underlying bulk water, leading to time constants essentially identical between surface and bulk and wavelength independent. On the contrary, when the surface is covered by a monolayer of lipids, the vibrational and thermal contact with bulk water is lost (or at least greatly slowed down) [79]. As a consequence, T_1 time constant varies strongly with the wavelength, and vibrational relaxation decays to a heated ground state, from which thermal relaxation takes much longer times. More precisely, T_1 raises with the vibrational energy in a way which matches models of hydrogen bonded water molecules: the stronger the H-bond, the lower the vibration frequency and the faster the vibrational relaxation. Such results confirm that water bound to a lipid monolayer is engaged in a privileged chemical relation with the lipid polar headgroups, and that vibrational relaxation relies only on its local H-bonding network. Inside the lipid layer, vibrational dynamics of CH_3 and CH_2 groups as a function of lipid structural organization and of polarization of the photon pump evidences a quicker relaxation for antisymmetric CH_3 than for symmetric CH_3 (respective 1.2 and 3.6 ps lifetimes) implying two distinct deexcitation pathways for both modes, contrary to the bulk situation, either by a preferential coupling of the antisymmetric CH_3 to CH_2 vibrations or as a consequence of a coupling to the rotational motion around the C–C bond). Such results hint at energy transfer to/from CH_3 excitations from/to other vibration modes [80]. In order to put such a phenomenon in evidence, it is necessary to perform two-color pump-probe SFG experiments.

Playing with colors and excited states: 2D-IR SFG, DR-SFG

Extracting multidimensional information on a system requires additional levels of complexity (technical, experimental, and theoretical). We have just seen that time could constitute one of the new dimensions added to surface-specific spectroscopy. However, it is possible to go beyond and mix in a single experiment time, infrared frequency-domain, and visible frequency-domain.

As a straightforward extension of the previous study on dynamics of CH_2 and CH_3 groups, it has been shown that the vibrational energy transfer between two moieties may be experimentally measured *in situ* [26, 80]. For that purpose, the same pump–probe scheme as before is employed, except that the pump and probe spectral ranges differ. Specifically, water molecules are pumped in a first place, and then SFG probes the CH groups from the lipids as a function of the delay between pump and probe. Vibrational energy acquired by water molecules is transferred to the lipid monolayer under the form of incoherent thermal excitation, which results in a decrease in the total SFG signal from CH moieties. Data are consistent with ultrafast energy transfer from water to lipids with time scale less than a picosecond (0.95 ps).

In this study, the infrared range was fixed and time varied in order to extract dynamic information on vibrational coupling. Inversely, it is possible to access full frequency-domain vibrational coupling by fixing the time-delay between pump and probe at an optimal value (e.g., for maximal bleaching) and scanning the narrow-bandwidth infrared pump frequency. The result is the surface-specific 2D-IR SFG spectroscopy similar to two-dimensional infrared absorption spectroscopy and multidimensional NMR spectroscopy. The decrease of the SFG spectrum is plotted as a function of the frequency of the pump in a 2D-IR SFG chart (Fig. 9). Diagonal peaks relate to the excitation of a vibration mode by light of its own frequency, whereas off-diagonal peaks illustrate the excitation of a vibration mode by absorption of a photon with another frequency, evidencing vibrational energy transfer from one mode to the other. At the time of writing, no such study was performed on biological material. However, a detailed study on a dodecanol monolayer on water has been the subject of several publications [81, 82]. Figure 9 shows its main result: the pump and probe band structures relate to the IR absorption spectrum (left) and SFG spectrum (top), respectively. The graphical chart illustrates the mutual vibrational coupling between antisymmetric/Fermi and symmetric stretching CH_3 modes, and an additional strong vibrational energy transfer from antisymmetric CH_2 to symmetric CH_3 vibration. It must be noted that the technique benefits from the differences in selection rules for both processes involved (IR absorption pump and SFG probe). As a consequence, it is possible to probe the coupling from CH_2 modes, which are essentially SFG inactive but dominate the IR spectrum, to CH_3 modes specific from the interface but undetectable by infrared absorption alone. These results complement the time-resolved study detailed in the previous section, which hinted at a transfer from antisymmetric CH_3 to CH_2 modes that could not be probed here as the CH_2 cross-section was too low.

As a summary, the recent literature evidences that the up-to-date SFG setups make it possible to access the dynamics of individual vibrations, as well as their specific one-to-one couplings, while maintaining the originality of SFG spectroscopy: sensitivity to small amounts of matter, surface/interface specificity. It seems reasonable to consider that such time-resolved techniques will lead to

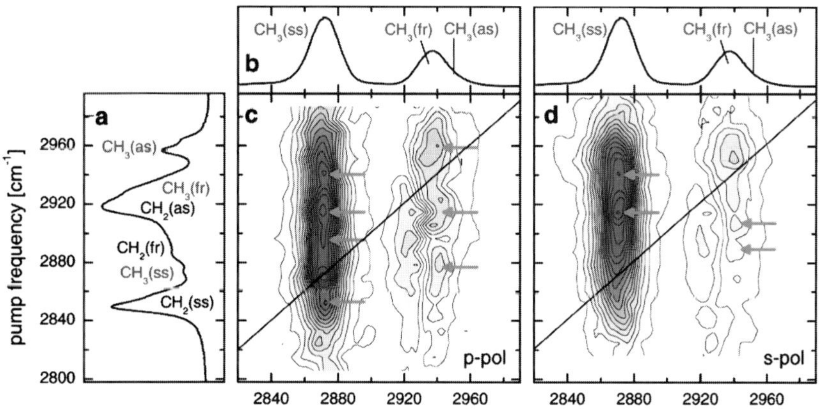

FIGURE 9 **2D-IR SFG spectra of a dodecanol monolayer on water. Dark denotes a pump-induced decrease in SFG intensity. (a) IR spectrum of crystalline dodecanol at 150 K. (b) SFG spectrum of the monolayer, polarizations: SFG/VIS/IR, s/s/p. (c) 2D-IR SFG spectrum, p-polarized pump, t = 0.7 ps. (d) Same as in panel c, with s-polarized pump, for which the in-plane CH_2 modes are efficiently excited, leading to the dominance in the 2D spectrum of the cross-peak between the CH_2 as and CH_3 ss modes. Lines indicate the diagonal; arrows point to off-diagonal peaks indicative of vibrational coupling. From [82].**

advances in complex processes involving energy transfers and ultrafast dynamics in biological material involving for example model lipid membranes and proteins.

As for the interplay between colors and excited states, we have focused so far on the infrared colors. The promising prospects in multidimensional SFG spectroscopy of biointerfaces also lie in another way to mix excitations by varying independently the visible and infrared frequencies of the two input beams. This becomes interesting when each beam excites a molecular transition, as it opens the door to experimental determination of couplings between excited states of different origins. As for the infrared beam, excitations relate as before to molecular vibrations. As for the visible frequency region (visible and SFG beams), several kinds of excited states and transitions may be at stake, either of molecular (e.g., electronic transitions) or extramolecular origin. When both the visible/SFG and infrared beams excite molecular transitions, the SFG process is called doubly resonant (DR) and leads to a strong enhancement of the measured signals. In such experiments, the SFG intensity may be drawn as above on a 2D graph as a function of the IR and visible/SFG frequencies to enlighten the couplings [83]. As outlined on Figure 2(c) (Theoretical Background section), the upper virtual states involved in the SFG process become real excited states, which belong to the vibrational structure of the electronic excited state. In fact, DR-SFG probes the molecular vibronic structure and allows in principle to extract electron-vibration coupling constants [84].

However, a full data analysis remains a rather delicate task. This, together with the low number of independently infrared and visible tunable SFG setups, accounts for the very few examples in the literature of such studies on biological samples (excluding the particular case of some optical chromophores commonly used in biological applications). We point out the evidence of a doubly resonant enhancement of the SFG process with the electronic structure of an adsorbed layer of Green Fluorescent Proteins mutant 2 (GFPMut2) on a platinum surface [85]. The GFP chromophore builds up the electronic excited states of the protein, among which only one of the two states active in absorption, related to the anionic native form, was reachable in the probed visible range (400–600 nm). An intermediate form of the chromophore, involved in photo-isomerization processes, is responsible for its fluorescent properties. Double resonance and enhancement of the SFG spectra from the chromophore in the CH_2 and CH_3 stretching region was evidenced, as a consequence of resonance not only with the electronic excited state related to the stable anionic form, but also with the intermediate fluorescent state. The rapid energy transfers between excited electronic states, associated with the picosecond (rather than fs) time scale of the SFG setup, account for these findings, which prove that doubly-resonant processes may even probe the electronic structures hardly reachable by the direct absorption of a photon.

This rapid overview shows the high potential of the very recently developed optical methods exposed here. In particular, extending the infrared range for time-resolved and 2D-IR SFG spectroscopy, and mixing time-resolved studies with doubly resonant processes should lead to a deeper understanding of molecular couplings in the vibrational and electronic structures, including dynamics, of biological materials.

SFG microscopy: toward high-resolution imaging of biointerfaces

Recent gains in sensitivity or resolution in time and frequency domains rely on heavy experimental developments. In parallel, this technological boost also has implications on the space domain. Reliable laser sources with high-quality profiles open the door to effective optical microscopy based on nonlinear processes. SHG microscopy is now well established, opening the way to SFG microscopy. Apart from the surface and interface sensitivity, SFG microscopy possesses another key asset in terms of lateral resolution for chemical imaging. Relying on the Rayleigh criterion, the resolution is limited to distances of the order of the wavelength of the probe radiation that is a few microns for IR microscopy. However, the resolution limit for SFG is imposed by the SFG wavelength, not the IR. In other words, as for Raman microscopies, resolutions below the micron, giving access to details inside entire cells, are reasonably accessible while still providing chemical information through infrared excitation of vibrational properties. A review of the various methods developed in the past to perform SFG microscopy can be found in [86]. In such microscopies, it is possible to mix imaging and spectral analysis on various points of the sample [87]. An example of SFG microscopic image of onion cells, with a resolution of the order of one micrometer (i.e., around three times below the Rayleigh limit), is presented in Figure 10.

The ultimate resolution of far-field SFG microspectrometers lies around 400 nm. However, near-field optics does not rely on the diffraction limit and, even if it is not possible to define a universal resolution for a given setup, its ability to separate two objects on an image goes far below the Rayleigh criterion. Near-field SFG microscopy is therefore a privileged way to mix high-resolution imaging and chemical fingerprint. Although the application to biological samples seems at present still out of reach, the

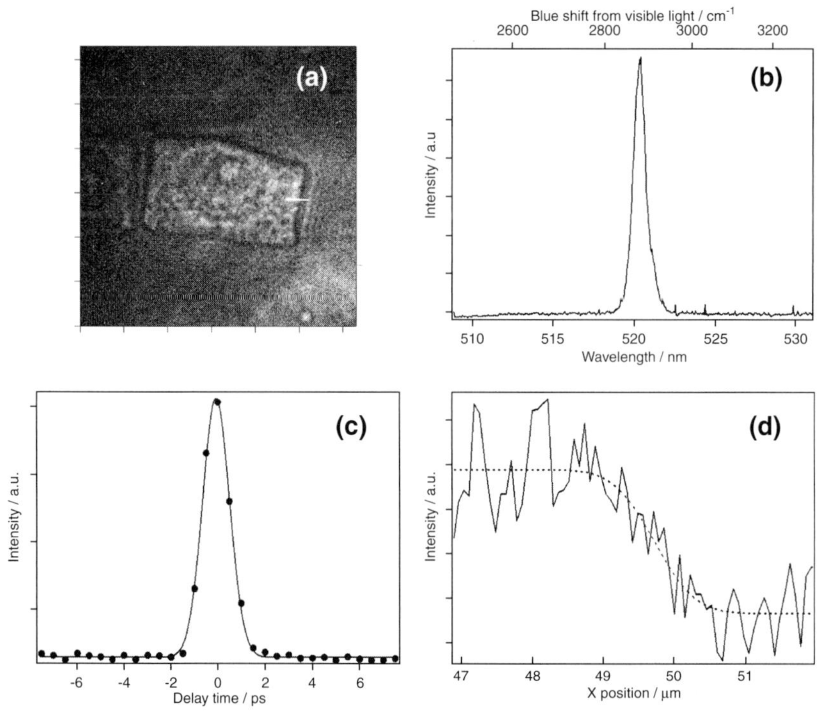

FIGURE 10 **(a) Vibrational (V) SFG image with 610 nm visible and 3472 nm (2880 cm^{-1}) IR beams. One scale division corresponds to 10 μm. (b) Emission spectrum of the VSFG signal of the cells shown in (a); the transverse axis indicates the blue shift from the incident visible beam. (c) Time evolution of the VSFG signal of the cells shown in (a) (filled circles). The delay time of the visible beam relative to the IR beam was scanned from -7 to 7 ps. The best fitted Gaussian curve is shown by a solid line. (d) The intensity profile of the VSFG signal along the white line in (a) (solid line). The fitting curve calculated by the convolution of a Heaviside-step function and a Gaussian point spread function is shown by the dashed line. FWHM of the point spread function was estimated to be 1.1± 0.3 μm. From [86].**

technological progress witnessed in the last decade in the field of nonlinear spectroscopy, as illustrated in [88], makes us confident that this goal is accessible.

CONCLUSIONS

In this presentation of infrared-visible SFG spectroscopy, it has been emphasized that this noninvasive optical probe could be considered as a versatile technique with various applications to biointerfaces. We have underlined that SFG is more than a complementary probe to classical IR and Raman spectroscopies. It possesses specific assets, which make it a fully original tool: sensitivity to orientation, order and symmetry; absolute intensity measurements; ultrashort temporal resolution; vibrational microscopy below the Rayleigh limit; possibility to study couplings between molecular levels at identical or different energy scales. Dedicated to analyze the vibrational fingerprint of interfaces, its peculiar and strict selection rules lead to submonolayer sensitivity from molecular to biological scale, in various kinds of environments and external constraints. However, SFG remains a spectroscopic tool among others, and we point out that original and reliable information often results from its coupling with IR spectroscopy, Raman spectroscopy, or chemical imaging.

While IR-visible SFG spectroscopy is a rather young optical probe of matter, it has proved mature enough to study complex interfaces and tackle delicate questions mainly related to *in vitro* biology. We have illustrated the most prominent to date in this chapter: structure, kinetics, and dynamics of model membranes; conformation of proteins at interfaces; interaction mechanisms between peptide/proteins and model membranes; biosensing at the molecular scale and the nanoscale including DNA hybridization; time, space, and energy mapping of vibrational and electronic properties of biological molecules and tissues.

Considering the speed at which this technique spreads worldwide and technically develops, it seems reasonable to believe that it should continuously evidence in the future new details on various mechanisms involving biological matter, while still experiencing new developments to improve its power of investigation. Scientific content described in the present chapter should therefore be considered as a beginning of a long relationship between SFG spectroscopy and biological materials.

REFERENCES

[1] Guyot-Sionnest P, Hunt JH, Shen YR. *Phys. Rev. Lett.* 1987;59:1597.
[2] Maiman TH. *Nature* 1960;187:493.
[3] Franken PA, Hill AE, Peters CW, Weinreich G. *Phys. Rev. Lett.* 1961;7:118.
[4] Bloembergen N. *Rev. Mod. Phys.* 1982;54:685.

[5] Chen JM, Bower JR, Wang CS, Lee CH. *Opt. Commun.* 1973;9:132.
[6] Olivier N, *et al. Science* 2010;329:967.
[7] Shen YR. *Nature* 1989;337:519.
[8] Braun R, Casson BD, Bain CD, van der Ham E. WM, Vrehen Q. HF, Eliel ER, Briggs AM, Davies PB. *J. Chem. Phys.* 1999;110:4634.
[9] Humbert C, Dreesen L, Mani AA, Caudano YY, Lemaire J-J, Thiry PA, Peremans A. *Surf. Sci.* 2002;203:502–503.
[10] Arnolds H, Bonn M. *Surf. Sci. Rep.* 2010;65:45.
[11] Boyd RW. *Nonlinear Optics,* 2nd edition. USA: Academic Press, Elsevier Ed; 2003.
[12] Shen YR. *The Principles of Nonlinear Optics.* USA: Wiley; 1984.
[13] Herzberg G. *Molecular Spectra and Molecular Structure* 1991;Vol. 2:USA: Krieger Publishing Company; 1991 Chapter III.
[14] Le Rille A, Tadjeddine A. *J. Electroanal. Chem.* 1999;467:238.
[15] Busson B, Tadjeddine A. *J. Phys. Chem. C* 2008;112:11813.
[16] Gan W, Wu B-H, Zheng Z, Guo Y, Wang H-F. *J. Phys. Chem. C* 2007;111:8716.
[17] Wang J, Clarke ML, Chen Z. *Anal. Chem.* 2004;76:2159.
[18] Zhuang X, Miranda PB, Kim D, Shen YR. *Phys. Rev. B* 1999;59:12632.
[19] Lin SH, Villaeys AA. *Phys. Rev. A* 1994;50:5134.
[20] Hirose C, Akamatsu N, Domen K. *Appl. Spectrosc.* 1992;46:1051.
[21] Wang J, Paszti Z, Clarke ML, Chen X, Chen Z. *J. Phys. Chem. B* 2007;111:6088.
[22] Gautam KS, Schwab AD, Dhinojwala A, Zhang D, Dougal SM, Yeganeh MS. *Phys. Rev. Lett.* 2000;85:3854.
[23] Lagutchev A, Lu GQ, Takeshita T, Dlott DD, Wieckowski A. *J. Chem. Phys.* 2006;125:154705.
[24] Ortega J-M, Glotin F, Prazeres R. *Infrared Phys. Technol.* 2006;49:133.
[25] Humbert C, Busson B, Six C, Gayral A, Gruselle M, Villain F, Tadjeddine A. *J. Electroanal. Chem.* 2008;621:314.
[26] Bonn M, Campen RK. *Surf. Sci.* 2009;603:1945.
[27] Harper KL, Allen HC. *Langmuir* 2007;23:8925.
[28] Sovago M, Vartiainen E, Bonn M. *J. Chem. Phys.* 2009;131:161107. Erratum: Sovago M, Vartiainen E, Bonn M. J. Chem. Phys. 2010;133:229901.
[29] Chen X, Hua W, Wang Z, Allen HC. *J. Am. Chem. Soc.* 2010;132:11336.
[30] Lis D, Guthmuller J, Champagne B, Humbert C, Busson B, Tadjeddine A, Peremans A, Cecchet F. *Chem. Phys. Lett.* 2010;489:12.
[31] Du Q, Superfine R, Freysz E, Shen YR. *Phys. Rev. Lett.* 1993;70:2313.
[32] Roke S, Schins J, Müller M, Bonn M. *Phys. Rev. Lett.* 2003;90:128101.
[33] Viswananth P, Aroti A, Motschmann M, Leontidis E. *J. Phys. Chem. B* 2009;113:14816.
[34] Sovago M, Wurpel G. WH, Smiths M, Müller M, Bonn M. *J. Am. Chem. Soc.* 2007;129:11079.
[35] Bonn M, Roke S, Berg O, Juurlink L. BF, Stamouli A, Müller M. *J. Phys. Chem. B* 2004;108:19083.
[36] Ma G, Allen HC. *Photochem. Photobiol.* 2006;82:1517.
[37] Nishida T, Johnson CM, Holman J, Osawa M, Davies PB, Ye S. *Phys. Rev. Lett.* 2006;96:077402.
[38] Anglin TC, Cooper MP, Li H, Chandler K, Conboy JC. *J. Phys. Chem. B* 2010;114:1903.
[39] Anglin TC, Brown KL, Conboy JC. *J. Struct. Biol.* 2009;168:37.
[40] Liu J, Conboy JC. *J. Am. Chem. Soc.* 2004;126:8894.
[41] Bourguignon B, Zheng W-Q, Carrez S, Ouvrard A, Fournier F, Dubost H. *Phys. Rev. B* 2009;79:125433.

[42] Anderson NA, Richter LJ, Stephenson JC, Briggman KA. *Langmuir* 2006;22:8333.
[43] Anderson NA, Richter LJ, Stephenson JC, Briggman KA. *J. Am. Chem. Soc.* 2009;127:2094.
[44] Levy D, Briggman KA. *Langmuir* 2007;23:7155.
[45] Wang J, Even MA, Chen X, Schmaier AH, Waite JH, Chen Z. *J. Am. Chem. Soc.* 2003;125:9914.
[46] Watry MR, Richmond GL. *J. Phys. Chem. B* 2002;106:12517.
[47] Ji N, Shen YR. *J. Chem. Phys.* 2004;120:7107.
[48] Chen X, Sagle LB, Cremer PS. *J. Am. Chem. Soc.* 2007;129:15104.
[49] Chen X, Wang J, Sniadecki JJ, Even MA, Chen Z. *Langmuir* 2005;21:2662.
[50] Wang J, Lee S-H, Chen Z. *J. Phys. Chem. B* 2008;112:2281.
[51] Shuji Y, Nguyen KT, Boughton AP, Mello CM, Chen Z. *Langmuir* 2010;26:6471.
[52] Wang J, Chen XY, Clarke ML, Chen Z. *Proc. Nat. Acad. Sci. U.S.A.* 2005;102:4978.
[53] Nguyen KT, King JT, Chen Z. *J. Phys. Chem. B* 2010;114:8291.
[54] Jung S-Y, Lin S-M, Albertorio F, Kim G, Gurau MC, Yang RD, Holden MA, Cremer PS. *J. Am. Chem. Soc.* 2003;125: 12782.
[55] Evans-Nguyen KM, Fuierer RR, Fitchett BD, Tolles LR, Conboy JC, Schoenfisch MH. *Langmuir* 2006;22:5115.
[56] Chen Z, Ward R, Tian Y, Malizia F, Gracias DH, Shen YR, Somorjai GA. *J. Biomed. Mater. Res.* 2002;62:254.
[57] York RL, Mermut O, Phillips DC, McCrea KR, Ward RS, Somorjai GA. *J. Phys. Chem. C* 2007;111:8866.
[58] Mermut O, Phillips DC, York RL, McCrea KR, Ward RS, Somorjai GA. *J. Am. Chem. Soc.* 2006;128:3598.
[59] Phillips DC, York RL, Mermut O, McCrea KR, Ward RS, Somorjai GA. *J. Phys. Chem. C* 2007;111:255.
[60] Shuji Y, Nguyen KT, Le Clair SV, Chen Z. *J. Struct. Biol.* 2009;168:61.
[61] Chen X, Chen Z. *Biochim. Biophys. Acta* 2006;1758:1257.
[62] Kim G, Gurau MC, Lim S-M, Cremer PS. *J. Phys. Chem. B* 2003;107:1403.
[63] Doyle AW, Fick J, Himmelhaus M, Eck W, Graziani I, Prudovsky I, Grunze M, Maciag T, Neivandt DJ. *Langmuir* 2004;20:8961.
[64] Chen X, Wang J, Boughton AP, Krystalin CB, Chen Z. *J. Am. Chem. Soc.* 2007;129:1420.
[65] Shuji Y, Nguyen KT, Chen Z. *J. Phys. Chem. B* 2010;114:3334.
[66] Yang L, Harroun TA, Weiss TM, Ding L, Huang HW. *Biophys. J.* 2001;81:1475.
[67] Chen X, Wang J, Krystalin CB, Chen Z. *Biophys. J.* 2007;93:866.
[68] Shai Y. *Biopolymers* 2002;66:236.
[69] Dreesen L, Sartenaer Y, Humbert C, Mani AA, Méthivier C, Pradier C-M, Thiry PA, Peremans A. *ChemPhysChem* 2004;5:1719.
[70] Pluchery O, Humbert C, Valamanesh M, Lacaze E, Busson B. *Phys. Chem. Chem. Phys.* 2009;11:7729.
[71] Tourillon G, Dreesen L, Volcke C, Sartenaer Y, Thiry PA, Peremans A. *J. Mater. Sci.* 2009;44:6805.
[72] Walter SR, Geiger FM. *J. Phys. Chem. Lett.* 2010;1:9.
[73] Sartenaer Y, Tourillon G, Dreesen L, Lis D, Mani AA, Thiry PA, Peremans A. *Biosens. Bioelectron.* 2007;22:2179.
[74] Asanuma H, Noguchi H, Uosaki K, Yu H-Z. *J. Am. Chem. Soc.* 2008;130:8016.
[75] Howell C, Schmidt R, Kurz V, Koelsch P. *Biointerphases* 2008;3:47.

[76] Stokes GY, Gibbs-Davies JM, Boman FC, Stepp BR, Condie AG, Nguyen ST, Geiger FM. *J. Am. Chem. Soc.* 2007;129:7492.
[77] Matranga C, Wehrenberg BL, Guyot-Sionnest P. *J. Phys. Chem. B* 2002;106:8172.
[78] Tong Y, Li N, Liu H, Ge A, Osawa M, Ye S. *Angewandte Chemie International Edition* 2010;49:2319.
[79] Gosh A, Smits M, Bredenbeck J, Bonn M. *J. Am. Chem. Soc.* 2007;129:9608.
[80] Smits M, Gosh A, Bredenbeck J, Yamamoto S, Müller M, Bonn M. *New J. Phys.* 2007;9:390.
[81] Bredenbeck J, Gosh A, Nienhuys H-K, Bonn M. *Accounts Chem. Res.* 2009;42:1332.
[82] Bredenbeck J, Gosh A, Smits M, Bonn M. *J. Am. Chem. Soc.* 2008;130:2152.
[83] Chou KC, Westerberg S, Shen YR, Ross PN, Somorjai GA. *Phys. Rev. B* 2004;69: 153413.
[84] Hayashi M, Lin SH, Raschke MB, Shen YR. *J. Phys. Chem. A* 2002;106.2271.
[85] Dreesen L, Humbert C, Sartenaer Y, Caudano Y, Volcke C, Mani AA, Peremans A, Thiry PA, Hanique S, Frère J-M. *Langmuir* 2004;20:7201.
[86] Inoue K, Fujii M, Sakai M. *Appl. Spectrosc.* 2010;64:275.
[87] Cimatu KA, Baldelli S. *J. Phys. Chem. C* 2009;113: 16575.
[88] Schaller RD, Johnson JC, Wilson KR, Lee LF, Haber LH, Saykally RJ. *J. Phys. Chem. B* 2002;106:5143.

Index